LIFE CYCLE APPROACHES TO SUSTAINABLE REGIONAL DEVELOPMENT

Edited by
Stefania Massari,
Guido Sonnemann and
Fritz Balkau

Routledge
Taylor & Francis Group
NEW YORK AND LONDON

earthscan
from Routledge

First published 2017 by Routledge

2 Park Square, Milton Park, Abingdon, Oxfordshire OX14 4RN
52 Vanderbilt Avenue, New York, NY 10017

Routledge is an imprint of the Taylor & Francis Group, an informa business

First issued in paperback 2020

Library of Congress Cataloging in Publication Data
Names: Massari, Stefania, 1971– editor. | Sonnemann, Guido, editor. | Balkau, Fritz, editor.
Title: Life cycle approaches to sustainable regional development / edited by Stefania Massari, Guido Sonnemann and Fritz Balkau.
Description: New York, NY: Routledge, 2017.
Identifiers: LCCN 2016022818 | ISBN 9781138940604 (hardback)
Subjects: LCSH: Regional planning – Environmental aspects. | Regional Economics – Environmental aspects. | Product life cycle – Environmental aspects. | Industrial ecology. | Sustainable development.
Classification: LCC HT391 .L54 2017 | DDC 307.1/2—dc23
LC record available at https://lccn.loc.gov/2016022818

ISBN: 978-1-138-94060-4 (hbk)
ISBN: 978-0-367-66991-1 (pbk)

Typeset in Corbel and Myriad Pro
by Florence Production Ltd, Stoodleigh, Devon, UK

LIFE CYCLE APPROACHES TO SUSTAINABLE REGIONAL DEVELOPMENT

Life Cycle Approaches to Sustainable Regional Development explains the ways life cycle methodologies and tools can be used to strengthen regional socio-economic planning and development in a more sustainable manner. The book advocates the adoption of systematic and long-term criteria for development decision-making, taking into account the full life cycle of materials and projects. It describes life cycle practices from both a scientific and a practitioner point of view, highlighting examples and case studies at regional level. The applications are relevant to key economic sectors, as well as for internal planning and administrative procedures. It concludes with a synthesis chapter that distils the key messages from the authors into practical guidance points on how best to use such approaches to enhance sustainability in regional development. The book is essential reading for regional and urban planners who are integrating life cycle thinking into their policy regimes, as well as for researchers working to further evolve life cycle methodologies.

Stefania Massari became full Professor in Commodity Sciences in 2004. She works at the Faculty of Economics at the University of Salento, Lecce, Italy. She is the Coordinator of the degree courses in the business area of the Faculty and she is the Director of the 'Life cycle approaches to sustainable regional development' summer school. She holds an MSc in Biology from the University of Bari. She did research at UNEP Paris and King's College London.

Guido Sonnemann is full Professor at the University of Bordeaux, France, where he heads a research group on life cycle assessment in the Institute of Molecular Sciences at the Department of Sciences and Technology. From 2002 to 2012, he worked as UNEP Programme Officer for Innovation and Life Cycle Management in Paris. In this function, he served as Science Focal Point for UNEP's Resource Efficiency subprogramme. Guido holds a PhD in Chemical Engineering from the University of Rovira and Virgili, Spain (2002), and an MSc in Environmental Chemistry from the University of Poitiers, France (1996), and he graduated as an environmental engineer from the Technical University of Berlin, Germany (1995).

Fritz Balkau is an independent international advisor on sustainable development, focusing on sustainable production and consumption practices, enhanced resource efficiency, and risk reduction in society. He has extensive environmental policy experience in government (Australia) and intergovernmental organisations, and has carried out professional consultancy projects for national governments and independent associations, as well as the OECD and UN.

'Regions present an ideal scope and scale for sustainable action: not too big, not too small – and politically capable of setting priorities and directions for economic, environmental and social goals. The power of life cycle thinking and sustainability approaches has proven effective when applied to goods and services; thus to extend the life cycle approach to the sustainable management of geographic areas is exciting and beneficial. This substantial volume will serve well as a guide for students of regional development and for practitioners seeking stimulation and new direction.'

Steven B. Young, Graduate Officer & SUSM Program Director, University of Waterloo, Canada

'Reorienting regional development towards a more inclusive, resource efficient and sustainable path is a critical part of implementing the sustainable development goals. This would require utilizing new sets of planning and decision making tools. *Life Cycle Approaches to Sustainable Regional Development* provides valuable insights and inputs to this process from a life cycle perspective.'

Desta Mebratu, Fellow, Stellenbosch Institute for Advanced Studies (STIAS)

CONTENTS

FIGURES

TABLES

BOXES

CONTRIBUTORS

Naeem Adibi has a multidisciplinary background in engineering, environmental science, and management. He has coordinated numerous life cycle assessment (LCA) projects in building construction, recycling, and waste treatment, and in a range of industry sectors. He is currently European project manager for Platform [avniR] in northern France, managed through the association cd2e.

Victor Barichello is Vice-President of *Empauer*, a global company that specialises in the creation and use of advanced software tools used to develop sustainable products and packaging for the consumer goods industry. His contribution to the book occurred while he was CEO of *Piqet*, an Australian company developing software tools for packaging design.

Riccardo Beltramo is Professor of Operations Management at the University of Turin, where he teaches a variety of subjects in industrial ecology, environmental management systems, and environmental technology. He is President of the Centre NatRisk, a network for research into risk management of natural disasters in the mountains.

Alberto Bezama is leader of the Bioeconomy and Biomass Resources working group at the Department of Bioenergy, focused on sustainability evaluation of bio-based technologies, development of LCA-based tools for enabling integrative sustainability evaluations, and development of regionalized LCA approaches.

Silu Bhochhibhoya is a PhD candidate at the University of Padua, Italy, working on research concerning environmental assessment of buildings in the Himalayan Region.

Keith Bradby helps lead Gondwana Link, a collaborative multi-stakeholder initiative that aims to restore ecological connectivity and ecosystem function across south-western Australia. It encompasses a wide range of ecologically focused activities that also provide social and economic co-benefits to local communities.

Ravina Brizmohun is a PhD student at the Faculty of Engineering, University of Mauritius, and works as Scientific Officer in the Ministry of Environment and Sustainable Development. Her research is based on life cycle assessment of electricity-generating systems in Mauritius.

Isabela Butnar works in the field of sustainability assessment of products and processes based on life cycle analysis and hybrid input–output (IO) LCA techniques, in research into environmental taxation based on IOLCA results, and estimation of pollution embodied in international trade.

Francesc Castells is Emeritus Professor in the Chemical Engineering Department of the University of Rovira and Virgili and led the AGA Group (environmental analysis and management R&D group), from its founding in 1995 until 2011. Current research fields are applications of life cycle assessment, eco-design, and sustainable development models.

Raffaele Cavalli is Professor of Forest Operations. His main lecturing and research activities deal with forest engineering, wood supply chain, and wood biomass utilization. He is Director of the Department of Land, Environment, Agriculture and Forestry, at the University of Padua, and is responsible for the laboratory for biofuel analysis (Laboratory ABC).

Marie Darul has a background in agriculture, environment, and food sciences. She is currently consultant for the company cd2e, focused on sustainable construction for the BuildavniR project and working with regional stakeholders in the building life chain. She has participated in and led a range of European projects in this sector concerning eco-design of products.

Federica De Leo is currently Assistant Professor of Commodity Science at the University of Salento. Her main research interests concern quality and safety in the food industry, environment and sustainable development, clean technologies, and the application and analysis of environmental pressure indicators.

Christelle Demaretz is the Circular Economy Project Manager in the environmental department of the Regional Council of Hauts-de-France, Nord-Pas-de-Calais, Picardie, in northern France. She is working on the mainstreaming of life cycle thinking and new business models. She has worked in different organizations such as Paris Airports, Renault, and CNRS.

Aldo Femia is senior researcher and leader of ISTAT's team on Environmental Accounts in Physical Units. He has also worked at the Wuppertal Institute and at the OECD. His research focuses on sustainability, well-being beyond GDP, material flows, and resource efficiency in the context of socio-economic metabolism in physical and monetary terms.

Giulia Ferrini is a technical advisor in UNEP's Energy Branch. She helps developing countries to build capacity in climate technologies, in the framework of the programmes of the Climate Technology Centre and Network. She also supports the Oil and Gas Methane Partnership, aimed at reducing methane emissions from oil and gas operations.

Annik Magerholm Fet is Professor of environmental management, systems engineering, and life cycle analysis, and is the Director of NTNU Sustainability. She is head of several national and international research projects in global production, the environment, and CSR. She is a national expert in life cycle environmental management, including product declarations and footprinting.

Matthias Finkbeiner is Professor of Sustainable Engineering and Managing Director of the Institute of Environmental Technology. He is the Chair of ISO TC207/SC5 Life Cycle Assessment, member of the Board of the UNEP/SETAC Life Cycle Initiative, and co-chair of its flagship on LCA of organizations. He chairs numerous committees concerning standardization and LCM.

Eskinder Demisse Gemechu is a life cycle assessment expert with a PhD and has trained in a number of European institutions. He has been working towards a methodological integration of LCA and resource criticality under the life cycle sustainability assessment framework. He is currently a freelance expert, also working at the University of Waterloo as a part-time researcher.

Ross Hamilton is a Director at the International Council on Mining and Metals (ICMM), based in London, responsible for environmental stewardship, climate change, and social progress programmes. Prior to joining ICMM in 2013, Ross held a number of senior sustainability-related roles within industry, including, most recently, with BHP Billiton in Australia.

Jakob Hildebrandt is a PhD candidate in the working group Bioeconomy and Biomass Resources, Department of Bioenergy. His scientific tasks include sustainability evaluation of bio-based technologies from a regional perspective and development of regionalized LCA approaches, particularly a Sustainability Index for assessing bioeconomy regions.

Atsushi Inaba is the Chair of the Japanese Mirror Committee for the ISO TC207/SC5 Life Cycle Assessment and the Japanese expert on the TC/207/SC3 Environmental Label and SC7/WG6 Carbon Footprint of Products. He is member of the Board of the UNEP/SETAC Life Cycle Initiative and is lead author of Chapter 12 of the *Fifth Assessment Report of IPCC WG-III*.

Harish Jeswani is currently working as a Research Fellow at the University of Manchester. For the past 7 years, he has worked on various research projects on sustainable production and consumption. His research interests include sustainable production and consumption, life cycle assessment, carbon and water footprinting, and corporate sustainability.

Bertrand Laratte is Assistant Professor in the Research Centre for Environmental Studies and Sustainability at the University of Technology of Troyes. His field of research concerns the integration of the temporal and spatial aspects of the inventory and the assessment phases of LCA. His research also encompasses dynamic material/substance flow analysis (MFA/SFA).

Agathe Laure is Programme Advisor to the Climate Technology Centre and Network (CTCN) hosted by UNEP in collaboration with UNIDO, providing technical assistance services and capacity-building activities in developing countries and supporting the deployment and transfer of climate technologies for mitigating and adapting to climate change.

Maria Llop is Associate Professor of Economic Analysis, researcher, and Director of the Chair for Energy and Economic Development at the University of Rovira and Virgili. Her research interests focus on economic impact analysis and environmental economics through the use of computable general equilibrium models and input–output models.

Eléonore Loiseau is a researcher at Irstea. She also belongs to the French research group ELSA for LCA and industrial ecology. While working on her PhD thesis, she developed 'territorial LCA' approaches for performing an environmental assessment of a territory as a whole. She continues her research in this field.

Massimo Lombardi is Sustainability Manager and CEO of Greener Italia, a company that aims to replicate in other companies the experience gained in Sabox in embedding sustainability into business routine and products. Greener Italia specializes in LCA analysis and promotes product carbon footprinting and the implementation of sustainability projects in public- and private-sector activities.

Philippe Loubet is an Assistant Professor at the Ecole Nationale Supérieure de Chimie, de Biologie et de Physique of Bordeaux. He undertakes research in LCA applied to green chemistry and materials. He has a background in process engineering and the water footprint of urban water systems, in collaboration with ELSA research group and Veolia.

Renato Marra Campanale is currently seconded national expert to the European Commission, DG Eurostat, from the Italian Institute for Environmental Protection and Research and has also worked at the OECD. His main expertise is in physical environmental accounting.

Julia Martínez-Blanco is an expert in the field of analysis and environmental assessment of sustainability, particularly in the agri-food sector, undertaking research related to the work groups of Sostenipra (UAB, Barcelona) and SEE (TU Berlin). She is currently the coordinator of the UNEP/SETAC Life Cycle Initiative's flagship project, LCA of Organizations.

Bernard Mazijn is Managing Director of the Institute for Sustainable Development, a not-for-profit organization in Bruges (Belgium), focusing on research and training, inter alia on life cycle sustainability assessment. The Institute also coordinates the Social LC Alliance. He has been Guest Professor since 1998 at Ghent University, teaching the Master's course in sustainable development.

Mesfin Mekonnen is a Post-doctoral Research Associate in the Robert B. Daugherty Water for Food Global Institute at the University of Nebraska

Pier Paolo Miglietta started his postgraduate research activities in the international PhD programme, the Economics and Management of Natural Resources, at the LUM Jean Monnet University, Bari. Since 2014, he has been a Research Fellow in Commodity Sciences at the University of Salento. He is author of research papers in national and international journals and conference proceedings.

Bruno Notarnicola is Professor and Head of the Ionian Department of Law, Economics and Environment at the University of Bari Aldo Moro. His main research is in industrial ecology and analytical and procedural instruments such as LCA, LCC, environmental input–output, and integrated product policy.

Asha Panwar is a PhD candidate at Indian Institute of Technology Bombay (India). Her doctoral research focuses on applications of life cycle thinking for waste management in India.

Andrew S. Parsons is leader of Gold Fields' sustainable development activities, including environmental management, mine closure, community relations, and climate change, and is responsible for advising the Board and other stakeholders on sustainable development. He has held positions in another mining company, the Chamber of Mines of South Africa, in UNEP and ICMM.

Vanessa Pasquet is project manager at [avniR], a collaborative platform in northern France, managed through the association cd2e, that supports the economic transition throughout new development models based on life cycle thinking. She also coordinates working groups on eco-design, LCA tools, and research in the building sector.

Donatella Porrini is Associate Professor of Political Economy at the Department of Management, Economics, Mathematics and Statistics, University of Salento. Her research activity focuses on economic analysis of legal approaches and deals with environmental policies and the role of insurance. She is author of textbooks, contributions in scientific books, and articles in international journals.

Maxime Pousse is LCA project manager for Nobatek, a French non-profit technological research centre contributing to innovation in sustainable development and energy efficiency in building and urban planning. Nobatek carries out both R&D activity and short-term assistance for private and public customers, aiming at collaboration on innovation and disseminating best practices.

Mark Radka has a background in engineering and public policy from universities in the USA and has extensive field experience in south-east Asia. As head of the UNEP Energy, Climate, and Technology Branch, he works to link the global energy and environment agendas, focusing on reducing emissions of greenhouse gases through greater use of renewable energy and improved energy efficiency, with a focus on developing countries.

Cristiana Reho works as an independent consultant, supporting certification bodies with the provision of training and audits. She also undertakes consultancy projects concerning management systems or the evaluation of environmental impacts of organisations and/or their products/services.

Pietro Alexander Renzulli is a full-time researcher in commodity sciences. Research topics include LCA, food LCA, industrial ecology, industrial symbiosis, energy management, and innovation of production cycles.

William Rivenbark is a Professor of Public Administration and Director of the Master of Public Administration Program at the University of North Carolina at Chapel Hill. His research and teaching involve local government administration, focusing primarily on the areas of performance and financial management, including budgeting and financial analysis.

Andrés Rolon is co-founder and project manager of Regenerativa, a company that aims to accelerate the transition to a circular economy, promoting sustainable development in companies and individual lifestyles. He provides strategic sustainability advice for private companies and public-sector bodies, and leads studies on packaging sustainability and climate change solutions.

Judit Roncz is a PhD candidate at the Enterprise Theory and Practice Doctoral School, University of Miskolc. Her research and publications are focused on the energy sector, economic development and sustainability, life cycle analysis, and environmental sustainability. Her research also touches the fields of regional development and spatial studies.

Philippe Roux has been working at Irstea since 1989. A research engineer in the field of life cycle assessment, Philippe is the team leader of the ELSA-Itap team and the co-founder of the ELSA research group. He has been involved in many research projects (European, French ANR, Chaire Elsa-Pact, etc.).

Marcello Ruberti is Aggregate Professor of Commodity Science at the University of Salento. He is author of numerous scientific publications, including articles, communications at conferences, and essays on issues concerning, energy, the environment, and agrifood.

Marta Schuhmacher is Professor of Environmental Engineering at the University of Rovira and Virgili. She is the Technical Manager of TecnAtox Research Group, a leader in developing new environmental tools and integrating techniques for environmental monitoring and analysis. She has directed national and international research projects and participated in various EU projects.

Anke Siebert is a PhD candidate in the Bioeconomy and Biomass Resources working group, in the Department of Bioenergy. Her scientific tasks include sustainability evaluation of bio-based technologies from a regional perspective and development of regionalized LCA approaches, particularly social LCA methods.

Fabrizio Striani is a tutor in Microeconomics at the University of Salento. He also teaches at the University of Basilicata. His research activity deals with consumer behavior during life cycle consumer selection, which was the topic of his PhD thesis.

Philip Strothmann is currently Chief Finance and Operations Officer of the Forum for Sustainability through Life Cycle Innovation e.V. He is also an independent consultant with experience in international sustainability projects focusing on natural resource management and life cycle thinking.

Giuseppe Tassielli is Professor of Commodity Sciences at the University of Bari. His main research interests are technological and economic analysis of productive processes, development of methodologies to assess the environmental quality of goods and productive processes, and application of analytical, procedural, and economic tools for the environmental assessment of companies.

Daniela Thrän is Head of the Department of Bioenergy at the Helmholtz Centre for Environmental Research, UFZ. She is also Head of the Bioenergy Systems Department at the Deutsches Biomasseforschungszentrum (DBFZ) in Leipzig and member of the German Bioeconomy Council.

Nelle Thyseen is a bio-engineer, working in Natuurinvest in Belgium, in charge of the FSC certification of the forests in Flanders, as well as coordinating an interregional project about innovation of forest management (exploitation and sale) in Flanders and Southern Netherlands. Earlier she was overseeing standardisation issues and LCA application in a Belgian institute Wood.be that contributes to sustainability in wood production and in the furniture sector.

Pierluigi Toma started his research in the Economics, Management and Quantitative Methods PhD programme at the University of Salento, conducting studies of non-parametric methods applied to environmental concerns. Since 2015, he has been a visiting scholar in the Institute of Statistics, Biostatistics and Actuarial Sciences in the Catholic University of Louvain-la-Neuve.

Klara Szita Tóthné is Professor and Head of the Doctorate School at the University of Miskolc. Her research field includes sustainability, LCA, and futures studies. She is a member of the Scientific Committee of Statistics and Futures Studies of the Hungarian Academy of Sciences and President of the Futures Studies Subcommittee. She is a past member of the SETAC Europe LCA Steering Committee.

Christian Traisnel has more than 30 years' background in industry. He is currently Director of cd2e and the head of the [avniR] international platform in northern France, working on new development models based on life cycle approaches. He is also General Director of TEAM2, a national cluster in France on recycling and circular economy.

Jean-François Vallés is Associate Director and co-founder of Synopter, a consulting firm in the field of sustainable performance of industrial estates and collaborative approaches.

Benjamin Walker is Director of the Institut Supérieur des Arts Appliqués in Strasbourg. He has extensive experience in industrial design through managing his own company, Touraine Design Studio, in France, and undertaking teaching and consultancy projects in eco-design for universities, companies, and international organizations.

Peter Whitbread-Abrutat is the owner of Future Terrains – a social enterprise working internationally to address the challenges of degraded lands. He is a committee member of the British Land Reclamation Society and a Churchill Fellow. He was intimately involved in the land redevelopment of the Eden Project in the UK.

Pia Wiche is a sustainability expert supporting decision-making by governments and organizations. She helps write ISO and national standards on life cycle tools and offers advice and consulting services on sustainable metrics. She is also a teacher of sustainable management and entrepreneurship at Pontificia Universidad Católica de Valparaíso.

Vanessa Zeller is a researcher in the Recycling and Environment research group at the Université Libre de Bruxelles. Her research activities focus on the environmental impact assessment of production and consumption activities on a regional scale.

Guoping Zhang is Programme Director of the Water Footprint Network, where he leads programmes to support water resources stakeholders in sustainable, efficient, and equitable water management through application of water footprint assessment. He has been lead expert for international projects on sustainable water resources management and water footprint assessment.

FOREWORD

This book is about sustainability, or, more precisely, it is about how to move towards sustainability through the use of life cycle methodologies and tools. The book has two chief axes of vision – the sustainability objectives of regions and of their stakeholders, and how to use life cycle approaches to achieve those objectives. It should be borne in mind that sustainable development is a journey as much as a destination (although it is that as well). Methodologies and tools are now available to measure progress and promote regional sustainability. At the same time, the methodologies continue to evolve.

Why the focus on regions? Because regions hold the key to many development decisions and have been given important roles in promoting their own industrial and natural resource development, and because they undertake land protection and enhance the social fabric. National governments are increasingly devolving administrative and regulatory responsibilities to their regions.

And why are life cycle approaches important? Because, although the full benefits and impacts of local products within a region will, of course, be central to our considerations, it is also a fact that the value chains of many goods and services now span the entire globe. Many of the products in our lives come into our region from distant places, have long life chains, and may contain forbidden substances or have hidden embedded impacts. Conversely, our own valuable exports to distant places may face scrutiny there, to identify any impacts unacceptable at their destination. Complex life cycle dynamics require a sophisticated management response that is coherent within both local and global development agendas.

A series of summer schools on 'Life Cycle Approaches to Sustainable Regional Development' – hosted by the University of Salento in 2012–14 and by the University of Bordeaux in 2015 – brought together life cycle specialists and graduates to explore some of the dynamics of global life cycles and the methodologies that can be used to manage them. This book draws inspiration and substance from those summer schools, augmented by commissioned papers from recognised experts.

The material presented reflects a specific moment in the journey to regional sustainability. Not all topics could be covered, some techniques are still evolving, and many subjects are not explored in depth. But there is, nonetheless, sufficient material to raise awareness, present the key ideas, and build understanding. The book presents valuable material to support immediate action, but also constitutes a basis for moving forward.

The different chapters give a glimpse of life cycle methods that could be usefully applied in and by regions and show how assessment tools provide a basis for new management approaches. Examples are given showing how a life cycle perspective can be applied in the important economic sectors that are present in many regions, and the book also looks at how infrastructure such as energy, buildings, and waste management can be more effective if planned around life cycle thinking.

Given the diversity of chapters, their interlinkages, and the large number of authors, some repetition of ideas and facts is inevitable. If anything, such repetition should reinforce the notion that the material is definitely mainstream and is now widely accepted as important background for decision-making.

Through the interpretive skills of the individual chapter authors, this work captures the essence of what others – specialists and managers, scientific societies and intergovernmental bodies, standards bodies and NGOs – have produced in the wide world of life cycle approaches and we consider relevant for the multiple actors at the regional level. We hope the book does justice to their efforts, efforts that are ongoing, because methodologies, needs, and policies continue to evolve. Key messages and learning points concerning life cycle applications in regions around the world are summarised in a synthesis chapter at the end of the book.

PREFACE

Regions everywhere now have increasing responsibilities for the pursuit of their own socio-economic development and for stimulating innovation. At the same time, they remain embedded in a national and international context of trade and demographic change that influences their development options. Decisions taken today will influence present and future generations and the viability of the regional industrial clusters they try to create.

Analysing the present and predicting the future remain essential skills if the development process is to be truly sustainable. Life cycle approaches are among the tools already available for these assessments and management decisions, even if many of them are still undergoing refinement and extension. It is essential that regions make use of such tools to avoid errors of the past where development processes were considered only from a short-term, single-issue perspective. These past errors are everywhere to be seen in our society that is afflicted with degraded land, abandoned industrial sites, massive unemployment, and lack of vitality in affected regions. Our current generation has the vision and the tools to do better and plan in a sustainable way.

This book assists the regional development process by outlining a range of life cycle approaches, such as life cycle analysis and eco-design for sustainable innovation, that can assist in improving decision-making by regions and business alike. In the Aquitaine region, we have given these tools a prominent role in our strategy towards a circular economy. No two regions are alike; the book, which I warmly welcome, does not offer an easy solution that solves problems with a simple wave of a magic wand. However, the synthesis provided is sound, the examples given are encouraging, and the direction forward is clearly described. The book can serve as both an inspiration and a guide to our regions as we gradually move forward to a more sustainable world by fostering sustainable innovation in multiple economic sectors, establishing a circular economy, and transforming the way we use and produce energy.

Alain Rousset
President, Region Nouvelle Aquitaine, France
Former President, Association of the Regions of France

ACKNOWLEDGEMENTS

This book has resulted from the combined efforts of many persons. The various authors and co-editors have freely given their time to contribute to this compilation, supported in some cases by researchers and assistants who searched out material and contributed text.

The book was inspired by the series of summer schools on 'Life Cycle Approaches to Sustainable Regional Development'. We would like to acknowledge the students and lecturers who attended the four editions of these summer schools since 2012. They have shown us that the topic identified is of interest also to a larger audience, at which this book is aimed.

Next, we appreciate that Routledge approached us and agreed to publish a book based on the summer school topic and its content. In particular, we thank Judith Newlin and Nicole Solano for accompanying us throughout the editing process to the point of publication.

A special thank you goes to Geoffrey Bird for the professional editing and Philip Strothmann from FSLCI, as well as Marcello Ruberti and Stefano Pastore from the University of Salento for helping us to manage the formatting of the chapters.

Finally, we would like to express our gratitude to our families and friends who provided support and advice during the preparation of the book. We would especially like to mention here Marisa, Marc, and Julia, who saw less of their husband and father while he worked on this compilation.

Stefania Massari, Guido Sonnemann, and Fritz Balkau
Editors

Part I

INTRODUCTION – PROVIDING THE CONTEXT

1

SUMMARY

This book is about sustainability, or, more precisely, it is about how to move towards sustainability at the regional level through the use of methodologies and tools that are based on life cycle thinking. The book has two chief axes of vision – the capacities of various regional actors to address the sustainability challenge more systematically and the applicability of various life cycle procedures and approaches to help achieve this goal.

The book is divided into six interrelated parts – Introduction, Methodologies, Implementation instruments, Regional applications, Economic sectors, and a Synthesis and Summary – giving a glimpse of how life cycle approaches can support sustainable, socio-economic development at regional level. No major attempt has been made to describe the sustainable development agenda in detail, nor has there been a lengthy description of regional frameworks and administration. These aspects are woven into the various chapters in a general way, and further details can be found in specialist literature on these subjects. Regions are very loosely defined as territorial entities that have a high degree of control over their economic, social, and environmental futures. The definition does not include clusters of nation states grouped into global regions, as in, for example, UN 'regions', even if some of the methodologies could be interesting also at that level.

The *Introductory part* explores some important issues that form the backdrop to regional sustainability. The editors – Balkau, Massari, and Sonnemann – describe how regional development is influenced by the globalised nature of the world economy, technology development, and demographic changes. Regions need to have an eye on both their own sustainability and that of the world as a whole, as regions cannot exist independent of their native countries or of the wider global context. Porrini and Striani describe the origins and actual situation regarding regional sustainability. They discuss the framework as well as the key drivers at the macro level, where the priorities and guidelines originate, their relationship with supranational frameworks in the UN (also mentioning policies at EU level), and how they cascade to lower levels in regions within nation states. As territorial development depends on the degree and conditions of financing, Balkau and Rivenbark explore the way some current financing principles and mechanisms can support sustainability policies in regions and in local government. They highlight that regional governments themselves need to exercise financial responsibility if they are to manage sustainability programmes that have positive longer-term outcomes in line with the strategic vision of the region. They describe some financing and investment mechanisms specifically designed to assist in the uptake of high-profile sustainability programmes such as biodiversity and energy. Sustainable investments consider the entire project life cycle, not just the productive phase of service delivery and profitability.

As business and industry will always be the motor of economic development, Sonnemann and Balkau review the principles of resource efficiency as an important factor in simultaneously reducing environmental (and social) impacts, while concurrently improving production efficiencies. Such an approach aims to achieve, to the maximum extent possible, the often talked about win–win strategies in development, in particular through innovation and optimisation of technologies and consumption patterns. They also highlight the importance of the current high level of resource and product consumption worldwide as a causal factor of the serious environmental stress that is compromising the future of many social and economic systems. In this context, they also introduce the concept of resource criticality, meaning a situation where an expected shortage of certain key resources – mineral and agricultural, for example – limits the possibility of technological solutions to adapt to future challenges. Life cycle management (LCM) refers to the efficient management of the entire span of the life of a product or process so as to properly take account of the upstream and downstream causes of both environmental and socio-economic impacts. LCM is thus indispensable in preventing point source intervention from merely shifting the problem up or down the life chain. Balkau and Sonnemann review the place of LCM in a regional framework, highlighting the experience so far in business and industry but with applicability also at regional level. They explain that, far from being an approach for product optimisation only, LCM can contribute meaningfully to more effective project planning and management, administration, and financing.

Part II examines the *Analytical basis for life cycle approaches*. Much of this part deals with various assessment techniques that provide better information for decision-making. Analytical techniques have been developed for a variety of purposes, many of them relevant to the activities that support regional development. Ruberti and Massari commence with a description of ecological footprinting as a way of understanding a region's current level of sustainability. Footprints may be calculated for specific issues of importance or for the entire sustainability agenda (the latter is more difficult to interpret). The footprint concept is useful at many levels, including that of an individual, but regional bodies will find that footprinting data are especially useful in highlighting action priorities for sustainability policies. Gemechu, Butnar, Llop, Castells, and Sonnemann continue by highlighting the value of environmentally extended input–output tables. Such tables allow a region to better understand the flow of materials and pollutants, thus helping to set priorities for intervention and education programmes. In a related chapter, Femia and Campanale expand on this concept by explaining how material flow accounts provide a detailed appreciation of substances and materials, and their embedded impacts, that flow through the region. Insofar as most imported materials in a region end up as local wastes that need to be disposed of properly, such studies contribute significantly to the improvement of waste management, with subsequent benefits for the long-term health of the population. In view of the complexity of individual materials and products now in common use in society, their proper management at national, regional, and local levels depends on a detailed understanding of the life cycle of such materials and of their impacts. Such knowledge is fundamental in preparing and understanding eco-labels and product declarations, as discussed by other authors in Part III.

Life cycle assessment (LCA) is described in some detail by Tassielli, Renzulli, and Notarnicola . LCA can be used to clarify the environmental and other impacts of selected materials and products that we purchase or approve for use, as well as which targets might be affected. In subsequent chapters, we see that the scope of LCA can be expanded beyond a purely product focus. In two of these, Jeswani, Loubet, and Sonnemann explain in more detail how LCA techniques can be used to understand regional carbon emissions, both direct and indirect, and also water use and release. Both these issues are major preoccupations for many regions, with a complex set of causes and effects that need careful evaluation. Whereas many LCA studies assume a static situation, Laratte and Sonnemann discuss the way in which dynamic and spatial modelling can contribute insights into how a product system functions and the dependency on time and space of such assessments. Mazijn takes the assessment methodologies to the next level by showing how additional issues can be incorporated into conventional assessment procedures to produce a 'total life cycle sustainability assessment'. This includes social and economic impacts, in addition to the traditional environmental assessments. The expanded method can thus be used to give a bigger picture of the sustainability situation.

The last chapter in this part looks beyond a purely materials flow picture. Martínez-Blanco, Inaba, and Finkbeiner expand the LCA horizon to describe the new procedure of organisational life cycle assessment (O-LCA), which allows calculation of an organisation's sustainability footprint. Although this calculation still includes a look at materials and products handled by the organisation, it specifically examines the functioning of the organisation itself – its consumption patterns and waste flows, its management of project cycles, and so on. In view of the broad strategic reach of regional administrations, this technique can be invaluable in identifying the success and challenges of meeting sustainability goals at the regional level. Taken together, the assessment methodologies described in this part can provide valuable information on sustainability hot spots to underpin longer-term regional planning.

Part III, on *Converting analysis into life cycle tools*, looks at the way in which intervention and action tools can be constructed to act on the data from the various assessment procedures previously mentioned. As a way of building confidence in the reliability of sustainability-oriented tools, Reho and Wiche describe the international systems of standardisation and certification that provide a common framework for such tools. They explain how the International Organization for Standardization (ISO) has provided a family of standards and recommendations designed to ensure compatible application of management tools and procedures around the world. Among the many intervention tools, eco-design is a potent concept to avoid the many impacts implicitly embedded in the various products with which we surround ourselves. Walker explains the principles of eco-design that can be applied at many levels, by a variety of actors. As an extension to this, Walker also describes the use of eco-labels as a recognition instrument for well-designed, responsibly manufactured products. We have already mentioned how eco-labels depend critically on a reliable application of LCA. It is pointed out that eco-labels have value for regions promoting their own sustainable products, linking local suppliers with distant customers who share the same values. The use of environmental product declarations, as described by Fet, takes labelling one step further through a formal procedure to reveal the embedded impacts in major products, allowing potential purchasers an insight into the total footprint of the items in question. The above play a fundamental role in (sustainable) purchasing decisions. Balkau describes the growth of sustainable purchasing around the world, as customers everywhere – individual and corporate – increasingly specify sustainability characteristics they want to see embedded in the products they purchase.

Sustainable purchasing links producers with 'green' end-users, including agencies that have in-house environmental objectives. Sustainable supply-chain management is already practised by major corporations. Green purchasing can also add sustainability factors to the criteria in common use by public bodies and civil society. Beyond helping to meet sustainability objectives, green purchasing ensures significant economic outcomes for regional organisations by achieving best value over the asset life cycle, rather than just lower initial cost. It also creates a basis for more innovative local industrial and commercial development.

Further chapters extend the focus beyond physical products. Many regions are responsible for industrial promotion and land management. Vallés explains how the clustering of companies in industrial zones creates significant opportunities for collaboration on sustainability (and of course also production) issues. Such an industrial ecology approach can facilitate meaningful service linkages with the surrounding districts and expand the development dimension, beyond the boundary of the zone and over a greater time period. The expected stability of an industrial zone over an extended time frame allows more permanent links to be established with the wider community, including local suppliers and service companies, all of whom have an interest in a longer cycle for industrial development.

Toma, Massari, and Miglietta describe the application of measures to enhance natural resource efficiency and economic productivity within industrial zones, but also for activities elsewhere. A variety of procedures and tools have been designed in recent years to achieve more with less, effectively reducing waste and other impacts and creating more profitable (and, hence, sustainable) companies. This chapter focuses on an economic analysis of resource efficiency that allows the design of more productive industrial systems and activities, linking also with the economic analysis in the later Chapter 25. Beltramo expands the life cycle vision to look at environmental and landscape management systems that allow regions a better appreciation of the visual quality of their land. This is a vital factor for economic activity sectors such as tourism and recreation, as well as enhancing a region's attractiveness for companies more generally.

When applied in a life cycle framework, the above constitutes a powerful set of instruments for regions to work towards their sustainability objectives and targets. The application of these life cycle approaches is described in more detail in the following two parts.

Part IV gives concrete examples of how *life cycle-based approaches can be used at regional levels*. Several chapters are based on actual cases studies where an assessment step preceded the intervention and programmatic phase. The ecological footprinting concept in particular is a powerful tool for analysis and communication to focus attention on issues and their priorities. Miglietta, De Leo, and Toma examine the policy perspective of ecological footprints with a view to a closer integration with regional development objectives. Such an analysis can also provide input to an economic analysis of regional sustainability, as described by Szita Tóthné and Roncz for the case of Hungary. Adibi et al. describe a case study from northern France where an initiative to revitalise the regional economy is taking place. This uses LCA applied to the building sector as a way to foster local innovation throughout the supply chain. This chapter is complemented by the analysis of Loiseau and Roux, who look at how LCA can take a geographical focus instead of a product focus. Sonnemann et al. use site-dependent life cycle impact assessment to identify health impacts from fixed industrial installations (in this case, a waste incinerator). Bezama describes a similar study in Chile to give decision-makers in remote regions improved in-depth, life cycle-oriented information about waste disposal options. Zhang and Mekonnen, and Jeswani present examples of footprint studies for water and energy, respectively, looking at allocation issues and fuel substitution options that are of direct interest to many regions. Mazijn expands the earlier description of sustainable procurement in Chapter 19 to show how life cycle costing can be an important and desirable element in purchasing decisions. Vallés also expands on his earlier chapter, describing actual examples of sustainable management of eco-parks in several different regions in Europe and elsewhere. Finally, Whitbread-Abrutat's description of the challenges and means of re-development of degraded areas (many regions have these) builds on the earlier chapter by Beltramo by showing how a consideration of the land's life cycle (and that of its occupiers) is important to achieving a satisfactory and sustainable outcome to such redevelopment.

Part V looks more closely at a number of selected *economic sectors* to see how the integrated use of life cycle approaches can help identify the most effective potential actions and make choices from among multiple sustainability options. Regional administrators and other stakeholders need this understanding if they are to adequately plan for and act within these sectors. Although individual companies and activities such as mineral extraction may have useful immediate economic outcomes, regions also need to be concerned with the longer-term consequences. And, depending on the products in question, their responsibilities for environmental impacts at different intervention points in the life cycle of a product need to be clarified. As global market forces are increasingly influenced by sustainability criteria, it is important for regions to evaluate the longer-term future of certain industrial operations concerning, for example, fossil fuel extraction, use of chemicals, and polluting processing operations. Regions need to remain abreast of such evolutions and assess for themselves the resource efficiency and environmental life cycle dimension of their socio-economic development choices. Accordingly, this section looks at nine examples where life cycle approaches have been used in relation to the industrial development in a region.

Notarnicola, Renzulli, and Tassielli look at agri-food chains from an LCA perspective to identify vulnerable components where regional intervention may be required. Bezama et al. describe attempts in Germany to develop a bioeconomy for the local wood sector, focusing especially on an analysis methodology for the socio-economic issues surrounding this sector. A further case study of a collaborative wood-based industry is described in the chapter by Thyssen and Zeller. They study the requirements for further consolidation along the life chain of wood to achieve improved economic conditions for this sector in Belgium. Hamilton and Parsons explain the manner in which long-term sustainability issues are now addressed in leading mining companies, with a gradual evolution away from an environmentally focused 'resource efficiency' approach to one of triple bottom line 'shared stewardship' with local and far-away stakeholders. This evolution has been triggered by the recognition that the range and complexity of sustainability issues in the mining sector make it difficult to succeed with a linear, top–down approach.

Radka et al. give a detailed description of the long-term life cycle issues inherent in the move to renewable sources of energy. They demonstrate how LCA gives valuable insights into issues that have often been overlooked in the past.

Lombardi describes the life cycle issues for packaging suppliers, showing how supply-chain and consumer linkages at regional level can make the entire packaging sector more viable. An example of how LCA can assist in better local planning for the building sector is given by Bhochhibhoya and Cavalli, who examine the relative performance of traditional and modern construction concepts in remote areas of Nepal. Panwar gives insights from India into the challenges of taking a life cycle perspective for waste management, building on the earlier chapters that also dealt with this topic. The last chapter returns to the issue of life cycle land management, illustrating along the way that grass-roots initiatives can also make effective use of life cycle thinking. Bradby describes a community-based regional initiative in Australia to combine conservation (biodiversity) with economic development (tourism) in a way that allows both to move ahead.

Chapter 45, on *Synthesis*, draws on common ideas from the various authors, distilling these into key points for moving forward. Regional agendas are typically focused on promoting development, expanding infrastructure, providing services, and social issues – each with its specific life cycles and timetables, each susceptible to different sustainability tools or ideas. Taken as a group, the chapters contain some useful principles and also outline similar challenges. The synthesis comments on the maturity of current life cycle tools in a regional context – clearly, there is still some evolution needed in some of them; others are ready to use in their current state. Some tools were originally developed for other purposes and have lately started to become more generalised. The human factor – the willingness of life cycle practitioners to reach out to new clients, and the ability of regional administrators to understand and absorb new ways to managing their sustainability agenda – is also important.

When one is applying life cycle concepts and procedures, it is important to retain a broad perspective. Scoping of the issues, objectives, and procedures is necessary to arrive at a practical and realistic outcome. A clear vision of the sustainability agenda itself is required: it is more than just environment, more than just managing a carbon budget, but it is also not 'everything'. A clear notion of what constitutes a workable 'life cycle' is needed. Although intellectually satisfying, a treatment of the entire cradle-to-cradle life of every material and every project quickly becomes unworkable.

The increasing use of life cycle approaches is most successful when it builds on existing experience. The use of case studies and evaluations can accelerate the introduction of successful approaches: there is no need for every region to experiment with the basics. Nevertheless, LCM is undergoing its own evolution, and the various tools intended for organisations (as for products) are evolving quickly. It is essential to find the right blend between lessons learned and leading-edge advances.

We stress to our readers that, in a vast discipline such as LCM, we have been able to present only a selection of key issues and examples, and, indeed, only in the form of short outlines. Readers are therefore encouraged to consult the literature for further details and additional examples, and to keep abreast of the numerous updates now pouring out of the research studies. This subject is both work in progress and a strong recommendation to 'do it now'.

2

SUSTAINABLE REGIONAL DEVELOPMENT IN A LIFE CYCLE CONTEXT

Fritz Balkau, Stefania Massari, and Guido Sonnemann

Introduction

Our societies are seeing a convergence of three important trends. First, local governments and regional administrations are being given increased responsibility for management of their development agenda, including fiscal and other instruments for implementation of policies. Second, heightened community awareness is leading administrators and politicians to reflect the sustainability agenda in their priorities and decisions. Third, organisational management methods are evolving to be able to take a wider, more holistic view of development, assisted by a range of life cycle assessment (LCA) tools in the context of the broader concepts of circular economy, industrial ecology, and resource stewardship.

For the moment, the various actors and stakeholders involved in the above are still tending to operate independently, even though the importance of 'co-ordination' is regularly mentioned in promotional information. This compartmentalisation is leading to many projects and initiatives that are narrowly focused and that do not always take into account a life cycle approach. This means going beyond the traditional focus on only the production facilities of companies. A product life cycle begins with the extraction of raw materials from natural resources in the ground and energy generation. Project life cycles begin at the conception and planning stages and extend to end of life and decommissioning. The focus is gradually changing as the various partners come to understand how the potential of modern management tools to address the full life cycle of their activities can connect with the different parts of their sustainability agendas. Nonetheless, there is still a general lack of appreciation of 'how to do it'.

Here, we briefly review these three elements in order to set the scene for the more detailed and specific chapters that follow.

The sustainable development dynamic

The 1987 Brundtland report, Our Common Future (UNCED 1987a), firmly established the basic components of the sustainable development agenda. The incorporation of three dimensions – social, environmental, and economic – into sustainable development is now clearly reflected in public policy in nearly all countries. It is also generally accepted that these three dimensions should be addressed concurrently, not in sequence, as was the practice in earlier times.

Less widely understood is the long-term time dimension of the outcomes implied in the word 'sustainable'. The term implies a higher degree of permanence and longer-term objectives than many past policies and programmes have aimed for. Much has certainly been achieved since the United Nations Conference on Environment and Development (UNCED) in sustainable development globally through major international programmes, some generously funded by both public- and private-sector contributions. But important questions remain about the efficacy of many initiatives, and we are still far from a consensus on objectives and methods.

How do we measure sustainability? Despite many attempts to compile a commonly accepted sustainability 'index', there is still no agreed standard metric for measuring progress towards sustainable development and its outcomes; only a few of the elements can be measured currently, and even fewer can be properly quantified. Some common metrics, such as gross domestic product (GDP), are universally agreed to be deficient in incorporating the notion of social welfare, but, as money flows are still the easiest to quantify, they continue to dominate our thinking. At the individual project level, it is difficult to determine on which criteria, and to what extent, many projects are 'sustainable'. The common practice of highlighting good results on a particular single sustainability factor often hides poor performance on others.

Beyond the confusion over terminology, we can see other conflicts in sustainable development policy. Perhaps the most important arise from a different perception of timescale. Conflicts of timescale can have several origins. Thus, the 'poverty trap' of disadvantaged communities means they are intrinsically unable to act to avoid deleterious long-term consequences. For the better off, the current high cost of borrowing money means that infrastructure and industrial project outcomes tend to focus on short-term returns. Short-termism also works against intergenerational equity. We have pushed off far too many end-of-life and remediation actions on to future generations without giving them a dedicated budget. Lastly, the renewability of our resource base, as well as more careful husbandry of non-renewables, is only possible if there is a long-term view of development.

In addition to the uncertainty of timescale and criteria, issue conflicts are also caused by geographical distribution. The delocalisation of production capacities to sites far from the point of consumption means that 'sustainable' at the local level does not necessarily imply sustainable at national or global levels. One example is the clean electric car, whose electricity supply (with its accompanying air pollution) is produced far away and so does not enter into local calculations. Blindness to geography has led major car manufacturers to produce vehicles that are promoted as having 'zero emissions'.

In any case, many stakeholders do not have the same objectives or decision-making frameworks. Although, for many actors, 'sustainability' is about ensuring environmental quality outcomes, for others, economic outcomes need to be maximised first in order to produce enough wealth to rehabilitate the environment degraded by, for example, extractive or industrial activity. Perhaps the most complex aspect of the sustainable development agenda has been how to achieve optimisation of performance in several dimensions simultaneously.

These internal contradictions in the development process have been appreciated for some time. Various concepts for a new model of development have been advocated over the years, often being displaced by other ideas as public consciousness evolves. These include societal models for a green economy, resource efficiency, and sustainable consumption and production, further explained in Chapter 5, which themselves are based on concepts such as Factor 10 (Schmidt-Bleek 2008), the Natural Step (TNS 2013), or Natural Capitalism (Hawken et al. 2010). On the social side, issues of stakeholder engagement, decent work, and human rights have become more visible, leading companies to increasingly embrace some form of social responsibility (*ISO 26000*; ISO 2010). In the economic sphere, approaches such as national resource accounting and efforts to quantify 'gross national happiness' have come to nuance our long-standing preoccupation with GDP as the only indicator of national performance. All these concepts and trends can (and should) have a strong influence on attempts to make regional development more sustainable. More fundamentally, most of them embody a life cycle perspective of the development process that has to be more effectively assessed via techniques such as LCA (see Chapter 10) and more adequately implemented by improved holistic management frameworks.

Holistic life cycle approaches in a regional development context

If social impacts and other longer-term consequences of development projects are ignored, the economic advantages often accrue only to a limited group of stakeholders of the current generation, leaving the wider, longer-term impacts to be resolved (and often funded) by others. In order to shed more light on the hidden or delayed impacts (and, hence, to avoid burden-shifting in the management response), a number of assessment instruments have been devised in recent years to provide more precise predictions of various non-economic consequences of development decision-making. Most major projects are now subject to environmental impact assessment (EIA) processes that predict likely environmental consequences and encourage modifications and improvements to be made. Although EIA often takes a wide view of possible impacts, it is not specifically structured around a holistic life cycle perspective, nor does it necessarily always incorporate social issues or specialised technical fields such as hazard evaluations. Other assessment tools such as social impact assessment, chemicals assessment, and industrial risk assessment, among others, have come to complement EIA in its mission. There is also now greater emphasis on tracking the flows of materials (most of which eventually become wastes in one way or another) through society and on estimating the 'footprint' of a product or a region as a further indirect way of assessing likely sustainability-related impacts. Life cycle costing can reveal the financial implications of products and services over the entire life cycle. Life cycle costing is discussed in Chapter 33, and descriptions of other related assessment tools are found in Parts II–IV of this book.

The assessment tools generate useful data for identifying problem areas and defining priorities, but some way still has to be found to translate the impact data into action. In particular, it is important to use a holistic framework of development decision-making that is not narrowly focused and considers upstream and downstream impacts in a systematic way. *Life cycle management* (LCM) has been developed as a framework to achieve this.

LCM is based on a variety of subsidiary concepts to achieve its goals. Among the most important are the concepts of a 'supply chain' for products and 'process life cycle' for projects. Sustainable supply chain management and green purchasing base decisions on what is happening upstream, whereas the downstream component is addressed through ecodesign, extended producer responsibility, product service systems and related tools (Balkau and Sonnemann 2011). Other tools – such as eco-labelling and environmental product declarations – are aimed at communication about more sustainable products and processes to enable better-informed consumer choice. Concepts such as circular economy and industrial ecology, which focus on the way resources flow through our societies, can guide both public- and private-sector policy-making. Each of the above approaches has a particular role to play and they are explained in Parts III and IV, with circular economy principles discussed in Chapter 26. A challenge for all of them is the number of sustainability aspects they can take into account simultaneously.

Despite the apparent complexity, it is a mistake to see all this as the domain of specialists. A life cycle approach is most effective when implemented as a mainstream management responsibility. Although authoritative *command and control* models are sometimes applied, LCM is most effective with co-ordination among the various actors and interest groups along the life chain, in line with a commonly agreed set of objectives. High-level political leadership for sustainability in development is usually required to ensure the necessary level of programme co-ordination and implementation.

Comments on the sustainable regional development process

Like the definition of 'region' itself, regional development is a broad concept – further elaborated in Chapter 3 – covering many aspects, subject to a range of aspirations, and faced with numerous obstacles. Some regions and municipalities have formally adopted sustainability objectives and programmes, perhaps as Local Agenda 21 (UNCED 1987b) or under other names. In some cases, these have been incorporated into legislation, such as, for example, Nova Scotia's Sustainable Prosperity Act (Nova Scotia 2012). Regrettably, many administrations lack the practical tools to effectively implement such objectives, in part owing to the inherent confusion and conflicts within the sustainable development agenda itself. A number of international bodies such as Local Governments for Sustainability (ICLEI) and the International Institute for Sustainable Development have established programmes to guide and assist them in this process. Numerous associations of regions or cities have been formed to allow sharing of experience, pooling of resources, and development of pilot programmes. Some of these initiatives are described as case studies elsewhere in this book; others can be located directly on the home pages of the relevant associations.

Regional development initiatives take many forms, both directly on the ground as well as in providing enabling frameworks. These include building projects, infrastructure development, expansion of different economic sectors, environment protection and direct intervention as, for example, in establishing industrial parks. In parallel with the physical planning process, normal ongoing community services, practices and programmes also contribute – or not – to the development process. These include maintenance and rehabilitation of community infrastructure, urban and rural waste services, transport networks and social services.

For many of the regional actors, and even within individual organisations, the timescales of their sustainability initiatives may differ significantly, from the short-term outcomes expected in industrial project development to the long-term objectives of research efforts, resource conservation and infrastructure services.

A region is no longer an isolated bubble in which development decisions are taken without reference to outside dynamics. Nowadays, regional objectives interface, not only with wider national policies, objectives, and practices, but also with global ones. Actually, each region is offering its products and services, often produced with local resources, to the world market. The regional development options inevitably have to reflect the realities of global trade patterns and resource flows, as well as of international finance mechanisms and corporate practices. This dependency will influence the choice of instruments and tools used to orientate the regional development process. Many of these tools are becoming standardised at international level. Those developed by the International Organization for Standardization are also in line with the rules of the World Trade Organization.

As all of the above may appear somewhat abstract, we can illustrate its relevance by listing some development functions where life cycle approaches applied by the relevant actors would assist regions in more easily meeting their sustainability objectives. These were all taken from actual examples around the world, and many are discussed in various places throughout the book, in particular in Parts IV and V:

- sustainable purchasing and procurement by regional authorities and institutions;
- certification of identified local products for export as 'sustainable';
- use of 'green funds' for financial placements and investments;
- pollution prevention and resource efficiency initiatives, regulations, and standards;
- resource recovery and recycling initiatives; circular economy;
- conception of more sustainable products in different sectors;
- sustainable building criteria and construction practices;
- natural resource conservation objectives and goals;
- sustainable agricultural practices and products;
- mobility solutions, including transport systems;
- sustainable tourism and tourism infrastructure;
- energy policy based on decentralised, renewable sources.

Many more areas of action can be identified at a local level, each with its own life cycle, its own relationship with sustainability criteria, and its own specific LCM toolbox.

Conclusions

A life cycle approach to regional development brings together three main considerations: regional development dynamics, sustainable development concepts and LCM tools. All three are evolving independently and bringing them together will be an ongoing, everyday challenge. Their successful convergence will require an exploration of the functional linkages, but also a further enhancement of some of the individual techniques that make up the holistic, LCM framework. Given the challenges our societies face in confronting an uncertain future, it is a journey well worth attempting.

References

Balkau, F., and Sonnemann, G. (2011) 'Addressing sustainability issues through enhanced supply-chain management', in *Supply Chain Management – New Perspectives*, ed. S. Renko, InTech, pp. 379–400. DOI: 10.5772/24579. Available from: http://www.intechopen.com/books/supply-chain-management-new-perspectives/addressing-sustainability-issues-through-enhanced-supply-chain-management

Hawken, P., Lovins, A.B., and Lovins, L.H. (2010) *Natural Capitalism: The Next Industrial Revolution*, 2nd edn, London: Earthscan. Available at: www.naturalcapitalism.info/ (accessed 22 July 2016).

ISO (2010) *ISO 26000. Guidance on Social Responsibility*, Geneva: ISO.

Nova Scotia (2012) Environmental Goals and Sustainable Prosperity Act, Halifax, Canada: Province of Nova Scotia. Available at: www.novascotia.ca/nse/egspa/ (accessed 22 July 2016).

Schmidt-Bleek, F. (2008) 'Factor 10: The future of stuff', *Sustainability: Science, Practice, & Policy*, 4 (1): 1–4.

TNS (2013) References relating to the Framework for Strategic Sustainable Development (FSSD): 'Backcasting from Sustainability Principles' or 'The Natural Step Framework', Stockholm: The Natural Step. Available at: www.thenaturalstep.org (accessed 22 July 2016).

UNCED (1987a) Our Common Future – Brundtland Report, New York: United Nations Conference on Environment and Development. Available at: www.un-documents.net/our-common-future.pdf (accessed 22 July 2016).

UNCED (1987b) Agenda 21, New York: United Nations Conference on Environment and Development. Available at: http://sustainabledevelopment.un.org/content/documents/Agenda21.pdf (accessed 22 July 2016).

3

REGIONAL DEVELOPMENT AND FRAMEWORKS FOR SUSTAINABLE DEVELOPMENT

Donatella Porrini and Fabrizio Striani

Introduction

As political and economic agendas become more complex, we see an increasing trend to regionalisation, with decision-making powers devolving to lower levels of government. This trend is not always accompanied by a transfer of managerial and budgetary capacities as well, with the result that many regions face serious challenges in implementing a coherent sustainability agenda and developing an economic development process that takes account of future social and environmental impacts.

The place of regions in a sustainable world

The Oxford dictionary defines a region as 'a tract of land, space or place having more or less definitively marked boundaries or characteristics'. The boundaries may be geographical, administrative, or political. The term 'region' can be used loosely to define a state, a province, or a cluster of provinces. At the international level, the UN has divided the world into six geographical regions; elsewhere, within defined national borders, a region may be quite small. In this book, we do not formally define 'region', intending only to refer to administered territories larger than a municipality and smaller than a country, even if many of the same principles and challenges apply to all of them. In this book, we assume in a general way that regions have a certain level of political and economic independence concerning the direction of their futures and that they can 'manage' their sustainability agenda.

Regions think about their future in different ways. 'Sustainable development' means different things to different persons, but the most frequently quoted definition is undoubtedly that of the Brundtland Commission as 'development that meets the needs of the present without compromising the ability of future generations to meet their own needs' (WCED 1987). The first global action plan for sustainable development was Agenda 21, prepared at the Rio Conference in 1992 as a programme to be implemented by governments, international agencies, local and regional administrators, organisations in the community and the general public (UN n.d.). Local government has a major role in providing information and education, mobilizing the general public, and drawing up its own Local Agenda 21 in close consultation with its citizens. Regional government is commonly associated with a 'local' Agenda 21. These initiatives have multiplied in recent years, becoming one of the main means of achieving the goals set in 1992 as part of a more general process of multilevel integration of sustainability policies pursued at regional, national and local levels. An example of a regional Agenda 21 programme is that of the Île de France (Île de France 2015).

Agenda 21 speaks prominently about holistic vision, a life cycle context, a prevention approach to decision-making, joining production and consumption into a single policy, resource stewardship and social justice, among other things. Local Agenda 21 provides a coherent framework for action, but has been less explicit than other approaches in prescribing how local actors should proceed. Consequently, although regions and local governments have been aware of the importance of sustainable development issues, they have spent much time searching for appropriate procedures to help them implement their sustainability policies.

Major international organisations continue to study how to assist the move forward on sustainability and, in particular, to support the position of regions. The report Regions Matter (OECD 2009) stresses the individual characteristics of regions and their 'place specificity', advocating a consideration that extends beyond 'one-size-fits-all' development approaches to instead build up region-specific solutions. The principle is that no region or country can tackle the sustainable development challenges alone. Integrated, holistic solutions must be found at local, national, regional, and global levels. Beyond simply adopting objectives, a local framework for sustainable development is needed to mobilize all stakeholders, explain the challenges, focus operational action at the right scale, and form a basis for a true partnership.

In this respect, the environmental action plans of the European Union already assign a central role to the consultation of, participation in, and sharing of, responsibilities between different levels of government. The preparation of sustainability strategies at all levels is encouraged, in collaboration and partnership with local authorities and all parties involved.

An important reference to the life cycle approach was included in the 7th Environment Action Programme that will guide European environment policy until 2020 (EU n.d.). Article 35 of the EAP establishes the need for a framework that sends appropriate signals to producers and consumers to promote resource efficiency and the circular economy. Specifically, Article 35 states:

> Measures will be taken to further improve the environmental performance of goods and services on the Union market over their whole life cycle, including measures to increase the supply of environmentally sustainable products and stimulate a significant shift in consumer demand for such products. This will be achieved using a balanced mix of incentives for consumers and businesses (including SMEs), market-based instruments and regulations to reduce the environmental impacts of their operations and products.

It is in this broader context that life cycle approaches to sustainability management find their importance. The impacts of products and materials occur along the entire value chain, often in distant places, invoking the principle of shared responsibility in the action needed to minimise them in the best way. Comparable notions of life cycle apply to natural resource exploitation, infrastructure development and social services. Regions are an intrinsic and inseparable part of such life cycles. Their use of holistic methodologies is thus an indispensible part of a wider national and global arrangement for sustainability management.

Regions for sustainable development

Encouraged by the movement towards 'sustainable cities', regional administrations everywhere are now increasing their involvement in, and action on, sustainable development. Individual regions are defining and adopting sustainability objectives, selecting priority action areas and monitoring results. For many of them, sustainability is not seen as a barrier; rather, it is their opportunity to define a new future for their citizens. As traditional industries go into decline, they are looking for renewal and innovation, as, for example, in the 'Third Industrial Revolution' of the French region of Hauts de France (see Chapter 27). The now-abandoned, all-too-visible industrial liabilities of previous industrial expansions have encouraged them to select more holistic models of development for the future, based on concepts of circular economy, industrial ecology, and bioeconomy, as described elsewhere in this book. Innovation, eco-design, life cycle partnerships, and resource-efficient processes are at the heart of many of these initiatives.

Few of these initiatives can be taken in isolation, and the advantages of collaboration are becoming clear to the regions. There are now numerous bodies and associations that share experience, pool resources and encourage capacity building on common issues, including sustainability. Thus, the Association des Régions de France (ARF 2015) has published a book on the experience of the national regions with initiatives to address climate change. At the pan-European level, the Assembly of European Regions (AER n.d.) has a diverse programme of information and exchange of experience dealing with priority issues under the 7th European Environmental Action Plan. In the Mediterranean region, local authorities, associations, and businesses are already adapting practices and innovating on sustainable development issues. They see these solutions as both preserving the climate on our planet and acting as boosters for economic development based on sustainable development (MEDCOP21 2015).

At the global level, the Forum of Global Associations of Regions (FOGAR n.d.) has the objective of supporting governments of a territory to promote sustainable development, with full respect for the environment. Associated with this is the objective of also promoting regional competitiveness and growth factors for social and territorial cohesion. Elsewhere, the Network of Regional Governments for Sustainable Development includes in its objectives the promotion of sustainable development at the level of subnational governments around the Globe (NRG4SD n.d.).

A number of independent capacity-building programmes have also evolved. The R20 initiative has the mission to help subnational governments around the world develop and communicate low-carbon, climate-resilient economic development projects. It is based on the observation that many clean energy projects are not currently happening, even though intergovernmental bodies, national governments, subnational governments, NGOs, academics, technology providers, and public and private financial partners share the same vision and want concrete action. This is owing to a number of factors:

- National and subnational politicians lack awareness of the long-term political, technical and financial solutions for a sustainable, safe, clean economy.
- Project developers do not know how to identify, design, and communicate the right information to project investors.

- Project investors are not sure how low-carbon investments fit into their existing investment instruments and portfolios.

R20 addresses these factors to help build an effective flow of '*green deals*' at subnational level (R20 n.d.). Although many of the above programmes are focused on sustainable energy, a similar set of challenges surrounds other sustainability issues and assistance programmes on these for regions would also be useful. To this end, the United Nations Environment Programme (UNEP) launched a partnership programme on cleaner production financing (see Chapter 4) to help address resource efficiency issues (UNEP 1999). This, like R20, identified regional lack of awareness on crucial issues as a major barrier to sustainable development.

These networks should also be seen in relation to the extensive networks of cities' associations and mechanisms working towards sustainability. The organisation Local Governments for Sustainability (ICLEI n.d.) has taken a part in, and sometimes leadership of, many initiatives aimed at helping local governments adopt and implement sustainable practices, whether under Agenda 21 or otherwise. Some of these networks, such as the Declaration of Mayors and Subnational Leaders on Climate Change (www.iclei.org/worldmayors summit.html) and the association of mayors (C40 n.d.), are focused on a specific priority issue (climate change, in this instance). Others, such as the European Green Capital Award, incorporate a broader range of sustainability issues in a composite programme. All of these initiatives share similar sustainability concerns and use similar programmes to address them.

Although programme concepts are important, it still remains to clarify how and where such approaches should be applied in the normal functioning of a region. It would be a mistake to see sustainable development and the methodologies that give it effect as an extra programme to be pursued when conditions allow. There is general consensus that the most effective approach is to integrate – '*mainstreaming*', to use some management jargon – sustainability into all day-to-day activities. The list below, which is not exhaustive, gives some examples of the regional functions where sustainability criteria and life cycle considerations should be integrated:

- region as a provider of infrastructure and services;
- region as a promoter of economic development and commerce;
- region as a regulator and enforcer of health and environmental standards;
- region as a protector of social values;
- region as a policymaker and planning body;
- region as an institution that consumes resources in its purchases and operations;
- region that spends, invests, receives taxes, and pays costs;
- region as a polluter in its own right (transport, waste, operations);
- region as an exploiter of natural resources (water, land, forests, minerals);
- region as an exporter of its wastes and other liabilities beyond its territorial frontiers;
- region as an investor in future development;
- region as a sponsor of intellectual advancement;
- region as a promoter of initiatives and ideas;
- region as a provider of subsidies.

Some of these issues are covered in various individual chapters in the book.

Conclusions

Being placed as they are at the interface between national and local government, regions share many of the preoccupations and challenges of both. The complexity of the national and global mechanisms to move forward on sustainability objectives, together with unfamiliarity with the methods to address these in a coherent manner, means that regions merit special attention in receiving guidance and capacity building. Agenda 21 provides an important framework for them to structure their sustainability programmes. Life cycle approaches are most effective in their implementation.

Regions are also an important partner in the promotion and implementation of sustainable development in a wider arena. Many sustainability issues play out at global level, requiring a development vision that extends beyond territorial boundaries and the employment of collaborative action methods. Although we have seen some important achievements, barriers to the broader implementation of life cycle thinking still point to a need for greater awareness of the benefits of such approaches.

At a regional level, it is now possible to construct a sustainable development process that fully considers the upstream and downstream trade-offs of goods and services. Such a process incorporates a 'cradle-to-grave' life cycle of materials, from the extraction of raw materials to the disposal of remaining wastes, without shifting the burdens of impacts such as climate change and carcinogenic effects to other sectors and regions and to future generations, as explained in Chapters 5 and 6.

The increasing number of regions adopting such programmes, assisted by a variety of associations and institutions, suggests that progress towards regional sustainability is now finally accelerating. It is important that sustainability programmes be integrated into the ongoing functions of the region, rather than seen as a separate, independent strand of activity.

References and further reading

AER (n.d.) Available at: http://aer.eu (accessed 22 July 2016).

ARF (2015) French Regions Positive Achievements and Prospects: Committing, Cooperating and Succeeding Together against Climate Change. Available at: www.arf.asso.fr/wp-content/uploads/2015/11/Cahier-des-Re%CC%81gions-de-France-en-anglais-BD.pdf (accessed 22 July 2016).

C40 (n.d.) Available at: www.c40.org/about (accessed 22 July 2016).

EU (n.d.) Environment Action Programme. Available at: http://ec.europa.eu/environment/action-programme/ (accessed 22 July 2016).

FOGAR (n.d.) Available at: www.regionsunies-fogar.org/en/about-oru/objectives-and-mission (accessed 22 July 2016).

ICLEI (n.d.) Available at: www.iclei.org (accessed 22 July 2016).

Île de France (2015) Agenda 21. Available at: www.iledefrance.fr/sites/default/files/brochure_a21_2014.pdf (accessed 22 July 2016).

MEDCOP21 (2015) Available at: www.medcop21.com/en (accessed 22 July 2016).

NRG4SD (n.d.) Available at: www.nrg4sd.org (accessed 22 July 2016).

OECD (2009) Regions Matter: Economic Recovery, Innovation and Sustainable Growth. Available at: www.oecd.org/gov/regional-policy/regionsmattereconomicrecoveryinnovationandsustainablegrowth.htm (accessed 22 July 2016).

R20 (n.d.) Available at: www.regions20.org (accessed 22 July 2016).

UN (n.d.) Agenda 21. Available at: www.unep.org/documents.multilingual/default.asp?documentid=52 (accessed 22 July 2016).

UNEP (1999) Financing Cleaner Production. Available at: www.financingcp.org (accessed 22 July 2016).

WCED (1987) Our Common Future. World Commission on Environment and Development, Oxford, UK: Oxford University Press.

World Bank (2009) Reshaping Economic Geography. Available at: https://openknowledge.worldbank.org/handle/10986/5991 (accessed 22 July 2016).

4

FINANCIAL INSTRUMENTS AND PROCEDURES FOR SUSTAINABLE REGIONAL DEVELOPMENT INVESTMENTS

William Rivenbark and Fritz Balkau

Introduction

The economic component of the sustainable development agenda has seen much change recently. Regions, cities and localities – still always seeking to capitalise on their natural resources, human skills and businesses' potential – are now increasingly favouring cleaner industries, renewable resources and low-impact activities, the so-called 'cleaner production' investments. The parameters for the financing of such longer-term, *'sustainable'* developments are not necessarily the same as those for conventional industries with their requirement for short-term profitability. Special investment mechanisms are in place for certain sustainability issues, to assist in the implementation of international environmental agreements and development of renewable energy sources.

This move has been assisted by growth in responsible financing processes in institutions and professions. There is also a change in the way natural resources are valued through new techniques such as natural resource accounting. More and more regional bodies are applying sustainability criteria to their own investment portfolios, with many public institutions retreating from investment in polluting, unsafe, or fossil fuel-based industries.

Many of the sustainable financing procedures and mechanisms are relatively recent. Taken together in the framework of longer-term, more life cycle-oriented development objectives for regions, their application requires new insights and skills that are not always found in textbooks.

Sustainable financial management

Although responsible budgeting by regional administrations remains a core condition for sustainability, it is increasingly taking a longer-term view of costs and incomes and investments and returns than in the past. The financial paradigm for sustainable regional development is evolving in two ways. The first involves a shift away from the traditional redistribution of wealth in the form of subsidies for economically deprived regions to an organic approach where all regions increase their competitiveness by capitalising on their inherent assets (OECD 2016). Various studies have shown that subsidies actually do little to create a stable climate of investment, employment and social improvements, leading instead to inefficient use of natural resources. The second concerns methods for calculating the return on investments from regional development initiatives. The tendency is to move from the myopic view that primarily concerns wealth generation to a more holistic view of economic prosperity, environmental health, and social vitality through tools such as life cycle costing and sustainable financing mechanisms. The Province of Nova Scotia in Canada refers to this as '*sustainable prosperity*'.

Common underlying factors in this evolution are the expanding responsibilities of local and regional government. Local governments face challenges in adding value to regional development initiatives if they are not financially sound within their own communities. To ensure that core services such as public security, transportation, and solid waste collection are sustainable in the broader context of long-term development, local officials must account for the financial inflows, outflows, and stocks of the organisation, but they must also understand the area of financial condition analysis to reduce financial risk. Although debt is an important mechanism by which local governments can provide regional growth initiatives, increasing the amount of annual debt service affects local government's ability to provide core services. A number of alternative financing mechanisms have been devised to overcome such limitations of accumulated debt, including some that are directly aimed at promoting sustainability outcomes. In a long-term sustainability context, with all financial instruments – traditional as well as novel – the life cycle of what is being proposed needs to be closely looked at. For example, in the past, decommissioning and end-of-life aspects of infrastructure were often omitted from the equation, with unknown costs simply passed to future generations.

Life cycle issues and financial sustainability

A life cycle view can be invaluable in improving financial efficiency. For example, public procurement was often based on the cheapest upfront purchase cost, because that was the way the accounting system was structured. It is increasingly being realised, however, that procurement based on value for money over the entire asset life is financially more efficient, as for example in the purchase of LED lighting systems rather than incandescent globes (see Chapters 19 and 33). Given that public-sector purchasing can represent up to 40 per cent of the national total, sustainability criteria in procurement are not a minor issue. In France, the central purchasing hub for local government (UGAP 2016) has adopted environmental criteria for its buying for the regions.

An important development in recent years has been the use of life cycle costing, an extension to traditional LCA that considers the full costs across the entire life cycle of a material or product and that could eventually be extended to projects. An illustration of this is given in Box 4.1. Chapter 33 gives further details on this subject.

BOX 4.1 LIFE CYCLE COSTING OF ENERGY INVESTMENTS AND CLIMATE DEBATE

Capital costs are often used by promoters of fossil fuels as evidence that coal and gas are, and will remain, cheaper than renewable energy sources such as wind and solar. But this focuses on the short-term aspects only – an approach used by opponents of climate action and clean energy who focus mainly on the upfront costs of policies.

Citigroup has published a detailed analysis of the costs of various energy sources and it concludes that, if all the costs of generation are included (known as the '*levellised*' cost of energy), then renewables turn out to be cheaper than fossil fuels and represent a benefit rather than a cost to society (RenewEconomy 2015).

Financing and sustainability

Traditional financing mechanisms have been much concerned with rates of return, without necessarily taking account of wider social or environmental issues ('externalities' in economic jargon) or assessing the diminished value of a community's natural resource assets that follows their exploitation. Similarly, regular budgeting procedures in common use frequently overlook environmental or social costs (or benefits) in organisations' accounts. Within business entities, substantial costs of environmental compliance are often invisible to senior managers, being buried deeply in maintenance budgets or similar. Likewise, purchase of clean technology is often viewed as an environmental cost rather than as an investment in productivity.

Numerous procedures have been advocated to better take environmental and social costs into account in both business and government (see www.financingcp.org). Many mainstream industries have now understood that 'pollution prevention pays', and that cleaner technologies are also more cost-efficient to operate. This understanding has been slower to infiltrate political and administrative circles, preoccupied as they often are with short-term outcomes.

This understanding of the value of a longer-term view is not yet universal. Some financing mechanisms are resistant to change. Of the many structures designed to deliver development outcomes in a framework of reduced public expenditures, public–private partnerships (PPIs) have become a popular concept in recent years. The leveraging of private-sector funds and expertise has many attractions, but also some risks. One of these is that it often leads to neglect of longer-term, life-chain consequences by business partners who have shorter objectives, or who are not domiciled in the region. Box 4.2 shows an extract from a recent discussion paper in India.

BOX 4.2 PPPS AND LONG-TERM VALUE

[There is still debate] . . . about the extent to which PPPs are delivering value for money in India. While traditional value for money, henceforth VfM, is interpreted to mean the lowest price alternative at the time of commissioning, in the context of sustainable development it is understood as value for money across the asset life cycle. This approach embodies the principles of 'total cost of ownership' and 'whole-life value' – accounting for the costs of planning, designing, building, operating and maintaining an asset. It can therefore be used to account for the medium and longer-term efficiency gains and cost reductions enabled by sustainable infrastructure.

Source: Taking Stock of PPP and PFI around the World (2012) London: Certified Accountants Educational Trust. Available at: www.accaglobal.com/content/dam/acca/global/PDF-technical/public-sector/rr-126-001.pdf (accessed 25 July 2016).

Greening of financial institutions

While attention is being paid to greater resource efficiency at the end-point of investments, financial institutions have themselves become more aware of the need to integrate sustainability criteria into their lending policy. The Equator Principles (www.equator-principles.com/) give firm support to the incorporation of sustainability criteria into infrastructure investments and have been endorsed by the World Bank and by the International Finance Corporation. Independent rating agencies now assess lending bodies for their practice (or not) of incorporating sustainability in regular operations, leading to some competition in adopting such criteria. Financial institutions are regularly encouraged to include sustainability criteria in their public reporting. The French government considered introduction of mandatory reporting of investors' carbon footprints in 2015. The Global Reporting Initiative (GRI 2016) has published a sector supplement to harmonise reporting practices for financial institutions. Major institutions have joined UNEP's Financing Initiative (UNEP FI 2016) in order to adopt a common approach to such integration. Within the private sector, some stock exchanges have adopted listing criteria concerning companies with unsustainable or socially undesirable activities and have launched the Sustainable Stock Exchange Initiative (www.sseinitiative.org/). Another recent initiative has created a network of mainstream investors and financing bodies to elaborate and adopt the United Nations Principles for Responsible Investment (www.unpri.org/). It is notable that a large number of national and regional pension funds have declared their support for, and are members of, this initiative.

With the financing suppliers increasingly ready to follow sustainability criteria, the challenge is now for regions to insist that such financing bodies be preferentially involved in infrastructure development in their region.

There is a final aspect of sustainable financing that is relevant to regions – the choices within their own investment portfolio (and, for example, pension funds) to favour green investments and to avoid exposure to portfolios based around unsustainable resource use (e.g. fossil fuels), environmentally damaging projects, or unhealthy products, to give just some examples.

Financing sustainability

To address the mismatch between short-term financing targets and long-term sustainability outcomes, a number of mechanisms have been devised to enhance the direct financing of, for example, renewable energy, clean technologies, greater resource efficiency, fair trade and decent work.

A major development in recent years in municipal financing for environmentally focused projects concerns green bonds. For example, the French region of Île de France (which incorporates Paris; Île de France 2016) has issued more than €1 billion in bonds for green energy projects, offering attractive rates of return. There is also innovation in tax-based measures such as 'property assessed clean energy financing', green financing via service bills and land taxes, as well as a range of indirect mechanisms such as land-value capture, end-user financing, energy efficiency performance contracting and solar leases, among others. Although the initial use of these instruments was at city level, many are also effective at regional scale. Figure 4.1 lists a number of innovative financing instruments for green infrastructure that regions can use.

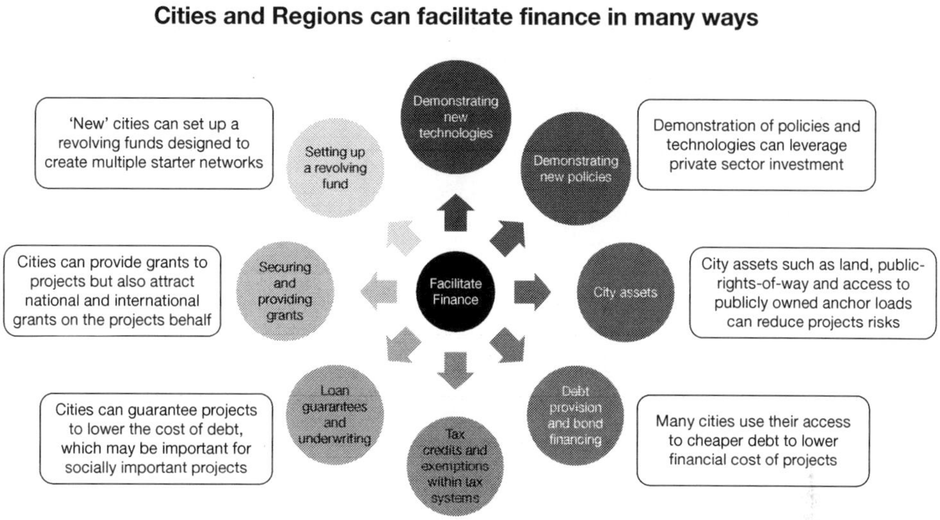

Figure 4.1
Green financing mechanisms for regions

Source: Reprinted with permission from the United Nations Environment Programme

Although most of the mechanisms are based on recovery of the cost of the initial investment, additional mechanisms have been devised to finance a part of selected development projects directly through supplementary funds. These may, for example, cover the additional costs of incorporating sustainability objectives in otherwise conventional (and presumably unsustainable) projects (see, for example, www.thegef.org/, www.multilateralfund.org/, https://cdm.unfccc.int/). Such funds are available especially to developing regions, to ensure that better account is taken of pollution, biodiversity, atmosphere (ozone) and climate issues. In some countries, national funds may be available to regions to deal with such issues. Some private-sector or NGO institutions also provide financial support to 'sustainable' regional development. The administration of all these funds requires sound budgeting and accounting practices within the region, as mentioned earlier.

Natural capital and resource accounting

As well as incorporating environmental, health, and social impacts into investment and accounting criteria, there is a deeper level of consideration that explicitly calculates the value of natural assets. The concept of 'ecosystem services' evaluates the hitherto unmeasured economic value of a healthy environment, that is, the intrinsic value of clean air, water and soil. The OECD refers to *natural resource accounting* as 'an accounting system that deals with stocks and stock changes of natural assets, comprising biota (produced or wild), subsoil assets (proved reserves), water and land with their aquatic and terrestrial ecosystems'. It refers to *environmental accounting* as 'physical and monetary accounts of environmental assets and the costs of their depletion and degradation'. We can, for example, think of abandoned, vacant, or park land in urban areas that, rather than being without monetary value, actually provides green services such as clean water and air and carbon dioxide removal worth millions of dollars (or other currency). The significance to regional institutions of such accounts is that they allow an estimation of the economic value of their current resource base that can be weighed against the economic impact of destructive and polluting development proposals. It is also a way of calculating the financial benefits of land before it is developed for other purposes. A number of countries are collaborating with the World Bank to explore the establishment of national (or regional) resource accounts as a way of influencing budget and development decisions at state level. It goes without saying that these accounts cover the entire life cycle of the resource assets being considered. Major financing institutions have joined in the *Natural Capital Declaration*. Box 4.3 is an extract from the Declaration.

BOX 4.3 NATURAL CAPITAL DECLARATION

[We the endorsing financial institutions wish to:]

> support the development of methodologies that can integrate natural capital considerations into the decision making process of all financial products and services . . . We aim to build on work undertaken through . . . the UN Principles for Responsible Investment, the Equator Principles, the UNEP Finance Initiative Principles for Sustainable Insurance, and The Economics of Ecosystems and Biodiversity (TEEB), so that we can develop methodologies to:
>
> a: apply a holistic approach to evaluating bonds and equities through the integration of natural capital considerations in environmental, social and governance (ESG) risk analysis . . . for investee companies;
> b: systematically value natural capital in the credit policies of specific sectors, including commodities, that may have a major impact on natural capital either directly or through the supply chain;
> c: systematically consider . . . natural capital in core insurance business strategies and operations.

(Natural Capital Declaration 2016)

As an extension of resource and environmental accounting that deals principally with inventories, we can also mention extended input–output tables and materials flow analysis (see Chapters 8 and 9, respectively), insofar as these techniques are able to put monetary value on basic resource flows through our societies from all sources. In some regions with substantial numbers of consumers and/or industries, such flows can be quite significant and can be usefully incorporated into planning and policy.

Environmental costs and revenues

Safeguarding environmental quality is not a cost: it is an investment – in the long term. In the short term, it is still useful to look at environmental costs, as these are often poorly accounted for, both within regional administrations and within companies. By way of anecdote, we can simply recall that, after land is negligently allowed to become polluted (saving money now), the eventual clean-up costs (spending it later) can be hugely superior to any sums that an initial upfront prevention action would have cost. Systematic life cycle costing is indispensable for regions taking a longer-term view of their development process.

In the meantime, environmental fees are a common mechanism to offset damage costs, or to encourage more ecological behaviour. Thus, the French eco-tax for road lorries was designed to discourage unecological transport, and the revenue is used to invest in cleaner energy sources. The proper setting of resource rents (water, land, forest, minerals, etc.) similarly acts both as a market-based influence on efficient behaviour and as a revenue stream for the region or state.

Finance, assets, and capital budgeting in regions

All of the above needs to be incorporated into a single financial system at regional level. There are several 'best practices' to guide local officials in capital budgeting and financial planning to encourage sound community and regional decisions (Marlow et al. 2009). Such processes must be oriented around the long-term goals of the region and its community, also reflecting their sustainability objectives in such areas as quality of life, public safety, economic development, environment, and infrastructure.

A transparent capital budgeting process identifies fixed assets via a rigorous benefit–cost analysis over their expected life cycle. A parallel process can be applied to natural resources and ecosystem services, which have been omitted from regional accounts in the past. The financial planning process for acquiring and maintaining the capital assets and resource base selects the appropriate financing source for each investment and asset. Debt financing can take on various forms, but local government inevitably enters into a long-term liability when using debt to finance capital assets. The debt service ratio can be used as a criterion to ensure that debt service obligations do not impede the ability to provide core services. Local governments also have the opportunity to address the benefit principle of public finance through intergenerational equity, where future taxpayers who benefit from the capital asset also pay for it. It has to be recognised, however, that future generations will be called upon to eventually pay for end-of-life liabilities and rehabilitation costs if these were not included at the outset.

Conclusions

To adequately foster sustainable regional development, local governments capitalise on and nurture their assets, both built and in the form of natural resources. To be effective in this role, they must understand and master their own financial capabilities. Local governments will only add value to development if they adopt best practices of capital budgeting and finance to deliver core services and public infrastructure, and exercise stewardship over their natural resource base.

Traditional financing instruments and budget processes were not always structured with the broader, long-term sustainability agenda in mind. New financing and revenue mechanisms have been developed that take a longer view of development than has often been the case in the past and that consider the life cycle of assets and projects more systematically. The budgeting process also needs to embrace a broader range of sustainability criteria in its thinking.

References

GRI (2016) Available at: www.globalreporting.org/reporting/sector-guidance/sector-guidance/financial-services (accessed 25 July 2016).

Île de France (2016) Available at: www.iledefrance.fr/sites/default/files/medias/2014/04/vigeo_seconde_opinion_green_sustainability_bond_idf_en.pdf (accessed 25 July 2016).

Marlow, J., Rivenbark, W.C., and Vogt, A.J. (2009) *Capital Budgeting and Finance: A Guide for Local Governments*, 2nd edn, Washington, DC: ICMA.

Natural Capital Declaration (2016) Available at: www.naturalcapitaldeclaration.org (accessed 25 July 2016).

OECD (2016) Available at: www.oecd.org (accessed 25 July 2016).

Reneweconomy (2015) Available at: reneweconomy.com.au (accessed 25 July 2016).

UGAP (2016) Available at: www.ugap.fr/actualites_113272.html (accessed 25 July 2016).

UNEP FI (2016) Available at: www.unepfi.org (accessed 25 July 2016).

5

LIFE CYCLE APPROACHES TO IMPROVE RESOURCE EFFICIENCY IN A REGIONAL CONTEXT

Guido Sonnemann, Philip Strothmann, and Fritz Balkau

Introduction

Although the term 'resource efficiency' is widely used in sustainability discussions, there is a need to explain its background, given that there is no universally accepted definition. 'Resource efficiency' in this book refers to the lowest possible use of natural resources to achieve an economic outcome with a minimum of environmental impact. This contributes to decoupling of economic growth and human welfare from resource use and environmental degradation. Life cycle approaches offer a way to enhance resource efficiency by identifying inefficiencies and impacts along the life cycle of products and services.

That resource efficiency and decoupling are needed more than ever has been highlighted in various reports by international organisations. The Millennium Ecosystem Assessment found that, over the past 50 years, human activity has changed ecosystems more extensively than in any comparable period in our history (Millennium Ecosystem Assessment 2005). These impacts mainly derive from extensive resource exploitation, agricultural expansion, and growth in manufacturing, and are unnecessarily reinforced by the inefficient use of resources at different scales.

In its Fifth Assessment Report, the Intergovernmental Panel on Climate Change (IPCC) outlines that greenhouse gas emissions are projected to grow and cause a rise in the global average temperature under every assessed scenario (IPCC 2014). This projection is based on the assumption that the global economy, and consequently resource use, will continue to grow. This is in line with conclusions from the OECD who found that, 'global extraction of material resources continues to grow, despite significant improvements in countries' resource productivity, i.e. the amount of material resources that is used per unit of output' (OECD 2008).

Examples of inefficient use of resources

Enhancing resource efficiency on a global level, however, remains a key priority, given that our present production and consumption system at a global level is still highly inefficient. For example, 'roughly one third of the food produced in the world for human consumption every year – approximately 1.3 billion tonnes – gets lost or wasted' (UNEP 2016).

Beyond the rather well-known example of food waste, resource efficiency is an issue of concern for manufactured goods, too. According to Hawken et al. (1999), as a global average, only 6–7 per cent of the initial resources are embedded in the final product, the remainder becoming waste at various points along the supply chain. An interesting example of low resource efficiency in manufacturing is also the production of silicon semiconductor microchips. The fabrication of a 2 g microchip requires more than 1.5 kg of chemical and fossil fuel input (Williams et al. 2002).

As Chapman (2005) outlines:

> Over 40 tonnes of waste are produced for each tonne of electronic products made, and users discard 98 per cent of these within just 6 months of purchase. By employing hazardous substances and energy-intensive processes, the waste this system generates is highly problematic.

Other examples of inefficiencies include ordinary light bulbs, which transform 90 per cent of their energy into heat, not light. And another good example is the amount of rock that is needed to produce one ounce of gold. As outlined in mining statistics for the first quarter in 2013 from Barrick Gold, it took the company anywhere from 2 t to 91 t of rock to produce just 1 oz of gold (Business Insider 2013).

New technologies, however, allow for inefficiencies to be better addressed. For example, 3D printing nowadays allows the aeronautical industry to manufacture aircraft parts by printing the metal components, rather than having to use the traditional wasteful practice of machining the wanted components out of blocks of solid metal, most of the block thus becoming expensive waste (PwC 2013). This applies also to the above-mentioned light bulbs being replaced by energy-efficient LED light bulbs. Nevertheless, it is system efficiency that really makes a difference, not improved efficiency at points along the life chain.

Taking the example of the light bulb and the related energy system, if we assume that the oil extraction is around 35 per cent efficient, electricity production is also around 35 per cent efficient and the old incandescent light bulb is less than 10 per cent efficient – in this case the overall system efficiency is just 1 per cent. The other 99 per cent of the original energy resource not only is wasted, but produces unwanted environmental and health impacts. Improving the efficiency of only the power station or the light bulb gives little overall improvement. In this case, replacing the light bulb with an energy-efficient LED with a typical energy efficiency of 80 per cent would still only result in an overall system efficiency of 10 per cent. To make major improvements, it is thus important to identify the key points in the system and address them simultaneously.

What are resource efficiency objectives in a sustainable development context?

On a global level, the issue is not new. Resource efficiency was addressed in the context of sustainable consumption and production (SCP) by the United Nations Conference on Environment and Development (UNCED) in 1987. The issue was subsequently included in concepts such as eco-efficiency (World Business Council for Sustainable Development 1999) and cleaner production (United Nations Environment Programme – UNEP), as well as industrial ecology (Graedel and Allenby 2002). In 2007, UNEP launched the International Resource Panel (IRP) to (a) provide independent, coherent, authoritative scientific assessments of policy relevant to the sustainable use of natural resources and the environmental impacts over their full life cycle, and (b) contribute to a better understanding of how to decouple economic growth from environmental degradation (IRP 2016).

In an effort to highlight the importance of resource efficiency and SCP at a policy level, the heads of state at the UN Conference on Sustainable Development (UNCSD; Rio+20) in 2012 adopted the 10-Year Framework of Programmes on Sustainable Consumption and Production (UNCSD 2012). Promoting SCP on a global level was also included in the UN's Sustainable Development Goals (UN DESA 2016). These global frameworks are structured to provide guidance in the first instance to governments, influencing their development planning, environment policy, and capacity-building programmes.

Nevertheless, they also provide guidance to companies and businesses, for which sustainability has become a megatrend in recent years according to the *Harvard Business Review* (Lubin and Esty 2010). We can observe that companies such as Unilever have integrated resource efficiency and SCP considerations into their decision-making processes:

> The Unilever Sustainable Living Plan is our blueprint for achieving our vision to double the size of the business, whilst reducing our environmental footprint and increasing our positive social impact. The Plan sets stretching targets to achieve by 2020 and includes how we source raw materials and how consumers use our brands.
>
> (Unilever 2016)

Still, these frameworks have the highest relevance for, and biggest impact on, policy-making. At the national level, many countries have incorporated resource efficiency and SCP policies into national action plans, as demonstrated in a survey conducted by the European Environmental Agency (EEA) on resource efficiency policies and approaches in the European Union (EEA 2011a). Beyond influencing national policies, they also have a direct impact on development planning, budget, and capacity-building programmes at the regional level, as discussed later in this chapter.

Life cycle approaches are available to foster resource efficiency

Implementing resource efficiency and SCP requires fundamental changes to established behaviour. It requires producers to change product design, production processes and marketing activities. Consumers will need to integrate environmental and social concerns into purchasing and consumption decisions, in addition to price, convenience, and quality.

Different stakeholders have applied a number of different tools to facilitate the application of resource efficiency strategies and SCP patterns. These have undergone a gradual evolution from single-point intervention tools to life cycle-based, holistic system solutions. They also use procedural and communication approaches to achieve improvements at relevant points in a value chain that could contribute to a greater level of resource efficiency.

Resource efficiency thus increasingly draws on life cycle thinking to link intervention activities across a product's entire life cycle, as introduced in Chapter 2. Life cycle thinking helps to avoid 'burden shifting' by ensuring that a reduction in the environmental impact at one stage in the life cycle does not increase the impact at other stages of consumption and production, in other geographic areas, in other products, or for future generations. Although it is nearly impossible to engage in consumption of anything without some environmental impact, life cycle thinking can help people make better alternative decisions towards mitigation of such impacts (European Commission 2011).

Much effort has gone into developing life cycle approaches to promote the application of SCP and resource efficiency concepts. Parts II and III of this book describe a number of tools that can be employed at different scales.

For analytical life cycle tools, we can distinguish those concerned with micro-level assessment – that is, the site or product level – from those focusing on the meso level – the sector and regional level – and those covering the macro economic level – the national or global level. Moreover, we can distinguish between those with a focus on resource and/or environmental impact questions only and those also covering the social and economic dimensions of sustainable development. Evidently, the notion of differentiation is an approximation; some tools can operate at various economic scales, if adequately applied.

At the macro level, we have the concept of ecological footprint, which is further discussed in Chapters 7 and 24. This is a measure of human demand for natural capital that can be compared with the planet's ecological capacity to regenerate. It represents the area of biologically productive land and sea necessary to supply the resources a human population consumes and to assimilate associated waste (Wackernagel 1994). Another tool used for environmental accounting purposes is environmentally extended input–output analysis, which reflects production and consumption structures within economies (Kitzes 2013) and is described in Chapter 8. A related tool, data envelopment analysis, as explained in Chapter 21, provides a direct estimation of productivity and resource efficiency.

The core tool at the meso level is material flow analysis (MFA), introduced in Chapter 9. This determines the main sources, flows, stocks, and emissions of man-made and natural materials (Brunner and Rechberger 2003). The principal concept underlying MFA is the interrelation between economy and environment, whereby the economy is an embedded subsystem of the environment, dependent on a constant throughput of materials and energy (Fischer-Kowalski et al. 2011).

The water footprint, covered in Chapters 12 and 31, is a geographically explicit indicator showing volumes of water use and pollution in a region or a country (Hoekstra and Chapagain, 2008). Allan (2011) demonstrated the 'virtual water' concept, which measures how water is embedded in the production and trade of food and other products.

At the micro level, we have site-specific instruments such as environmental impact and risk assessments applied to entire projects. The most widely used tool for assessing resource use and environmental impacts of products is life cycle assessment (LCA), explained in Chapter 10.

Looking beyond assessment, we need also to examine the 'intervention tools' at our disposal. There are, for example, procedures for developing more sustainable, resource-efficient products: the so-called 'design for sustainability' approaches (UNEP 2008). Eco-design and product service systems are among the elements of this approach. Sustainable purchasing is a powerful way of influencing the supply chain to provide more efficient, sustainable products.

Finally, a number of communication-orientated life cycle approaches, such as eco-labelling and certification (Frankl and Rubik 2005), can be used to foster resource efficiency and SCP. All these life cycle approaches that have been developed to move the marketplace towards greater resource efficiency and SCP are described in Part III.

Regional use of life cycle approaches in resource efficiency strategies

The different approaches above provide a useful package to allow regions to foster resource efficiency and SCP. In particular, increasingly mature analytical tools are available to measure how resources are used. But, beyond the individual potential actions by companies and consumers, a holistic view of the system itself is still required to optimise each stage of the life cycle and prioritise the most effective action.

Attention also needs to be put on the downstream user phase, as this is one of the major sources of inefficient energy and materials use in modern society. As we will see in Chapter 6, life cycle management serves as a framework to bind the actions of the various stakeholders into a coherent package. A toolbox of life cycle approaches can be built up at the regional level as part of a resource efficiency strategy (see Box 5.1) to increase competiveness, economic productivity, and social inclusiveness while minimising resource use and environmental impact.

BOX 5.1 THE CIRCULAR ECONOMY ROADMAP OF THE AQUITAINE REGION, FRANCE

One of the examples of how regions use life cycle approaches as part of SCP policies is the Aquitaine region in the south west of France, whose capital is Bordeaux. The region has adopted a circular economy roadmap, in which circular economy is defined as an economic system of consumption and production that, at all stages of the product life cycle, aims to increase the efficiency of resource use and reduce the impact on the environment. The circular economy is generally seeking to drastically reduce resource wastage and to decouple resource consumption from economic growth, while ensuring the reduction of environmental impacts. It aims to do more and better with less. The circular economy is opposed to the traditional model of the linear economy, in which the resources extracted and used in products are consumed and discarded in their so-called end-of-life phase. The circular economy is based on six key elements:

1. moderate and most efficient use of non-renewable resources;
2. use of regional renewable resources within the ecological carrying capacity;
3. eco-design and clean production;
4. environment-friendly consumption;
5. recovery of waste as a resource; and
6. treatment of waste without harm.

A multistakeholder committee has elaborated twenty proposals for concrete actions, within three areas of intervention that are (1) waste and recycling, (2) industrial ecology and territorial ecology, and (3) eco-design and functional economy, which particularly makes reference to LCA as a crucial approach for measuring progress towards a circular economy. The most relevant proposals making reference to the use of life cycle approaches are:

- creation of a virtual platform and annual conference on regional circular economy;
- fostering of university education on life cycle approaches;
- promotion of lifelong learning on topics related to the circular economy;
- reinforcement of the diffusion of existing expertise throughout the region;
- setting-up of a knowledge platform on regional material flows;
- development of circular economy-oriented industrial areas;
- positioning of operational life cycle tools in companies;
- mainstreaming the use of life cycle management in SMEs using a sector-based approach;
- provision of incentives for the market uptake of eco-designed products;
- financial support for research and investment in more sustainable products and services.

This example illustrates the important role life cycle approaches have gained for putting resource efficiency and SCP into practice at the regional level. The driver is not just protection the environment; it is to do more with less, to foster sustainable innovation in order to strengthen competiveness and ensure regional economic development, using the natural resources available in the region within its ecological carrying capacity.

Reference

Conseil Régional d'Aquitaine (2014) Aquitaine Développement Economie Circulaire, Bordeaux, France.

Prepared by Guido Sonnemann (University of Bordeaux)

In 2006, the government of the Flanders region in Belgium introduced the 'Flanders in Action' plan and the resulting 'Pact 2020'. Among other objectives, the plan focuses 'on the improvement of eco-efficiency of the Flemish economy'. This is to be achieved by an increase in material and energy efficiency within different sectors, as based on several indicators mentioned below (EEA 2011b). With a similar objective of enhancing regional development, the roadmap of the Aquitaine region, France (see Box 5.1), lays out ways to increase the resource efficiency of that region, making direct reference to the IRP and promoting the use of eco-design and LCA.

Public administration has multiple intervention points, ranging from economic stimulation and employment to research and innovation, as well as action on environment and climate change, supported by communication and education policies. Intervention can take place through regulation, subsidies, preferential loans, labelling, communication campaigns, etc. As outlined above, all these tools and interventions need to be applied in a simultaneous and coordinated manner to ensure that efficiency gains are achieved on a system level.

Conclusions

International frameworks on resource efficiency and SCP define a way forward for governments and regions, also setting new conditions under which private-sector companies operate.

Resource efficiency can be regarded from a local level and from a system level. For the former, we can often point to levels of efficiency improved through measures such as cleaner production processes and more sustainable patterns of consumption. A systems view takes into account all the stages in the life cycle of a product and establishes a co-ordinated intervention at priority points along the value chain. Regions need to look at both local-level and system improvements in resource efficiency.

In a regional context, the challenge is to use a variety of analytical life cycle approaches to pinpoint priority resource efficiency 'hot spots' and then to employ an LCM framework to make the sequence of co-ordinated local changes along the value chain. This process uses a situation-specific life cycle toolbox, as described in Chapter 6 and illustrated with examples in Parts IV and V.

References

Allan, T. (2011) *Virtual Water: Tackling the Threat to Our Planet's Most Precious Resource*, London: I.B. Tauris.

Brunner, P.H. and Rechberger, H. (2003) *Practical Handbook of Material Flow Analysis*, Boca Raton, FL: Lewis.

Business Insider (2013) Here's How Many Tons of Rock You Have to Mine just for an Ounce of Gold. Available at: www.businessinsider.com/tons-of-rock-for-an-ounce-of-gold-2013-4?IR=T (accessed 15 January 2016).

Chapman, J. (2005) *Emotionally Durable Design: Objects, Experiences and Empathy*, London: Earthscan.

European Commission (2011) Life Cycle Thinking and Assessment. Available at: http://lct.jrc.ec.europa.eu/index_jrc (accessed 18 November 2015).

EEA (2011a) Survey of Resource Efficiency Policies and Approaches. Available at: www.eea.europa.eu/themes/economy/resource-efficiency/resource-efficiency-policies-country-profiles (accessed 15 January 2016).

EEA (2011b) 2011 Survey of Resource Efficiency Policies in EEA Member and Cooperating Countries. Country Profile: Belgium. Available at: www.eea.europa.eu/themes/economy/resource-efficiency/belgium-country-profile-on-resource (accessed 15 January 2016).

Fischer-Kowalski, M., Krausmann, F., Giljum, S., Lutter, S., Mayer, A., Bringezu, S., Moriguchi, Y., Schütz, H., Schandl, H., and Weisz, H. (2011) 'Methodology and indicators of economy-wide material flow accounting', *Journal Industrial Ecology*, 15: 855–76.

Frankl, P., and Rubik, F. (2005) *The Future of Eco-labelling: Making Environmental Product Information Systems Effective*, Sheffield, UK: Greenleaf.

Graedel, T.E., and Allenby, B.R. (2002) *Industrial Ecology*, Upper Saddle River, NJ: Prentice Hall.

Hawken, P., Lovins, A.B., and Lovins, L.H. (1999) *Natural Capitalism: The Next Industrial Revolution*, London: Earthscan.

Hoekstra, A.Y., and Chapagain, A.K. (2008) *Globalization of Water: Sharing the Planet's Freshwater Resources*, Oxford, UK: Blackwell.

IPCC (2014) Climate Change 2014: Synthesis Report. Contribution of Working Groups I, II and III to the Fifth Assessment Report of the Intergovernmental Panel on Climate Change, Geneva: IPCC.

IRP (2016) Mission & Vision. Available at: www.unep.org/resourcepanel/Whoweare/IRPMission/tabid/133308/Default.aspx (accessed 15 January 2016).

Lubin, D.A., and Esty, D.C. (2010) 'The sustainability imperative', *Harvard Business Review*, May.

Millennium Ecosystem Assessment (2005) *Ecosystems and Human Well-Being: Synthesis*, Washington, DC: Island Press.

OECD (2008) Resource Productivity in the G8 and the OECD – A Report in the Framework of the Kobe 3R Action Plan. Available at: www.oecd.org/env/waste/resourceproductivityintheg8andtheoecd.htm (accessed 15 January 2016).

PwC (2013) 3D Printing: A Potential Game Changer for Aerospace and Defense. Available at: www.pwc.com/us/en/industrial-products/publications/assets/pwc-gaining-altitude-issue-7-3d-printing.pdf (accessed 15 January 2016).

UN DESA (2016) Sustainable Development Goals. Available at: https://sustainabledevelopment.un.org/?menu=1300 (accessed 15 January 2016).

UNCSD (2012) 10-Year Framework of Programmes on Sustainable Consumption and Production Patterns, A/CONF.216/5 adopted document, Rio de Janeiro: UNCSD.

UNEP (2008) *Design for Sustainability: A Global Guide*, Paris: UNEP.

UNEP (2016) Food Waste Facts. Available at: http://staging.unep.org/wed/2013/quickfacts/ (accessed 26 July 2016).

Unilever (2016) Sustainable Living. Available at: www.unilever.com/sustainable-living/ (accessed 15 January 2016).

Wackernagel, M. (1994) *Ecological Footprint and Appropriated Carrying Capacity: A Tool for Planning Toward Sustainability* (PhD thesis). Vancouver, BC: School of Community and Regional Planning, University of British Columbia.

Williams, E.D., Ayres, R.U., and Heller, M. (2002) 'The 1.7 Kilogram Microchip: Energy and material use in the production of semiconductor devices', *Environment, Science & Technology*, 36: 5504–10.

World Business Council for Sustainable Development (1999) Eco-efficiency Indicators and Reporting, Working Group on Eco-efficiency Metrics & Reporting, Geneva: World Business Council for Sustainable Development.

6

LIFE CYCLE MANAGEMENT FOR REGIONS

Fritz Balkau and Guido Sonnemann

Introduction

We live on a smaller planet than we like to believe. Expanding international trade, globalising industrial production, and increasing consumerism have combined to accelerate environmental degradation, putting the world's ecosystems and our resource base under increasing pressure. Along the way, it has become clear that the older decision-making tools in common use are insufficient for the new development agenda of the twenty-first century. As already mentioned in Chapter 5, many 'solutions' simply moved the impacts to other environmental or geographical compartments and had a short-term perspective. Interventions often addressed the symptoms rather than the cause (e.g. the waste mountain that has to be 'safely disposed of'). The situation called for a more rigorous approach to development planning based on upstream action, multi-criteria objectives, and a longer-term perspective. Life cycle management (LCM; Sonnemann and Margni 2015) promises to provide such a framework for action on many of our urgent sustainability challenges.

The adoption of life cycle approaches is as relevant to regions as it is to companies. Both suffer the consequences of, and thus need to manage to the best extent possible, decisions made both locally and along the supply chains of their goods and materials.

The suite of assessment procedures now available (see Part II) can show the upstream and downstream sustainability impacts from a given material or product. Approaches such as ecological footprinting, materials flow analysis and life cycle assessment (LCA) can reveal how a region or a company is situated with respect to its potential impacts. The range of intervention (management) procedures presented in Part III has been developed to capitalise on the data flowing from such assessments. Some, such as eco-labels and environmental product declarations (see Chapters 18 and 23), have been standardised to improve purchasing decisions by organisations and individuals. Others, such as supply-chain management, product service systems, social responsibility, and sustainability reporting, remain more flexible to allow application to different situations.

All these approaches can be combined for effective use within the so-called LCM toolbox (UNEP/SETAC 2007).

Sustainable development goals as a basis for LCM

The release in 1987 of the Brundtland Commission report, Our Common Future, was a defining moment for more holistic thinking (UNCED 1987a). The earlier development focus, based purely on economic outcomes, was supplemented by two additional objectives – social and environmental. Subsequent meetings refined the objectives and attempted to set fixed targets for achievement. The latest, in New York in 2015, adopted an updated set of sustainable development goals (UN 2015).

However, there has been less success in constructing a universal decision-making framework for actually achieving them. With the sustainable development goals coming out of the international processes initiated by UNCED, life cycle thinking now has the potential to help in the implementation of these goals. Recent methodological evolution based on LCA (see Chapter 10) is now also grappling with areas such as social impacts and economic consequences. Attempts are being made to deal with biodiversity and landscape assessment is gradually coming into wider use. Natural resource accounting is being used to look at the monetary value of the natural environment. Even the actors are now being assessed: large companies are evaluated by their peers and their stakeholders on a range of sustainability criteria, and sustainability reporting is gaining ground among public-sector organisations, as exemplified by the recent public report by Yarracity (2013) in Australia.

What is still missing is a management or decision-making framework that moves from comprehensive assessment to systematic action. LCM has already come into use in corporations to improve the sustainability performance of their products and services (UNEP/SETAC 2009). But LCM can also be used by public authorities, as many of the objects of such management – products, materials, services – are the same. In fact, many authorities are already making use of other corporate management instruments such as environmental management systems (EMSs), supply-chain management, and sustainability reporting; however, this is still on a piecemeal basis. An evolution to a more holistic management approach, as used by companies, would provide many advantages in addressing complex sustainability issues.

In this, it is important to also understand the differences. In particular, the sustainability perspective of regional administrations is typically broader and longer term than that of companies. Many regions have developed ambitious sustainability objectives, whether under Local Agenda 21 programmes (UNCED 1987b) or independently developed. As regions are constrained by national legislation in their obligations and in the services they provide, the boundaries of the management they exercise are thus different from those of the private sector. This will influence the LCM framework they adopt and, in particular, the selection of the assessment and intervention tools they decide to use.

What is LCM and how is it applied?

LCM is still an evolving art. It is generally understood to be a broad, non-standardised concept that builds a coherent action along the life cycle of projects, products, and services to achieve certain defined (sustainability) outcomes.

The UNEP/SETAC Life Cycle Initiative has defined LCM as an organisational tool to make value chains more sustainable by decreasing the total footprint of products and materials:

> Life cycle management is a [. . .] management approach that can be used by . . . businesses and other organisations to improve products and the sustainability performance of the companies and their value-chains.
>
> (UNEP/SETAC 2007)

Although this may seem to focus exclusively on business performance, LCM also has major advantages for the public sector and for consumers. The importance of the phrase 'and other organisations' in the above definition should not be overlooked.

Thus, in the business sector, (sustainable) supply-chain management is a major corporate preoccupation, dealing with the upstream life cycle of products and materials being purchased (Balkau and Sonnemann 2011). It increasingly reaches into the downstream area of product responsibility, sometimes termed extended producer responsibility. Public administrations, as major consumers in their own right, are increasingly looking at their procurement practices. Procurement policy is not only about reducing negative impacts; it can also be used to stimulate sustainable local economic development and product innovation that gives added value along the entire production chain. Chapters 19 and 33 deal with procurement issues at greater length.

Non-material factors such as infrastructure projects, financial planning, or social programmes also have a life cycle whose impacts can be managed for sustainability outcomes. Similarly, land management and sustainable exploitation of natural resources benefit from a long-term life cycle view.

For LCM to be successful, its purpose needs to be clearly established, and the sustainability criteria that have become relevant at regional level must be taken into account. In most cases, these reflect the global Agenda 21 or the more recent sustainable development goals. Several urgent issues, such as climate, water, and biodiversity, are at the top of the agenda for public bodies. There is also regular reference to the need to address pressing social issues of employment, equity, gender, social freedoms and health and safety. Each of these issues plays out differently along the life cycles of various products and administrative procedures. An important function of the LCM concept is to ensure that action to address the issues in one point in time and place do not have the perverse effect of exacerbating the problem somewhere else.

Despite the huge potential, the application of LCM to such issues on the ground is not always self-evident. Although the application of life cycle approaches to product design is increasing fast, its contribution to programmatic and thematic development is less visible. In part, this is owing to lack of sufficient technical expertise and easy-to-use data sets. But arguably, the biggest obstacle is the lack of understanding of non-technical actors such as public administrators and project managers of their potential utility.

In this respect, some progress is being made by ISO to address this gap. The recent ISO Standard 14072 on organisational LCA (O-LCA; see Chapter 15) describes the application of the LCA method to an organisation rather than its products (the latter are included indirectly in the organisation's report). By describing the organisation's activities in this way, key life cycle factors can be revealed. O-LCA does not explicitly invoke LCM intervention, but the possibilities for further action will be clear from the assessment.

It is important to appreciate that LCM is not a tool as such: it is a management framework. LCM builds a specific toolbox for each application or sector from the components that are available. One of the strengths of the LCM framework is that it incorporates the ISO 9000 quality management standard steps of 'plan–do–check–act', to gradually build up a system that functions effectively. In this, it fits within current concepts of EMSs such as ISO 14001 or similar. The specific tools in the box will include those described in various ISO standards, technical handbooks, and various guides published by recognised international or national organisations.

An example is the toolbox for low-emissions cities: the Urban Low Emissions Strategies (LEDS) project, coordinated by ICLEI and UN Habitat (2016) with funding from the European Union, regroups a number of cities worldwide into the project. It is based on vertical integration between different levels of government – that is, all those implicated somewhere along the life cycle of the energy agenda – to achieve a coherent approach to a low-energy future for cities. It relies on a variety of emission accounting tools and a reporting platform to enhance credible action, both mitigation and adaptation.

LCM and regions

Regions can be viewed from a number of perspectives. At the simplest level, we can consider public authorities as consumers and providers of goods and services. LCA can shed light on the world behind individual products that are consumed by the region and its consumers. Moreover, analytical life cycle tools such as ecological footprinting, materials flow analysis, and environmentally extended input–output tables can provide a sustainability profile of the region by mapping materials and energy flows and highlighting areas of potential environmental impact of a region's economic sectors. These tools are described in Part II. Results of these different assessments tools should, in each case, be viewed in the context of regional objectives on sustainability.

LCM tools such as procurement, extended producer responsibility, product service systems and eco-design, as explained in Part III, can be employed to assist in best meeting these objectives. For example, the buy/not buy option in procurement is temptingly simple, but cooperating directly with suppliers and clients in developing mutually beneficial solutions is more effective in the long term and more efficient in avoiding burden-shifting.

Public authorities are also service providers. They plan, build, and maintain infrastructure over extended periods of time. LCM has great relevance here. The issues are, therefore, not confined to physical objects and resources. Full-cost accounting, life cycle costing, cleaner production financing, green investment, as mentioned in Chapter 4 – all are components of a more life cycle vision of financing and budgeting, encouraging the valuing of assets over their entire lifetime rather than only the production part of the cycle.

Insofar as regions foster further economic development and product innovation, they are able to apply life cycle tools to encourage the most appropriate sectors, companies, services, and products that fit into the sustainability vision of the region. This also applies to resource development at regional level.

An example of this is the French region of Hauts de France, which is using an LCM framework for its industrial revitalisation programme – the so-called 'Third Industrial Revolution'. The building sector is one of the focus areas where the actors in the value chain have accepted an agreed set of sustainability objectives, subsequently building a common approach to management practices based on LCA of the materials employed. Outreach and capacity building for the life chain actors, especially smaller companies and municipalities, are important components of the efforts in promoting LCM. Further details are given in Chapter 27.

Importance of clear objectives

Good data and clear objectives underlie any systematic management approach. Regions are obviously obliged to respect national commitments to environmental, economic, and social treaties. However, a region that has espoused its own principles of, for example, environmental integrity, social justice, and equality should also protect such values in the territory of its trading partners through its purchasing and export practices. This avoids the frequent reproach that the developed world 'exports' its pollution through the purchase of foreign goods that are manufactured under unsafe conditions elsewhere.

We have remarked elsewhere in this book that LCM, and the LCAs on which it depends, have often had a narrow focus on a small set of objectives. This reflects partly the specific priorities of the companies that have been the pioneers in using those approaches and partly the intrinsic limitations of the quantitative approach to LCA that is a consequence of its traditional preoccupation with energy and waste issues. Only recently, for instance, has a component on water been added.

Unlike companies, public authorities cannot so easily set aside the wider scope of sustainable development, with its multiple objectives derived from the sustainable development goals. The use of LCM by regions will require a response to a broader set of objectives than has been common in the past. This, in turn, will encourage further development of more comprehensive quantitative assessment tools by the LCA community. In this sense, the use of territorial LCA, as described in Chapter 28, is a welcome opportunity for regions.

Conclusions

The sustainability objectives, timetables, and organisational means of public institutions do not always mirror those of companies. But there are also similarities. Public authorities deal with, or encourage the production and consumption of, goods and services. They are major purchasers of products and infrastructure that have significant sustainability footprints. They manage resources that, under wise stewardship, have indefinite lifetimes, or, if non-renewable, should be managed to give the greatest benefit over the longest time frame. Much regional infrastructure has an extended lifetime, with significant upstream and downstream costs and social implications.

Accordingly, regions have, and are surrounded by, important life cycle dynamics that have to be managed in a holistic fashion. It is always useful to build upon what has already been accomplished. The LCM frameworks and tools developed for companies can also be applied by public administrations. To more visibly extend its relevance into the public sector, the current UNEP/SETAC definition of LCM could thus usefully be paraphrased as also including 'the sustainability performance of regional authorities and their value chains'.

LCM frameworks, although not standardised into any particular structure, draw extensively for their toolbox on existing procedural and managerial standards, whether under ISO or independent. There is already good experience with conventional LCM in many sectors, including corporations, to demonstrate the potential of its wider application in the public arena. The collaborative LCM procedures used in sector-specific schemes for forest products, the textile trade, or certain high-tech electronic products are also useful models for public authorities. In all cases, it must be remembered that LCM is an extension of current managerial processes, not something radically new, nor simply an extension of LCA. Its application is most effective where it is deeply integrated into the organisation (in the form of 'mainstreaming'), rather than constituting an 'add-on' function to an existing structure.

LCM remains a challenge for smaller organisations – both private- and public-sector – that often lack the analytical and managerial capacity for such an exercise. Here, it will be important to establish new partnerships of appropriate institutions and individuals who can provide the relevant advisory services. The example of the French region of Hauts de France above shows that this is possible.

The specifics of the different regional actors need to be taken into account. The various partners in a region have significantly different timetables over which they intend to realise their sustainability objectives, and also have different options for action. In general, public authorities take a longer view of the development process than companies. As well as regions having different timescales, their sustainability objectives are usually broader, and their stakeholders more diverse. The scope of their LCM exercises will thus be wider, and possibly more complex.

Overall, the recent maturing of both analytical and intervention-orientated life cycle approaches means that LCM now gives regions a new management framework through which to advance sustainable development.

References

Balkau, F., and Sonnemann, G. (2011) 'Addressing sustainability issues through enhanced supply-chain management', in *Supply Chain Management – New Perspectives*, ed. S. Renko, InTech, pp. 379–400.

ICLEI and UN Habitat (2016) Urban LEDS Project – Low Emission Development Strategies in Emerging Economy Countries. Local Governments for Sustainability (ICLEI) and United Nations Human Settlements Programme (UN Habitat). Available at: http://urbanleds.iclei.org/ (accessed 15 January 2016).

Sonnemann, G., and Margni, M. (2015) *Life Cycle Management, Book Series LCA Compendium – The Complete World of Life Cycle Assessment*, Dordrecht, Netherlands: Springer.

UN (2015) Transforming Our World: The 2030 Agenda for Sustainable Development. United Nations document A/RES/70/1. Available at: https://sustainabledevelopment.un.org/post2015/transformingourworld/publication (accessed 15 January 2016).

UNCED (1987a) Our Common Future – Brundtland Report, New York: United Nations Conference on Environment and Development. Available at: www.un-documents.net/our-common-future.pdf (accessed 15 January 2016).

UNCED (1987b) Agenda 21, New York: United Nations Conference on Environment and Development. Available at: http://sustainabledevelopment.un.org/content/documents/Agenda21.pdf (accessed 15 January 2016).

UNEP/SETAC (2007) Life Cycle Management: A Business Guide to Sustainability. UNEP/SETAC Life Cycle Initiative. Available at: www.unep.org/pdf/dtie/DTI0889PA.pdf (15 January 2016).

UNEP/SETAC (2009) Life Cycle Management: How Business Uses it to Decrease Footprint, Create Opportunities and Make Value Chains More Sustainable. Available at: www.unep.org/pdf/DTIE_PDFS/DTIx1208xPA-LifeCycleApproach-Howbusinessusesit.pdf (accessed 15 January 2016).

Yarracity (2013) The Yarra Environmental Strategy 2013–2017. Available at: www.yarracity.vic.gov.au/environment/Environment-Strategy/ (accessed 15 January 2016).

Part II

ANALYTICAL BASIS FOR LIFE CYCLE APPROACHES

7

ECOLOGICAL FOOTPRINT

Principles and methodology

Marcello Ruberti and Stefania Massari

Introduction

According to the UN Department of Economic and Social Affairs, Population Division (2013), the world population is steadily increasing and could reach, by 2050, the threshold of 9.6 billion people. This continuous population growth forces mankind to find new sources of food, materials, and energy and to produce more and more. This implies depletion of natural resources and ecological damage, such as deforestation, air pollution, acidification, nitrification and erosion of soil, loss of agricultural land, global fresh-water degradation, eutrophication, biodiversity loss and so on.

Measuring sustainability, and in particular environmental sustainability, using different tools could be a valid action to provide a basis for the public and policymakers, both governments and non-governmental organizations, for defining objectives and targets useful to rethink our current lifestyles and the patterns for human welfare.

It is widely recognized, in fact, that objectives of economic growth, applied to national planning and regional development policies and measured by the misleading GDP, are no longer the best choice, because they involve over-exploitation of natural resources and indiscriminate production of waste.

It is necessary to promote, not only the economic welfare of a community, but also its quality of life and the state of the environment in which the same population lives, reconciling the needs of present and future generations and taking into account the physical limits of our planet.

In particular, for 'environmental sustainability', one of the eight Millennium Development Goals established in 2000 by the UN, it is necessary to provide a set of suitable tools to measure the global availability of resources – in particular those at risk of depletion – and to understand and, possibly, avoid the environmental changes resulting from human pressure (Hoekstra and Wiedmann 2014). In this chapter, we'll give a brief overview of the ecological footprint (EF) indicator – a useful tool to assess the sustainability of our living patterns – describing its boundaries and calculation methods.

Environmental sustainability assessment by environmental footprints

The first concepts dealing with the limits of the natural environment, in terms of pollution absorption and resources availability, have been the Framework for

Strategic Sustainable Development and Factor X. The former, developed by the Swedish Natural Step organization, which was founded by Karl-Henrik Robèrt in 1989, is a comprehensive method for planning human activities taking into account the Earth's system conditions, as established by the laws of thermodynamics, for our survival on this planet (Lindahl et al. 2014).

Factor X, initially termed Factor 10, was originally proposed by authors of the Wuppertal Institute in the early 1990s. It has the aim to minimize the current global resource turnover by a factor *x*, preserving the planet's carrying capacity and allowing satisfactory living standards for future generations (Global Development Research Center 2015).

Since then, many very different instruments have been proposed and used for the assessment of environmental impact and sustainability, such as indicators, methodologies, standards, and so on (as described in Parts II and III of this book).

Among the most developed instruments of recent years, there are the footprint indicators, which represent quantitative measures of the pressure on natural resources caused by mankind. The concept of footprinting in general easily recalls, in a visual way, the pressure exerted on a surface by an individual and it is easy to understand, as the dimensions and the depth of the footprint depend on the weight and dimensions of the person.

The EF, developed on the basis of the pre-existing biological concept of 'carrying capacity', by William Rees (1992) of British Columbia (Canada) and, later, by Mathis Wackernagel (1994) and other scholars, is currently widely adopted as an indicator for assessing environmental sustainability. It is easy to understand, in fact, that environmental sustainability is ensured when, in a defined period of time, the consumption of renewable resources is less than or equal to the regenerative capacity of the biosphere, and the amount of waste produced is equal to or less than that that nature is able to assimilate.

Many other indicators, in addition to the EF, have been developed based on the concept of footprinting from a life cycle perspective, such as: carbon, water, energy, nitrogen, phosphorus, biodiversity, land footprints, and so on. Quite recently, the European Commission defined the Product Environmental Footprint, a new tool for measuring the environmental performance of a product or service throughout its life cycle, as better described in Chapter 23 of this book.

Some of these footprint indicators are in the early stages, and so they do not have a well standardized and widely recognized calculation methodology yet (Cucek et al. 2015).

Environmental footprints are aggregate, concise indicators, conceptually easy to communicate and to understand. According to a literature review, the most used and developed indicators to assess environmental sustainability are the ecological, carbon, and water footprints, which, jointly, form the so-called 'footprint family' (Galli et al. 2012). These footprints together capture the main sustainability issues of mankind: food, water and climate.

The EF, in particular, is a good accounting tool that – even with its criticized simplifications – illustrates, much better than other indicators, the *minimum criteria* of environmental sustainability, because it can provide a clear picture of the gap existing between the real boundaries of a certain area and the limits of the virtual area that would be needed instead to support the life of the population living in that area (Kitzes et al. 2009).

EF is obtained through physical accounting of resources using, as common 'currency', the land required to produce these resources in a renewable and sustainable way and to assimilate the waste produced. Adopting this common unit of account, it is possible to carry out time and spatial comparisons, at a local, regional, or global level and identify areas with ecological deficits or surplus.

Ecological footprint: Concept, boundaries, and calculation method

The EF measures the total area of biologically productive land and water ecosystems, expressed in global hectares (gha) per capita (equivalent hectares of productive space having the average biological productivity of the planet), required to produce the consumed resources and to assimilate the emissions (CO_2 sequestration) produced by a given population (GFN 2010). The calculations can be made at different scales, ranging from single products or individuals to humanity as a whole.

The calculation method, reported in the 2012 edition of the National Footprint Accounts, is based on the methodology described in Borucke et al. (2013). In 2006, the Global Footprint Network (GFN) also released international operation standards for the EF calculation for national and subnational products and organizations levels, which were updated in 2009 (GFN 2009).

To calculate the EF for a specific area, it is appropriate to divide the demands of the particular community into five categories (food, housing, transport, consumer goods and services) and to consider the flows of energy and materials into six different production ecological systems. The EF of each person is in fact the sum of the following six different components of bioproductive land, which correspond to the ecosystem services provided to human activities: the area of cultivated land needed to produce food and natural resources, the grazing area needed to breed animals and produce animal products, the forest area needed to produce wood and paper, the marine area needed to produce fish and seafood, the land area occupied by infrastructures, and the forested area needed to absorb carbon dioxide emissions associated with energy consumption. Each category of consumption corresponds to an impact on the specific ecological system.

EF, as a consumption-based indicator, is calculated as follows (Galli et al. 2014):

$$EF_C = EF_P + EF_I - EF_E = \sum_{i=1}^{n} C_i q_i \tag{7.1}$$

where EF_C is the EF of consumption of the area under investigation, EF_P is the EF of production, EF_I and EF_E are the EFs embodied in imported or exported commodity flows, C_i is the consumption of the product i, and q_i is the inverse of the average productivity for the product i. Thereafter, the EF per capita is calculated by dividing EF_C for the resident population N.

Instead, the basic (simplified) equation of EF of production for a specific primary product is (Ewing et al. 2010):

$$EF_P = \frac{P}{Y_N} \cdot YF \cdot EQF \tag{7.2}$$

where P is the quantity of product, Y_N is the national average yield for P, and YF and EQF are, respectively, the yield factor and equivalence factor for the given land-use type. YF is the ratio between the national and the world average productivity for P. EQF is a coefficient converting the local area used into global average units of bioproductive land (Monfreda et al. 2004).

Like all the other indicators, the EF has some deficiencies, recognized by the authors themselves. Most of them arise because not all the environmental implications of human activities are accounted for (Galli et al. 2012). For example, the EF does not quantify, except in an incomplete and indirect way, use of fresh water; loss of biodiversity; pollution by radioactive, toxic, or non-biodegradable substances; or waste for which the biosphere has a low assimilation capacity (Kitzes et al. 2007). Further, land for energy, calculated by converting CO_2 emissions from fossil fuels into wooded area, represents on average more than 50 per cent of EF value (Fiala 2008). In any case, it must be underlined that EF is an accounting and assessment tool that can be used for decision-making.

Despite these limitations, the EF represents a good compromise between complexity and simplicity. Considering the clarity of its unambiguous message, its relatively easy method of calculation, and its clear diagnostic validity (Moffatt 2000), it has been adopted by many international organizations.

National and regional footprint analysis

The EF works very well at a global level because, owing to the globalization of markets, the limitations resulting from the specific productivity of local ecosystems are greatly expanded thanks to the worldwide commercial flows. However, because it is at a local level that environmental damages take place, and it is at a local level that is necessary to have concise information to support decisions, the determination of EF at a regional level, although not exhaustive, can be very useful, as shown in Chapter 24. In Italy, for example, many regions have already calculated their own EF and used it as support for decision-making.

In this regard, in the literature, there are several methods to calculate the EF, although the following two are basically used most: the 'compound method' (top–down) and the 'component-based method' (bottom–up; Wackernagel and Monfreda 2004).

The compound method is well suited when the unit of analysis is a nation. First, there is the calculation of the consumption (food and non-food) of a population, divided into different entries, starting from the production data, adding imports, and subtracting exports. Then, using the UN Food and Agriculture Organization estimates of the average productivity of the global cultivable areas, the above consumption is translated into the corresponding ecologically productive land. Adding the land for energy, we get the EF per capita. For the EF calculation, it is possible to use suitable matrices, in the form of Excel sheets, wherein each category of consumption is associated, through appropriate conversion factors, to the impact on the different land categories.

The component-based calculation method is useful for the EF of specific activities, in a particular region or country, using different sources of information:

for example, the data based on LCA of products. Adding the several EFs of different activities or categories of consumption, we get the aggregate EF of an individual, an organization, or a region.

These two methods, having different sensitivity and reliability levels, are suitable for use in different contexts: the first method, unlike the second one, has the advantage of the availability of statistical data, but it does not provide detailed information on the different processes and, therefore, it is of little use for political regional governance (Wackernagel et al. 2005).

One of the main difficulties with the calculation of regional EF lies mainly in the fact that the data relating to trade and consumption are not always available at a regional level, and so it is necessary to conduct appropriate approximations and evaluations, using aggregated statistics and establishing, on the basis of appropriate criteria, in what proportion the national data differ from the regional ones, which can be based on different lifestyles and consumption patterns.

Conclusions

The EF allows estimation of the 'ecological deficit' (or level of sustainability) of any country or region compared with the biological availability and productivity of the same territory, highlighting the dependence of the specific area on the biocapacity of external territories. The total amount of these deficits provides an estimate of the 'overload' of the carrying capacity of the ecosystems by the human population. Because of the physical limits of our planet, the GFN and WWF in fact warn that, by as early as 2030, we will need, paradoxically, two planets to support our lifestyles (WWF 2012). By comparing the EF with biocapacity, the GFN also calculates, annually, the 'overshoot day', that is, the annual marker of when the population under study begins living beyond the means provided by nature in a given year.

Even with all its limitations, and although the indicator was born of a global approach, the determination of the EF at a regional scale represents a good way of raising awareness of the unsustainability of the current standards of consumption and of directing really effective, sustainable development policies for governments and organizations, as shown in Chapter 24.

References

Borucke, M., Moore, D., Cranston, G., Gracey, K., Iha, K., Larson, J., Lazarus, E., Morales, J.C., Wackernagel, M., and Galli, A. (2013) 'Accounting for demand and supply of the biosphere's regenerative capacity: The National Footprint Accounts' underlying methodology and framework', *Ecological Indicators*, 24: 518–33.

Cucek, L., Klemeš, J.J., and Kravanja, Z. (2015) 'Overview of environmental footprints', in *Assessing and Measuring Environmental Impact and Sustainability*, ed. J.J. Klemeš, Oxford,UK: Butterworth-Heinemann, Chap. 5, pp. 131–93.

Ewing, B., Reed, A., Galli, A., Kitzes, J., and Wackernagel, M. (2010) *Calculation Methodology for the National Footprint Accounts*, Oakland, CA: Global Footprint Network.

Fiala, N. (2008) 'Measuring sustainability: Why the Ecological Footprint is bad economics and bad environmental science', *Ecological Economics*, 67: 519–25.

Galli, A., Wackernagel, M., Iha, K., and Lazarus, E. (2014) 'Ecological Footprint: Implications for biodiversity', *Biological Conservation*, 173: 121–32.

Galli, A., Wiedmann, T.O., Ercin, A.E., Knoblauch, D., Ewing, B.R., and Giljum, S. (2012) 'Integrating ecological, carbon and water footprint into a "footprint family" of indicators: Definition and role in tracking human pressure on the planet', *Ecological Indicators*, 16: 100–12.

Global Development Research Center (2015) Sustainability Concepts. Factor 10. Available at: www.gdrc.org/sustdev/concepts/11-f10.html (accessed 8 September 2015).

GFN (2009) Ecological Footprint Standards 2009. Available at: www.footprintstandards.org (accessed 5 September 2015).

GFN (2010) Footprint Basics Overview. Available at: www.footprintnetwork.org/en/index.php/GFN/page/footprint_basics_overview (accessed 21 February 2014).

Hoekstra, A.Y., and Wiedmann, T.O. (2014) 'Humanity's unsustainable environmental footprint', *Science*, 344: 1114–17.

Kitzes, J., Moran, D., Galli, A., Wada, Y., and Wackernagel, M. (2009) 'Interpretation and application of the ecological footprint: A reply to Fiala (2008)', *Ecological Economics*, 68: 929–30.

Kitzes, J., Peller, A., Goldfinger, S., and Wackernagel, M. (2007) 'Current methods for calculating national Ecological Footprint Accounts', *Science for Environment & Sustainable Society*, 4: 1–9.

Lindahl, P., Robèrt, K.H., Ny, H., and Broman, G. (2014) 'Strategic sustainability considerations in materials management', *Journal of Cleaner Production*, 64: 98–103.

Moffatt, I. (2000) 'Ecological footprint and sustainable development', *Ecological Economics*, 32: 359–62.

Monfreda, C., Wackernagel, M., and Deumling, D. (2004) 'Establishing national natural capital accounts based on detailed Ecological Footprint and biological capacity assessments', *Land Use Policy*, 21: 231–46.

Rees, W.E. (1992) 'Ecological footprints and appropriated carrying capacity: What urban economics leaves out', *Environment & Urbanization*, 4: 121–30.

UN, Department of Economic and Social Affairs, Population Division (2013) World Population Prospects: The 2012 Revision, Vol. I: Comprehensive Tables ST/ESA/SER.A/336, New York: UN.

Wackernagel, M. (1994) *Ecological Footprint and Appropriated Carrying Capacity: A Tool for Planning Toward Sustainability* (PhD thesis). Vancouver, BC: School of Community and Regional Planning, University of British Columbia.

Wackernagel, M., and Monfreda, C. (2004) 'Ecological Footprints and energy', *Encyclopedia of Energy*, 2: pp. 1–11. Available at: www.weizmann.ac.il/plants/Milo/images/Ecological Footprints_energy.pdf (accessed 12 May 2015).

Wackernagel, M., Monfreda, C., Moran, D., Wermer, P., Goldfnger, S., Deumling, D., and Murray, M. (2005) National Footprint and Biocapacity Accounts 2005: The Underlying Calculation Method. Global Footprint Network. Available at: www.footprintnetwork.org/download.php?id=5 (accessed 10 May 2015).

WWF (2012) Living Planet Report 2012. Available at: http://d2ouvy59podg6k.cloudfront.net/downloads/1_lpr_2012_online_full_size_single_pages_final_120516.pdf [12 May 2015].

8

ENVIRONMENTALLY EXTENDED INPUT–OUTPUT ANALYSIS FOR SUSTAINABLE REGIONAL DEVELOPMENT

Eskinder Demisse Gemechu, Isabela Butnar, Maria Llop, Francesc Castells, and Guido Sonnemann

Introduction

The ever-increasing need for responses from governments, global companies, and individuals to reduce their potential environmental impacts has motivated and contributed to the development and improvement of already existing methods for quantifying resource use and emissions and calculating different related environmental impacts. Life cycle assessment (LCA) is one of the most widely used tools in this regard (Stewart and Weidema 2006; Finnveden et al. 2009; Guinée et al. 2002; ISO 2006a; Sonnemann et al. 2003) and is described in detail in Chapter 10. LCA is a systematic tool for the evaluation of the environmental impacts of a product system, taking into account all the processes in its life cycle from the extraction of raw materials, manufacturing process, production, and use to end-of-life treatments (ISO 2006a, 2006b). The environmental impacts of a product system can be modelled by using either the process-based LCA approach or economic input–output LCA (EIO-LCA), based on environmentally extended input–output (EEIO) analysis (Hendrickson et al. 2006), or a combination of both approaches, which is referred to as hybrid IOLCA (Suh and Huppes 2005). Despite the fact that there is variation in modelling among the different methods introduced, they are all based on a life cycle approach, quantify resource use and emissions and calculate environmental impacts.

This chapter aims to elaborate the methodological aspects of EEIO analysis as a life cycle approach and its relevance for sustainable regional development. EEIO analysis is a top–down approach used to account for resource consumption and emissions release based on economic IO tables (Kitzes 2013; Hendrickson et al. 2006). The approach uses generic data at international, national, or regional level to evaluate the raw material and energy resource flows and the associated environmental impacts from different sectors in a given economy.

The method of EEIO analysis

The EEIO model is derived from the Leontief's IO table (Leontief 1941). The simplified scheme of an IO table is illustrated in Table 8.1. It is symmetrical in nature as it is based on a one-to-one economic sector and product relationship, that is, each industry is assumed to produce only one product, and each product is produced by only one industry. Each row in the IO table is read as the use of

Table 8.1 A simplified scheme of economic IO table framework

	Industry	Final use		Total
Industry	Intermediate use	Domestic demand	Export demand	Total use
	Import intermediate	Import final demand		
	Value added			
	Total supply			

products by the corresponding economic sector, either as intermediate input by the production system or as a final demand. The final demand side comprises demand by household, government, and export; often they are classified as domestic and foreign demands.

The sum across the row refers to the total sectoral output. Each column in the IO table contains information about the purchase of products by the corresponding industry. It comprises purchases from the production system, from payment sectors, and imports. The sum along the column is the total input, which is equal to the total output (Miller and Blair 2009).

The formulation of the IO model is based on two sets of data: data on transactions among sectors and data on the purchase of goods and services by final consumers. Using the two data sets, the output from a given sector can be distributed to all sectors in the economy and final consumers. The IO model relates the total output from a given sector to its intermediate supplies and final demand. The simple relationship can be mathematically represented in matrix notation as follows:

$$\boldsymbol{x} = \boldsymbol{Z}i + \boldsymbol{y} \tag{8.1}$$

where $\boldsymbol{x}$ is the vector of sectoral total output. Each element x_i represents the total output of sector i. $\boldsymbol{Z}$ is a transaction matrix in which each element z_{ij} stands for the amount of sale by sector i to sector j. $\boldsymbol{y}$ is a vector of final demand, which comprises demand by households, government, and foreign trades. i is a column vector of ones.

Equation 8.1 can also be expressed to reflect the dependence of intermediate flows on the total output of each sector using the following equation:

$$\boldsymbol{x} = \boldsymbol{A}\boldsymbol{x} + \boldsymbol{y} \tag{8.2}$$

where $\boldsymbol{A}$ is a matrix of technological coefficients (the terms IO coefficient or direct input coefficient could also be used, alternatively). It is derived from the transaction matrix, $\boldsymbol{Z}$, and the total output vector $\boldsymbol{x}$ as:

$$\boldsymbol{A} = \boldsymbol{Z}\hat{\boldsymbol{x}}^{-1}$$

Each element a_{ij} represents the amount of input requirement by industry j from industry i in order for industry j to produce a unit monetary output for final demand. Here, the definition of the technology coefficient and then the Leontief IO model are based on two assumptions: the constant returns to scale and fixed proportion. For any change in total production of industry j, the amount requirement from industry i is in the same proportion as the change in the total output; this implies that the technological coefficient a_{ij} keeps a fixed relationship between the total output of an industry and its inputs, which ignores the economies of scale in production. On the other hand, the model considers that, for any change in total output of a given industry, the proportions of inputs to the industries remain the same, which means that the input proportions are fixed with the technological coefficients (Miller and Blair 2009).

The linearity assumption of the IO model allows expression of the output of a given sector for an arbitrary demand ***y***, which could be demand from household or government or any demand on a specific sector, and so on. Therefore, Equation 8.2 can be formulated as:

$$\boldsymbol{x} = \left(\boldsymbol{I} - \boldsymbol{A}\right)^{-1} \boldsymbol{y} \tag{8.3}$$

where $\boldsymbol{I}$ is an identity matrix with 1s in the main diagonal and 0s elsewhere. $(\boldsymbol{I} - \boldsymbol{A})^{-1}$ is the Leontief inverse or the total requirement matrix, which shows the dependency of total output on the final demand.

The economic IO model represented in Equation 8.3 needs to be extended with environmental data, such as emissions, resource consumptions, or waste generation, in order to be used as a tool for environmental assessments. The impacts associated with any arbitrary final demand can be calculated using the equation below:

$$\boldsymbol{n} = \boldsymbol{e}'\left(\boldsymbol{I} - \boldsymbol{A}\right)^{-1} \boldsymbol{y} \tag{8.4}$$

where $\boldsymbol{e}$ is a column vector of environmental impacts per total output. Each element e_j represents the impact associated with a unit output from industry j.

Two broad classes of data set are required to perform EIO-LCA: the economic IO tables and environmental extension data. The IO tables are constructed based on data from international, national, or regional accounts, and this provides consistent, complete data sets for measuring the economic activity of the economic area studied. In most cases, IO tables are built on the national accounts that are habitually available at country level, whereas, for international and regional purposes, they have to be derived and include aggregated statistics, such as GDP, investment, and spending and saving of the economic system. They are compiled using an internationally standardized industry classification provided by the UN. In most countries, IO tables are not published in every activity year; therefore, they need to be constructed using economic tables (use and supply tables) from the national accounts. Detailed information on how to construct IO tables can be found in Miller and Blair (2009).

Environmental extension data are used in environmental accounts that are also provided by the national statistical office or by researchers (Kitzes 2013). In most cases, they are supplied with the same industry classification as the economic IO tables. When it is not the case, additional effort may be required to aggregate or disaggregate the sectors in the environmental accounts in order to have the same level as the IO tables. For example, the greenhouse gas (GHG) emissions data from the Spanish Satellite Atmospheric Emissions Account are aggregated to twenty-nine industries, whereas the symmetric IO table classifies the economy into seventy-three industries (INE 2008).

EEIO analysis for sustainable regional development

EEIO analysis has important features that make it a relevant life cycle approach for regional sustainable development. Whereas process-based LCA modelling focuses on the use of product- or plant-specific information for companies, most decisions made by regional governments concern the economic sectors and related regional industrial activities; therefore, they need an assessment at this level, which is provided by EEIO analysis.

EEIO models link all economic sectors in a given region and, hence, allow the interdependence of industries to be studied. They are able to cover all economic activities in a defined region and, hence, provide a complete, economy-wide environmental impact assessment of different production and consumption decisions with regard to a system boundary definition. They are also able to assess both direct and indirect environmental impacts explicitly, which allows them to cover impacts associated with the final demand of a given product. Through path analysis, they also allow a detailed tracing of the main sources and drivers behind each environmental impact.

To focus Danish product-oriented environmental policy, a study applying EEIO analysis was performed, identifying the most important product groups from an environmental perspective. The environmental impacts were analysed from different perspectives, differing primarily in their system delimitation. The top ten environmentally most important product groups (out of 138 industry products and ninety-eight final consumption groups) were listed for each of the perspectives, using both total environmental impact and environmental impact intensity as ranking principles. The study covered all substances that contribute significantly to the environmental impact categories of global warming, ozone depletion, acidification, nutrient enrichment, photochemical ozone formation, ecotoxicity, human toxicity and nature occupation. The differences in results between the perspectives were elaborated, and their policy relevance was discussed. The top ten product groups accounted for a surprisingly large share of the total environmental impact of Danish production and consumption (up to 45 per cent, depending upon the perspective). This implied that product-oriented environmental policy might achieve large improvements by focusing on a rather small number of product groups. Especially noticeable were the export of meat and ship transportation. This led to the recommendation to include specific policy measures targeting both foreign producers and foreign markets. The results confirm results of other, similar studies, but are more detailed and have lower uncertainty, owing to a number of improvements in data and methodology and the focus being limited to Denmark, a relatively small country with fewer than 6 million inhabitants (Weidema et al. 2006).

The use of EEIO analysis was also demonstrated for quantifying the overall sustainability impacts of forest industries in the Finnish economy in 2005. Direct GHG emission, land use, employment, and import impacts of economic sectors were transformed into impact intensities of products. The intensities were used to construct a final demand-based emission inventory demonstrating the relative importance of export, investment, and consumption activities in causing environmental and social impacts (Mattila et al. 2011).

A special form of EEIO analysis focusing just on materials is the approach to building material flow accounts, as described in Chapter 9. The construction of a basic set of economy-wide material flow accounts is mandatory for EU countries.

Hence, these different examples show that countries that are relatively small in size and population, which we would consider regions in larger countries, also successfully use EEIO analysis, including for material flow accounts. An application of a similar approach to regional sustainability is presented in Chapter 25.

Remarks and conclusion

Despite these important uses of EEIO analysis thanks to its unique features, it is criticized for its poor level of resolution for application at the product level. A collection of industries that have quite heterogeneous processes and outputs is usually aggregated as a single industry with a single output (e.g. mineral water, soft drinks, and alcoholic drinks are aggregated into manufacture of beverages). The environmental intervention from this industry is then approximated to be the average of interventions from all industries under this umbrella. This is one of the potential sources of uncertainty in the use of EEIO analysis, specifically when used in the analysis of the environmental impacts of a specific product, for example, mineral water. However, some authors point out that the errors due to industrial aggregation are still lower than the truncation error from cut-off in a process-based LCA approach (Bullard et al. 1978; Lenzen 2000).

The treatment of imported goods is also another limitation of single-region EEIO analysis. When dealing with only with a single region, EEIO models usually assume that imported products are produced with the same technology as in the importing countries and, therefore, they are approximated to have the same environmental burden as corresponding products in the domestic economy. This could be adjusted by further extending the model and considering specific production processes for imported products, by constructing multiregional input–output model (Gemechu et al. 2015; Peters and Hertwich 2009).

In conclusion, EEIO analysis is a powerful life cycle approach for assessing the environmental sustainability of new developments at regional level. By interlinking all economic activities within a defined regional system, the method allows decision-makers to see the consequence of any change in the production and consumption structure of a given sector for the economy. Such an assessment can be done with relatively limited time and labour investment if the responsible statistical agency provides the input–output data at the regional level. Overall, it can be said that EEIO analysis is able to provide a smarter overview of relevant resource use and environmental impacts of production and consumption activities at the regional level than multiple process-based LCA studies, which each time demand an exhaustive inventory to model the resource use and emissions of a specific product system.

References

Bullard, C.W., Penner, P.S., and Pilati, D.A. (1978) 'Net energy analysis: Handbook for combining process and input-output analysis', *Resources & Energy*, 1(3): 267–313.

Finnveden, G., Hauschild, M.Z., Ekvall, T., Guinée, J., Heijungs, R., Hellweg, S., and Suh, S. (2009) 'Recent developments in Life Cycle Assessment', *Journal of Environmental Management*, 91(1): 1–21.

Gemechu, E.D., Butnar, I., Llop, M., Castells, F., and Sonnemann, G. (2015) 'CO_2 emissions flow due to international trade: Multi-regional input–output approach for Spain', *Greenhouse Gas Measurement & Management*, 4(2–4): 201–14.

Guinée, J.B., Gorrée, M., Heijungs, R., Huppes, G., Kleijn, R., de Koning, A., and Huijbregts, M.A.J. (2002) *Handbook on Life Cycle Assessment: Operational Guide to the ISO Standards*, Dordrecht, Netherlands, Boston, MA, and London: Kluwer Academic.

Hendrickson, C.T., Lave, L.B., and Matthews, H.S. (2006) *Environmental Life Cycle Assessment of Goods and Services: An Input–Output Approach*, London: Routledge.

Kitzes, J. (2013) 'An introduction to environmentally-extended input–output analysis', *Resources*, 2: 489–503.

INE (2008) *2008 Statistical Yearbook of Spain*, Madrid: Instituto Nacional de Estadística (INE).

ISO (2006a) *ISO14040: Environmental Management – Life Cycle Assessment – Principles and Framework*, Geneva: ISO.

ISO (2006b) *ISO14040: Environmental Management – Life Cycle Assessment – Requirements and Guidelines*, Geneva, ISO.

Lenzen, M. (2000) 'Errors in conventional and input–output-based life–cycle inventories', *Journal of Industrial Ecology*, 4(4): 127–48.

Leontief, W. (1941) *The Structure of American Economy, 1919–1939: An Empirical Application of Equilibrium Analysis*, Oxford, UK: Oxford University Press.

Mattila, T., Leskinen, P., Mäenpää, I., and Seppälä J. (2011) 'An environmentally extended input–output analysis to support sustainable use of forest resources', *The Open Forest Science Journal*, 4(1): 15–23.

Miller, R.E., and Blair, P.D. (2009) *Input–Output Analysis: Foundations and Extensions*, 2nd edn, New York: Cambridge University Press.

Peters, G.P., and Hertwich, E.G. (2009) 'The application of multi-regional input–output analysis to industrial ecology', in *Handbook of Input-Output Economics in Industrial Ecology*, ed. A. Tukker, Dordrecht, Netherlands: Springer, vol. 23, pp. 847–63.

Sonnemann, G., Castells, F., and Schuhmacher, M. (2003) *Integrated Life-Cycle and Risk Assessment for Industrial Processes*, Boca Raton, FL: Lewish.

Stewart, M., and Weidema, B. (2006) 'LCA methodology: A consistent framework for assessing the impacts from resource use – a focus on resource functionality', *The International Journal of Life Cycle Assessment*, 10(4): 240–7.

Suh, S., and Huppes, G. (2005) 'Methods for life cycle inventory of a product', *Journal of Cleaner Production*, 13(7): 687–97.

Suh, S., Lenzen, M., Treloar, G.J., Hondo, H., Horvath, A., Huppes, G., and Norris, G. (2004) 'System boundary selection in life-cycle inventories using hybrid approaches', *Environmental Science & Technology*, 38(3): 657–64.

Weidema, B., Suh, S., and Notten, P. (2006) 'Setting priorities within product-oriented environmental policy', *Journal of Industrial Ecology*, 10(3): 73–87.

9

MATERIAL FLOW ACCOUNTS

Aldo Femia and Renato Marra Campanale

Introduction

Material flow analysis (MFA) is a wide family of tools. These have in common the systematic quantitative assessment, in physical units, of the materials that flow within a system. That system is clearly defined in space and time (i.e. between its subsystems) and between the system and its environment. The system analysed may be a plant, a whole firm, a specific substance (substance flow analysis), an industrial material (e.g. steel, cotton, gypsum), or a product (in which case, MFA overlaps with life cycle assessment (LCA), which will be introduced in Chapter 10, and in particular with the life cycle inventory phase), or it may be entire national or regional economies (OECD 2008).

Relevance of MFA

To the extent that nature constitutes a system's environment, assessment of the material flows between the system and its environment is of immediate environmental relevance, as the inputs and outputs are, respectively, outputs from and inputs to the natural environment. Both of these kinds of flow influence the structure, quality, and functioning of the natural environment.

Moreover, as all materials used in human systems come from, and eventually end up in, the natural environment, any assessment of material flows may be considered relevant, at least indirectly, in environmental terms, as such assessment may help to optimise the way materials are handled in the pursuit of their intended purpose. In other words, MFA may help avoid, reduce, or substitute the intake of materials into the analysed system and/or the restitution of residues to its environment. Reducing the intake means (directly or indirectly) reducing the demand for materials that have to be taken from nature. Reducing the residues means – immediately or eventually – reducing flows of degraded matter from the human system to the natural environment.

Economy-wide material flow accounting

Economy-wide material flow accounting (EW-MFA) is the most aggregated tool in the MFA family. EW-MFA uses a standardized methodology (Eurostat 2001, 2013, forthcoming) to provide an overview of the material inputs and outputs of an entire socio-economic system, both at national (or supranational) and at regional (or local) level. It quantifies the system's material flows, excluding only the inputs of water and air not embodied in products or residuals. A nation's or a region's socio-economic system is viewed as a 'black box' in these accounts: only what comes in and what goes out are monitored. Internal flows remain invisible.

Knowing the quantities of resources used in an entire socio-economic system means knowing the absolute size of its metabolism. Measures of this size provide basic information on the system's potential to generate pressures through the handling of materials. This is of primary importance for sustainability, which can be approached only by placing this potential under control. The focus of EW-MFA is the absolute limits placed by nature on the expansion of human activities in general, rather than specific, well-circumscribed environmental issues. In other words, we are concerned with the very long run of socio-metabolic patterns and transitions (Fischer-Kowalski and Haberl 1993), rather than the day-to-day management of marginal policy changes or management of specific substances.

The quality and origin of the flows are also addressed within EW-MFA, but to a minimum extent. Completely different kinds of tool – such as LCA and also environmental risk assessment, belonging to the same wide family of system analytical MFA-based tools – are necessary to deal with low-volume but high-risk substances such as toxic chemicals, soil contaminants, endocrine disruptors, ozone depleting substances, or nano-products.

The construction of a basic set of EW-MFA is mandatory for EU countries: it is included in Regulation (EU) No 538/2014 on environmental economic accounts. This means that countries of small geographical size and population – which would be considered regions in larger countries – also prepare basic sets of EW-MFA. Their data production cycle is well established, and derived indicators are frequently used by policymakers. For instance, domestic material consumption (DMC) is the leading indicator of EU resource efficiency policy and resource productivity, measured as GDP/DMC, and features in the EU's set of sustainable development indicators as the headline indicator for sustainable production and consumption. With this approach, the basic task of policy is to promote qualitatively and quantitatively responsible lifestyles and to make it easier for citizens to adopt them.

The information required by the European regulation covers domestic extraction of used materials, imports, and exports and is quite detailed concerning the classification to be used for reporting. This allows the building of linkages to the economic sectors from which the flows stem – either as direct intake from nature or as natural resources transformed into raw materials and products. The first level of classification identifies four kinds of material: biomass, ores, fossil fuels, and other minerals. Biomass may be cultivated crops, used residuals from cultivation, wood and other gatherings from forests, fish and other wild animals. Ores are ferrous and non-ferrous, detailed according to the kind of metal they contain. The term fossil fuels covers all kinds of fuel: solid, liquid, and gaseous. Non-metal, non-energy minerals are detailed according to ten categories, based on a mix of mineralogy (e.g. limestone, clay, basalt, feldspar) and possible forms and uses (e.g. blocks, stones, loose materials; ornamental, for construction, for chemistry).

Many important flows – relating to the output side, unused materials and indirect extraction – are beyond the scope of the European regulation but are covered by EW-MFA methodology and are included in many practical applications. These flows are outlined below.

Flows on the output side, other than exports, include (1) additions of matter to man-made useful stocks, such as buildings, infrastructure, machinery, live domesticated animals, durables, human beings themselves; (2) additions of matter to man-made non-useful stocks, such as waste that is landfilled or stockpiled; and (3) emissions of all kinds to soil, water, and air.

The environmental relevance of these flows is obvious, but only the third kind is very well (though not fully) covered by environmental statistics. Air emissions are also the specific target of another piece of the same European regulation as EW-MFA. In that case, the focus is on the linkage with economic activities, resulting in much detail both for the emitted substances and the determinants underlying the flows. But, insofar as they are covered by environmental statistics and accounts, residual flows are dealt with individually, not as part of an organic, coherent system where all parts are interdependent. Only counting all of these flows, next to those included in the Regulation and to ancillary air and water flows, will enable formulation of an economy-wide balance of the matter throughput (Boulding 1968) on which an assessment of an economy's efficiency in transforming materials can be based.

Flows of unused materials include those that are not embodied in marketable products but that have to be removed from their natural site in order to gain access to useful materials (e.g. mining overburden), those that are needed to carry out an activity (e.g. soil excavated for building foundations or for road construction), or those that are an unintended by-product of primary activities (such as the non-usable parts of cultivated plants). The environmental relevance of these flows is not always obvious. However, they are, by definition, wastes that need to be taken care of as soon as they are extracted from the natural environment.

Indirect flows of used and unused materials – that is, materials whose removal from nature is necessary in order that a certain product can be obtained but that are not physically embodied in that good or service, for example the part of the ore that remains in the metal producing country as smelting waste, while the metal is exported to another country.

The material footprint

Together with the actual flows, these define the concept of material footprint, which – like MFA in general – can be applied to a product or organisation, or to an economy's or region's activities. This concept coincides with that of raw material equivalents (RMEs) if we consider only used materials. There are basically two methods to determine indirect flows, material footprints, and RMEs: (1) technical coefficients derived from the materials' use inventories of LCA (see Chapter 10 for more details) and (2) the Leontievian input–output (IO) analysis of materials' intake from nature. The latter methodology can be seen as a specific form of environmentally extended input–out (EEIO) analysis explained in Chapter 8. It is usually carried out making use of monetary IO tables. It may be based on a 'domestic technology assumption' – leading to a determination of avoided flows rather than actual flows – or on interconnection of several IO tables (multiregional IO tables) so as to mimic the actual functioning of the world economy and deliver estimates of actual flows.

Both approaches have shortcomings and limitations, both methodological and practical (the latter especially due to data availability). The first is very detailed, but incomplete, and suffers from the cut-off criteria applied in LCA. The second is complete but lacks detail and is imprecise, as it suffers from high aggregation of products and activities (especially for the most environmentally relevant ones) and low appropriateness of some monetary flows in representing the physical 'activation' relations. The practitioners' response to this challenge is the development of hybrid approaches, that is, the application of various mixes of higher disaggregation of IO tables and modular formulation of the problem, with parts and/or layers of the calculations based on LCA and parts/layers based on EEIO analysis.

A further extension of EW-MFA concerns flows internal to the economic system, with the development of supply/use and IO tables where the inter-industry and production-to-final-use flows are expressed in physical units (physical supply use table and physical IO table). This development may also be partial, that is, limited to products and activities for which the relevance of monetary relations is known to be low. The use of these hybrid tables in EEIO analysis (i.e. using IO tables that include rows and columns in physical units) constitutes another avenue of improvement for the calculation of material footprints, which, mixed with LCA and disaggregation, is important in the ongoing search for the best feasible model.

MFA and policy

The system perspective adopted by EW-MFA allows for a full analysis of the environmental releases associated with different stages of a particular material flow (e.g. releases from extraction or production versus use or disposal) or the flows of different materials. These accounts improve understanding of the driving forces behind the environmental burden. Greater insight will then help decision-makers prevent environmental problems, reduce inefficiencies in materials use, and improve resource productivity.

Several EW-MFA potential policy uses (frequently in conjunction with other tools) could be identified for tackling various issues at different levels, including the regional (or local) levels. An OECD report (OECD 2008) provides examples of existing applications.

First of all, environmental policies could use the EW-MFA detailed breakdown of materials to map the flows of nutrients or contaminants in a region and to identify whether, where, and to what extent environmental degradation downstream stems from these flows. EW-MFA, in conjunction with substance flow analysis and MFA (Brunner and Rechberger 2003), can help assess environmental pressures from metal ore extraction and metal production and the part due to inefficiencies in production technologies, together with the benefits that could be gained from new technologies and from improved recovery and recycling. EW-MFA can monitor and help understand indirect and unused material flows and their effects on the environment, in domestic as well as in foreign economies.

As regards waste, the output side of EW-MFA can be used to analyse trends and how they affect opportunities for resource conservation, resource productivity, and material recovery and recycling. There is a high policy interest in adding, or at least relating, circular material flows (also known as secondary raw material or recycled material input) to the existing EW-MFA. These circular material flows may be set against the usual material inputs (domestic extractions, imports, direct material input, DMC) in order to monitor progress in resource productivity due to increased efforts of recycling.

As long as economic and environmental policies are fully integrated, EW-MFA can (1) measure aspects of the physical performance of the economy and relate it to its economic performance; (2) measure the degree of decoupling between direct and indirect environmental pressures (pollution, waste, primary resource inputs) and economic growth; (3) support structural analysis of the global economy in physical terms: that is, analyse the effects of globalisation on international material flows, study the substitution of domestic raw materials with imported ones, and examine the interaction between production and consumption patterns.

Structural effects of trade and environmental measures on international materials markets and on flows of environmentally significant materials (e.g. hazardous materials, secondary raw materials, recyclable materials) can also be monitored by EW-MFA, along with the environmental implications of changes in international material flows, including environmental pressures from indirect flows abroad associated with trade and environmentally significant materials embedded in imported goods.

Conclusion

If business as usual is not an option to ensure sustainable resource use, the possible alternatives include strengthening existing good practices in material use and setting coherent international, national, and local targets for resource use (Dittrich et al. 2012). However, targets – such as those proposed by Giljum et al. 2014, for instance – should be designed in such a way that they enable a reduction of problems in various categories of natural resource use. Shifting problems from one resource category to another could thereby be avoided, and positive outcomes in environmental, economic and social terms could be allowed instead, which would then avoid the maximisation of the outcome in one dimension while deteriorating the situation elsewhere.

Overall, we can say that, in order to move towards more sustainable resource use, natural resource accounting by means of material flow accounts is crucial, and, hence, we can observe an increase in its use at the national and international level. A next step, among others, is to implement material flow accounting at the regional level, using, as far as possible, good practices of accounting for different materials (biomass, metal, non-metallic minerals, and fossil energy carriers) and calculating per capita consumption levels.

References

Boulding, K.E. (1968) 'The economics of the coming spaceship Earth', in *Beyond Economics*, Ann Arbor, MI: University of Michigan Press, pp. 275–87.

Brunner, P.H., and Rechberger, H. (2003) *Practical Handbook of Material Flow Analysis*, Boca Raton, FL: CRC Press.

Dittrich, M., Giljum, S., Lutter, S., and Polzin, C. (2012) *Green Economies around the World? Implications of Resource Use for Development and the Environment*, Vienna: Sustainable Europe Research Institute.

Eurostat (2001) *Economy-Wide Material Flow Accounts and Derived Indicators. A Methodological Guide*, Luxembourg: European Communities.

Eurostat (2013) *Economy-Wide Material Flow Accounts (EW-MFA), Compilation Guide 2013*, Luxembourg: European Communities.

Eurostat (forthcoming) *Economy-wide Material Flow Accounts (EW-MFA) Manual*.

Fischer-Kowalski, M., and Haberl, H. (1993) *Metabolism and Colonisation – Modes of Production and the Physical Exchange between Societies and Nature*, Vienna: Schriftenreihe Soziale Ökologie, Band 26.

Giljum, S., Hartwig, F., Hinterberger, F., and de Schutter, L. (2014) *Perspectives and Assumptions for Setting Resource Targets Report of IntRESS WP 2a*, Vienna: Sustainable Europe Research Institute.

OECD (2008) *Measuring Material Flows and Resource Productivity*, Paris: OECD.

10

LIFE CYCLE ASSESSMENT

Giuseppe Tassielli, Pietro Alexander Renzulli, and Bruno Notarnicola

Introduction

The exponential increase in the production of goods for human consumption and development that has occurred in recent decades has inevitably made mankind aware of the corresponding potential risks for the environment and, hence, highlighted the need to clearly understand the real value of such goods in terms of their interaction with the biosphere (e.g. resource use and interference with air, ground, and water).

This interaction with the environment can only be understood if the processes used to make these products are known and if there is information on raw material and energy use and on emissions and by-products throughout production, as well as during the product's use and disposal phases. In other words, a holistic method is needed to study the whole 'history' of the life cycle of a product. This approach, called life cycle thinking, goes beyond the traditional focus on production sites and manufacturing processes to include the environmental, and even social and economic, impacts of a product over its entire life cycle.

The technical instrument used to assess the environmental impacts in such a systemic kind of study is known as life cycle assessment (LCA). The initial LCA ideas and schemes were developed at the end of the 1960s (Hunt and Franklin 1996). Since then, LCA has attracted the attention of policymakers, academia, industry, and consumers, giving life to a new, standardised approach for the management of environmental issues related to product systems and services.

The LCA methodology according to ISO standards

LCA was first standardised as a process for environmental management in 1997, under the ISO 14040 and ISO 14041–42–43 standards, published between 1998 and 2000. The first two standards were then revised and substituted by ISO 14040:2006 and 14044:2006 (ISO 2006a, 2006b). The first describes the principles and framework of LCA, and the latter specifies requirements and provides guidelines for an LCA study. The methodology described in these standards is the backbone for any LCA-based study, applying equally to those that consider only single-issue approaches – such as carbon footprint (CF; see Chapter 11) and water footprint (WF; Chapter 12) – and to those addressing broader life cycle sustainability assessment, covering life cycle costing (LCC) and social LCA (Chapter 14), as well as organisational LCA (Chapter 15).

In summary, there are four phases in an LCA study. The first is the goal and scope definition phase, stipulating the intended application, the reasons for carrying out the LCA study, the intended audience, and how the results are to be used. The product system/service is also described in this phase, and the life cycle phases are clearly identified, together with the system boundaries. The functional unit is also defined (providing a reference to which the input and output data are normalised), as are the approaches that will be used to allocate the environmental burden resulting from the different by-products (if any) of the product system under analysis. Lastly, the impact assessment methodology is defined, together with assumptions, limitations, data requirements, type of critical review, and other information regarding the type and format of the report.

The second phase is the life cycle inventory (LCI), in which quantitative and qualitative data regarding each unit process included within the system boundary are collected and/or calculated. This brings together a large quantity of input–output data on the system being studied, concerning amounts of natural resources and energy used, products (or by-products), and emissions entering the biosphere. This is usually the most complicated, time-consuming, and expensive phase of an LCA. There are cases where the goal of an LCA can be attained by the performance of an inventory analysis only and by interpretation its results. This is usually referred to as an LCI study.

Next is the life cycle impact assessment (LCIA) phase, in which all the above-mentioned input and output inventory data are classified and grouped based on their potential contribution to environmental impact categories. The potential contribution of these data to the impact categories is then calculated via characterisation models. The impact categories are numerous and are chosen based on the goal and scope of the study. Examples of impact categories can be found in the ISO/TR 14047 standard. The above-mentioned steps of the LCIA are mandatory for all LCAs. Other optional steps for the impact assessment concern calculation of the magnitude of the impact category indicator results relative to reference information (normalisation); sorting, and possible ranking, of the impact categories (grouping); and convertion, and possible aggregation, of indicator results across impact categories using numerical factors based on value choices (weighting). The implementation of the above-mentioned LCIA steps will provide an indication of the sustainability, in terms of potential impacts, of the life cycle of the product system or service under analysis.

The final stage of an LCA study is the interpretation phase. This includes: identification of the significant issues based on the results of the LCI and LCIA phases; an evaluation of results that considers completeness, sensitivity, uncertainty and consistency checks; and the drawing of conclusions, establishment of limitations, and making of recommendations. The appropriateness of the definitions of the system functions, of the functional unit, and of the system boundary are also considered in this last phase, together with limitations identified by a data quality assessment and a sensitivity analysis.

Other life cycle approaches

As mentioned in the previous section, LCA is also the backbone for a series of other life cycle approaches and tools used to evaluate the sustainability, in economic, environmental, and social terms, of product–service systems and also of organisations or even of entire sectors. These are all similar and they all form a part of a common basket of techniques belonging to the so-called life cycle management toolbox (see Chapter 6), which involves a flexible, integrated management framework of concepts, techniques, and procedures to address environmental, economic, and social aspects of products, procedures, and organisations.

Single-issue approaches

Some LCA-based approaches address the environmental performance of product systems by focusing on a particular environmental aspect. Such methods, typically called footprints, have recently become popular and focus on environmental burdens currently considered of paramount importance (e.g. global warming). However, it should be pointed out that such approaches give only a partial view of the environmental profile of a product system and may not reveal other useful information or may even give indications for reducing one type of impact that inadvertently increase the burden of others.

The CF (see Chapter 11) is a very common LCA-based approach that considers the effects of a product system on global warming. There are various methodologies, such as that defined by ISO 14067:2013, PAS 2050 (a public initiative of the British Carbon Trust), the WBCSD/WRIGreenhouse Protocol, and numerous other international public initiatives.

The WF (see Chapter 12) has been developed as a means of accounting for the direct and indirect use of water arising from the consumption/production of goods and services or use at the level of entire regions. There are several initiatives concerning WF specifications, including the recent ISO 14046:2014.

Economic, social, and integrated approaches

LCC (see Chapters 14 and 33), with an underlying framework similar to that of traditional environmental LCA, essentially evaluates the costs of a product system from a life cycle perspective. Specifically, LCC summarises all of the costs associated with a product's life cycle (relating to actual money flows) covered by one or more actors in the life cycle. It may also include the costs associated with environmental externalities (e.g. social costs borne by those subjected to the adverse environmental impacts associated with production), in which case the assessment is intended as an environmental LCC.

Current research efforts are concentrating on incorporation of a process-oriented costing method into LCA (Settanni et al. 2011). At present, however, LCC is typically used in combination with LCA, as separate but consistent tools without any formal integration, as the cost information on material flows that is collected, organised, and computed is typically of poorer quality than the data needed for an LCA. Another limitation of LCC is that it can only be used as a tool to support investment decision-making concerning durable goods; it is less suited to non-durable assets, which can, on the other hand, be assessed via LCA.

Social life cycle assessment (SLCA; see Chapter 14) is intended as a tool to effectively and efficiently evaluate social, sociological, and socio-economic conditions of production and consumption. Like all LCA-based approaches, it is based on the ISO 14040 framework and complements environmental LCA and LCC. It can be applied either on its own or in combination with an environmental LCA. Social and socio-economic aspects assessed with SLCA are those that may affect stakeholders directly throughout a product's life cycle. These aspects may be linked to the behaviour of enterprises, to socio-economic processes, or to impacts on social capital. The results of an SLCA may be used for decision-making, prompting dialogue on the social and socio-economic aspects of production and consumption of goods, with the aim of improving organisations' performance in line with organisational LCA (Chapter 15) and, above all, for the well-being of stakeholders. Even though there are some guidelines, such as those from UNEP for SLCA of products, the tool is by no means fully standardised, and its methodological development is still a subject of keen interest for the academic research community.

The increasing interest in developing methods to better understand and address the sustainability of products along their life cycle has also recently brought into existence an integrated framework called life cycle sustainability assessment (LCSA; Zamagni 2012). This involves the unified use of LCA, LCC, and SLCA as a means of simultaneously evaluating the sustainability of products and services from the environmental, social, and economic points of view (see Chapter 14). This framework is still under development as, for the economic and especially for the social dimensions, there is still a need for consistent and robust indicators and methods.

Main applications of LCA and related approaches

LCA is a key sustainable innovation tool for more sustainable product design (design for environment or eco-design), as described in detail in Chapter 17. It is also used for the identification of significant environmental aspects related to environmental management system regulations.

LCA is currently often used for marketing purposes by private companies, for green public procurement, and for environmental labelling of products. In fact, growing awareness of the effects of their lifestyle on the environment on the part of consumers has created a need for information on the sustainability of products and services. This has led to 'eco-labelling' or 'environmental labelling' systems providing environmental information on a product/service and encouraging supply and demand for products and services that cause less damage to the environment, thereby stimulating a continuous environmental improvement process managed by the market. To this end, voluntary environmental labels/declarations or eco-labels have been standardised according to the ISO 14020 series. These labelling schemes, and the uses of LCA to support them, are explained further in Chapter 18.

LCA use at regional level and as a means for sustainable regional planning

The use of LCA is important at the regional level to support companies and research organisations in their efforts in sustainable innovation for the products and services that are produced and used in a given area. In this way, LCA and the other approaches based on it are factors for increased competitiveness of a region.

Even though LCA was originally developed for the environmental evaluation of products, it can also be applied to systems at a macro level, in order to evaluate the resource efficiency of economic sectors. Typically, this involves the evaluation of a sector defined as a particular system implemented at regional level (for example, energy, transport, or waste management systems), in terms of its impacts, and comparison with possible future solutions involving different technologies (Kewmmel et al. 1997). In this context, environmentally extended input–output analysis, as introduced in Chapter 8, is also a relevant life cycle approach that uses LCA elements in the impact assessment phase. Overall, with these different approaches, it is possible to compare future scenarios with different management systems and technological set-ups, whose lifespan may reach 100 years, and that can help regional decision-makers in their choices regarding the future.

Lastly, planning sustainable development at a regional scale may also include the evaluation of different scenarios involving financial savings, as well as reductions in the environmental impact of purchases made by public authorities (which are major consumers of goods and services). This type of process, named green public procurement, involves the combined use of LCA and LCC, as explained in Chapters 19 and 33, on behalf of public authorities, to try to identify supplies, services, or works that are most environmentally friendly and economically advantageous, not only in terms of purchasing price but also in terms of overall costs of an asset throughout its life cycle.

Methodological issues and future perspectives

There are many methodological issues that may have to be overcome when an LCA is implemented. An indicative, non-exhaustive description of some of the major issues is given below (other methodological issues can be found, for instance, in Chapter 36).

Typically, during the inventory phase, primary data may not be easily accessible. This can occur when companies or organisations are unwilling to disclose information, or when data collection is either too expensive or too time-consuming. In such cases, secondary data from the literature and from commercial databases are often used, either as direct substitutes or as a source for calculating approximations. It is important to carefully verify the data from databases and the literature for dependency on time and on a particular site that may render them inadequate for use in the case study (see also Chapter 13 for further details on dynamic and spatial modelling).

LCA is intended as a means for evaluating a product–service system in terms of potential impacts, independently of where and when the product or service is produced, consumed, and disposed of. This means that the evaluated impact does not represent the real, effective damage that may be caused to the environment. Furthermore, there are many impact categories and numerous methodologies used for their calculation (e.g. human and environmental toxicity) that make the evaluation of the potential damage uncertain and difficult. There is currently no generally accepted methodology for assessing the impacts on biodiversity caused by land use. Such aspects are strongly site- and time-dependent and are also very complex, whereas current LCA practices for evaluating changes in biodiversity tend to simplify impact estimation. A framework for the evaluation of the impact of land use has been proposed by the UNEP/SETAC Life Cycle Initiative working group (Milà i Canals et al. 2007), using biomes as a spatial unit for biogeographic differentiation. This approach distinguishes land-use impacts between transformation impacts, occupation impacts, and permanent irreversible impacts on ecosystems.

Conclusions

In conclusion, LCA is a powerful tool for the assessment of the environmental sustainability aspects of products and services. Currently, a broad range of LCA studies is available for product systems from different economic sectors. Up to now, those assessments have mostly not considered the regional context of development. However, a potential to do so exists.

References

Bare, J.C., Hofstetter, P., Pennington, D.W., and Udo de Haes, H.A. (2000) 'Midpoints versus endpoints: The Sacrifices and benefits', *International Journal of Life Cycle Assessment*, 5(6): 319–26.

Hunt, R., and Franklin, W. (1996) 'LCA – how it came about – personal reflections on the origin and the development of LCA in the USA', *International Journal of Life Cycle Assessment*, 1(1): 4–7.

ISO (2006a) *ISO 14040: Environmental Management – Life Cycle Assessment – Principles and Framework*, Geneva: ISO.

ISO (2006b) *ISO 14044: Life Cycle Assessment – Requirements and Guidelines*, Geneva: ISO.

Kewmmel, B., Nielsen, S.K., and Sorensen, B. (1997) *Life Cycle Analysis of Energy Systems*, Frederiksberg, Denmark: Roskilde University Press.

Milà i Canals, L., Bauer, C., Depestele, J., Dubreuil, A., Freiermuth Knuchel, R., Gaillard, G., Michelsen, O., Müller-Wenk, R., and Rydgren, B. (2007) 'Key elements in a framework for land use impact assessment within LCA', *International Journal of Life Cycle Assessment*, 12(1): 5–15.

Settanni, E., Tassielli, G., and Notarnicola, B. (2011) 'An input–output technological model of life cycle costing: Computational aspects and implementation issues in a generalised supply chain perspective', in *Environmental Management Accounting and Supply Chain Management*, eds R. Burritt, S. Schaltegger, M. Bennett, T. Pohjola, and M. Csutora, New York: Springer, pp. 55–109.

Zamagni, A. (2012) 'Life cycle sustainability assessment (editorial)', *International Journal of Life Cycle Assessment*, 17: 373–6.

11

CARBON FOOTPRINT PRINCIPLES

Harish Jeswani

Introduction

Climate change represents a significant threat to the environment and livelihoods of large proportions of humanity. It is widely accepted that greenhouse gas (GHG) emissions, caused by anthropogenic activities, are responsible for global warming and the resulting dramatic changes in our climate. To prevent projected and unforeseen effects of climate change, international agreements and scientific assessments call for a concerted action for substantial reductions in GHG emissions (IPCC 2013). In order to quantify and manage GHG emissions, carbon footprint as an indicator is being widely used. It has also become an increasingly popular concept for communicating climate change impacts among industries, consumers, and policymakers. In a policy context, carbon footprint is being used for knowledge-based decision-making for sustainable consumption and production policies. To increase transparency and consistency in measuring and reporting carbon footprints, several guidelines and standards have been developed. This chapter gives an overview of the carbon footprint concept and discusses the standards and approaches for assessing carbon footprints. Some applications of the carbon footprints and its relevance for regional authorities are also discussed.

Carbon footprint concept

The term 'carbon footprint', which was inspired by ecological footprint, is commonly used to describe the impact of activities on the global climate. As described in Chapter 7, ecological footprint is a resource accounting tool to measure human demand on the Earth's ecosystem, and carbon footprint represents the net emissions of CO_2 and other GHGs over the full life cycle of a product, process, service, or organisation (Wiedmann and Minx 2007). Carbon footprint analysis is also known as GHG accounting or GHG inventory. Other terms, such as embodied carbon, carbon content, carbon flows, virtual carbon, GHG footprint and climate footprint are often used interchangeably with carbon footprint (Pandey et al. 2011; Peters 2010).

The life cycle concept of the carbon footprint means all direct (on-site, internal) and indirect (off-site, external, embodied, upstream and downstream) emission sources, sinks, and storage, within the specified spatial and temporal system boundary, need to be taken into account. Normally, it is expressed as a CO_2 equivalent (usually in kilograms or tonnes per unit of analysis), and as such it is equivalent to the usual LCA impact category of global warming potential (GWP). It is calculated as the sum of emissions of GHGs multiplied by their respective GWP factors. The GWP factors are available for time horizons of 20, 100, and 500 years; however, GWP factors for a 100-year time horizon

Table 11.1 100-year GWP of six main GHG gases

GHG	Second Assessment Report (IPPC, 1995)	Third Assessment Report (IPCC, 2001)	Fourth Assessment Report (IPCC, 2007)	Fifth Assessment Report (IPCC, 2013)
Carbon dioxide (CO_2)	1	1	1	1
Methane (CH_4):				
– Biogenic	21	23	25	28
– Fossil	21	23	25	30
Nitrous oxide (N_2O)	310	296	298	265
Hydrofluorocarbons	140–11,700	97–12,000	92–14,800	116–12,400
Perfluorocarbons:				
– Hexafluoroethane ($C2F_6$)	9,200	11,900	12,200	11,100
– Carbon tetrafluoride (CF_4)	6,500	5,700	7,390	6,630
Sulphur hexafluoride (SF_6)	23,900	22,200	22,800	23,500

have been accepted for calculating carbon footprint by various standards. The Intergovernmental Panel on Climate Change (IPCC), the primary authority on climate change science, has updated the GWP values several times over the years to reflect the advances in scientific understanding, as shown in Table 11.1. The latest updated GWP values for GHG gases were published in the Fifth Assessment Report (IPCC 2013). However, GWP values published in the Fourth Assessment Report (IPCC 2007) are still widely used.

Standards and protocols

Standards and guidance are available for GHG accounting depending on the scale of analysis. The GHG protocol of the World Resource Institute (WRI)/World Business Council on Sustainable Development (WBCSD) has developed standards for GHG accounting and reporting for organisations, cities, projects, and products. ISO, BSI, and several other organisations have also developed carbon footprinting standards at the product level. These standards are discussed below.

GHG Protocol for cities

The purpose of the GHG Protocol for cities is to assist local governments to develop a comprehensive GHG inventory, through the use of standardised approaches and principles, to support strategies to tackle climate change and to enable benchmarking with other cities (WRI/WBCSD 2014). It builds upon the corporate standard (WRI/WBCSD 2004) and is based on the following five principles, which are intended to underpin all aspects of GHG accounting and reporting:

- *Relevance*: Information provided in the GHG inventory appropriately should reflect the GHG emission occurring as a result of activities and consumption patterns of the city and can be used for planning and decision-making.
- *Completeness*: All GHG emission sources and activities within the chosen inventory boundary should be accounted for and reported.
- *Consistency*: Consistent methodologies should be used to allow for meaningful comparisons of emissions over time, as well as within and across cities.

- *Transparency*: To allow for a clear audit trail, all relevant assumptions, data sources used, and calculation procedures should be clearly documented.
- *Accuracy*: GHG emissions should be quantified with sufficient accuracy to ensure the integrity of the reported information, and any uncertainties should be reduced as much as possible.

The standard covers the accounting of the seven GHGs in the Kyoto Protocol: carbon dioxide (CO_2), methane (CH_4), nitrous oxide (N_2O), hydrofluorocarbons (HFCs), perfluorocarbons (PFCs), sulphur hexafluoride (SF_6), and nitrogen trifluoride (NF_3). Reporting of other GHG emissions is optional. For GHG accounting, emissions are classified in three scopes (Scopes 1, 2, and 3), which are widely adopted in corporate inventories. Under the standard, cities are required to report on Scope 1, Scope 2 and Scope 3 emissions separately. CO_2 emissions from biogenic sources are reported separately to avoid double counting.

Scope 1 emissions are also termed 'territorial' emissions, because they are produced from sources solely within the territory defined by the geographic boundary. Scope 2 refers to GHG emissions from the use of grid electricity, heat, steam, and cooling within the city boundary. All other GHG emissions that occur outside the city boundary as a result of activities taking place within the city boundary are referred as Scope 3. All scope emissions are reported for the six sectors: stationary energy; transportation; waste; industrial processes and product use; and agriculture, forestry, and other land use. The use of region-specific emission factors and models is recommended for estimating emissions from Scope 1 and Scope 2 activities. The standard allows both production- and consumption-based (environmentally extended input–output; EEIO) approaches to be used, as appropriate. Considering that these emissions are typically estimated based on generic data and varying methodologies, they are thus associated with considerably higher uncertainties. As the protocol does not provide step-by-step approaches for estimating GHG emissions, the inventories would be highly influenced by subjective choices and assumptions, which, hence, would affect the robustness of the analytical outcome.

Product carbon footprint standards

At the product level, several standards, calculation methods, and approaches for carbon footprinting have been proposed and are used. The widely used standards are Publicly Available Specifications (PASs) 2050 (BSI 2011), GHG Protocol – Product Life Cycle Accounting and Reporting Standard (WRI/WBCSD 2011), and ISO 14067 (ISO 2013).

These standards are based on the LCA approach as established by ISO 14040 and 14044 standards for LCA (ISO 2006a, 2006b; as described in Chapter 10) and provide requirements and guidelines for several common methodological issues, such as goal and scope definition and allocation, as well as more specific issues, including land-use change, biogenic carbon emissions, soil carbon change, carbon storage in products, and delayed emissions. PAS 2050 and the GHG Protocol provide more detailed requirements and guidance, with less room for interpretation, whereas ISO 14067 is the more general standard. However, ISO 14067 provides more detailed guidelines for communicating products' carbon footprints through labelling, declaring, and performance reports. PAS 2050, one of the earliest standards, was first published in 2008 and was revised in 2011. The revised version is in alignment with the GHG Protocol Product Standard regarding key topics such as biogenic carbon, recycling, land-use change, and delayed emissions. Although these standards are largely in alignment, the existence of a large number of standards and initiatives can be overwhelming and confusing to companies.

Methodological approaches

The task of calculating carbon footprints can be approached methodologically using three approaches: 'top–down', based on EEIO analysis; 'bottom–up', based on process analysis; and a hybrid approach (Wiedmann and Minx 2007; Matthews et al. 2008).

The top–down approach makes use of an economic input–output model extended with environment-related information for each sector, such as its emissions and natural resource use. EEIO analysis, as described in Chapter 8, is suitable for estimating carbon footprints at the macro (countries, regions, cities) and meso (industrial sector) levels. Another physical accounting method, economy-wide material flow accounting (MFA; see Chapter 9), which is widely compatible with the accounting rules, can be used to provide data for environmental analysis at regional and sector level. However, owing to aggregated data at sector levels, neither EEIO nor MFA is suitable for micro-level analysis of processes, products, or services.

The bottom–up approach, which is based on LCA, is mainly used for the estimation of the carbon footprints of products and small entities. The results based on this approach are more detailed and accurate and, hence, useful for organisations to identify improvement opportunities and compare products. However, the collection of data for complex products can become time-consuming and costly. The accuracy of results depends on the quality of data for the foreground (direct) and background (indirect) emission sources. There are various software tools available, such as Gabi, SimaPro, Umberto, and CCaLC, for assessing carbon footprints and other LCA impacts. These tools also contain emission databases for a wide range of industrial activities, processes, and products.

The hybrid approach to carbon footprinting combines process analysis (using LCA) with EEIO (Lenzen and Crawford 2009). In the hybrid approach, lower-order process data are collected for the product or service, and higher-order requirements are covered by input–output analysis (Wiedmann and Minx 2007). This approach retains the detail and accuracy of the bottom–up approach and the robustness of EEIO analysis, thus increasing the completeness and reliability of estimates.

Carbon footprinting in regional development

Carbon footprinting or GHG accounting is a very relevant and important indicator for businesses and regional authorities for several reasons. At a regional level, such a type of analysis of the key impact areas, such as transport, housing, waste management, and provision of goods and services, could help with better planning of services and infrastructure. Also, by understanding the key drivers behind the carbon footprint of regions, authorities can prepare effective policies and legislation. Sensing a carbon-constrained economy in the near future, the corporate sector has embraced the concept to take competitive advantage. Calculating a carbon footprint is only the beginning of carbon management, that is, taking actions to reduce emissions and improve efficiency. In that regard, carbon footprinting can assist businesses to focus on carbon 'hot spots' and cost-effective reduction opportunities and, hence, can be a useful exercise as part of a complete environmental management system (see Box 11.1).

BOX 11.1 CARBON FOOTPRINT FOR A TOYOTA VEHICLE

To reduce the carbon footprint of its vehicles, in 2009 Toyota Australia embarked on an action programme covering the life cycle of its vehicles and implicating the entire regional network of local suppliers and dealers.

LCA studies revealed the major embedded impacts (energy/carbon) – see Figure 11.1. The main areas of the life chain amenable to direct action were subsequently addressed. Even sources of minor contribution were included as a way of engaging the entire life chain and motivate actors there to consider further action where feasible. (Note that minor for the production cycle does not necessarily mean minor for the relevant actor.)

Some of initiatives to reduce the carbon footprint were:

- *supplier actions* to survey their electricity and gas use and undertake energy efficiency initiatives;
- *logistics actions*, including eco-driving and load management;
- *manufacturing actions*, such as 'just-in-time' delivery of molten aluminium for the casting plant to avoid having to reheat delivered solid metal, as was done previously;

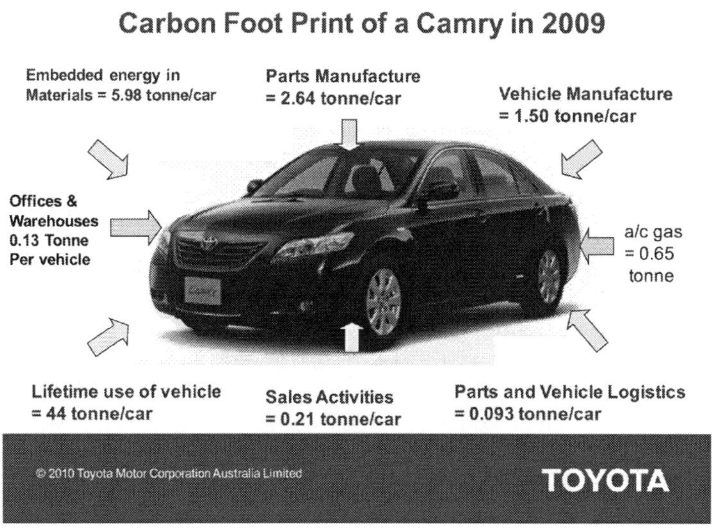

Figure 11.1
Carbon footprint of a Camry in 2009

Source: Prepared by F. Balkau, with permission from Toyota Australia

- *office actions*, including lower-impact, onsite power generation rather than grid power, and a 4-star energy-efficient building envelope;
- *dealer actions* – The Toyota Environmental Dealer (TED) Program.

It needs to be understood that 80 per cent of the lifetime energy footprint of a vehicle is on the road. How it is driven is really the key to overall efficiency. Although some engineering design features (weight, engine efficiency, etc.) can improve energy performance, the major factor of driving habits is not so amenable to direct intervention, except through educational and advisory programmes such as already found in national road programmes.

Carbon mitigation projects are often linked with energy-efficiency measures, and therefore reducing the carbon footprint provides the opportunity for cost reductions. Businesses and regional authorities can use this information to promote lower-carbon alternatives and to set policies and standards for developing low-carbon products. Regional authorities and local governments can also use carbon footprinting as one of the indicators to measure progress towards their sustainability targets and communicate the performance of their regions. Moreover, by using this as a public communication tool, local authorities can encourage their citizens to alter their behaviours and adopt sustainable consumption practices.

In various countries, there are legislative requirements for companies to measure, report, and reduce Scope 1 and Scope 2 GHG emissions. Moreover, to address carbon emissions embedded in goods and services, many organisations need to assess the carbon footprint of their supply chains (Scope 3). Because of its ease of conveying information about the GHG intensity of a variety of products and activities among the general public, carbon footprinting also offers a simple mode of communication to different stakeholders. As a result, carbon labelling of products has become popular in the recent past in the various industrial sectors, as also described in Chapter 18.

At sector level, carbon footprinting can assist in identifying mitigation options and best practice. For example, comparison of the carbon footprint performance of various waste management options (such as recycling, incineration, landfilling, composting) is helpful in making decisions on different options in integrated waste management plans, as discussed in Chapter 43. Similarly, as discussed in Chapters 32 and 39, promotion of renewable technologies and biofuels largely depends on their performance with respect to savings in GHG emissions compared with traditional energy sources. More examples from other sectors, including building construction, agri-food systems, packaging, and tourism, are discussed in detail in Parts IV and V of this book.

Conclusions

Quantification of current performance as well as future possible GHG reduction strategies is key to preventing climate change, and, therefore, carbon footprinting is a very relevant, important indicator for businesses and regional authorities. However, there are some technical and financial barriers, such as the lack and poor quality of input data and high costs, which could affect the adoption of the methodology at a regional level. Several standards, guidelines, software, and databases are available to facilitate GHG accounting in a transparent and consistent way, but there is considerable scope to harmonise GHG estimation methodologies and improve databases to reduce uncertainty.

Assessing the carbon footprint is only the first step in tackling climate change. It is, therefore, important that this tool is widely used in regional development to promote lower-carbon alternatives and to develop GHG emission reduction strategies.

References

BSI (2011) *Publicly Available Specification 2050:2011: Specification for the Assessment of the Life Cycle Greenhouse Gas Emissions of Goods and Services*, London: British Standards Institute.

IPCC (1995) IPCC Second Assessment Report, Climate Change 1995, Cambridge, UK: IPCC.

IPCC (2001) Climate Change 2001: The Scientific Basis, Contribution of Working Group I to the First Assessment Report of the Intergovernmental Panel on Climate Change, Cambridge, UK: IPCC.

IPCC (2007) IPCC Fourth Assessment Report: Climate Change 2007, Working Group I: The Physical Science Basis, Cambridge, UK: IPCC.

IPCC (2013) Climate Change 2013: The Physical Science Basis. Contribution of Working Group I to the Fifth Assessment Report of the Intergovernmental Panel on Climate Change, Cambridge, UK: IPCC.

ISO (2006a) *ISO 14040: Environmental Management–Life Cycle Assessment – Principles and Framework*, Geneva: ISO.

ISO (2006b) *ISO 14044: Environmental Management–Life Cycle Assessment – Requirements and Guidelines*, Geneva: ISO.

ISO (2013) *Greenhouse Gases – Carbon Footprint of Products – Requirements and Guidelines for Quantification and Communication*, Geneva: ISO.

Lenzen, M., and Crawford, R. (2009) 'The path exchange method for hybrid LCA', *Environmental Science & Technololgy*, 43: 8251–6.

Matthews, S.C., Hendrickson, C.T., and Weber, C.L. (2008) 'Estimating Carbon footprints with input output models', presented at *International Input Output Meeting on Managing the Environment*, 9–11 July, Seville, Spain.

Pandey, D., Agrawal, M., and Pandey, J.S. (2011) 'Carbon footprints: Current methods of estimation', *Environmental Monititoring & Assessment*, 178: 135–60.

Peters, G.P. (2010) 'Carbon footprints and embodied carbon at multiple scales', *Current Opinion in Environmental Sustainability*, 2: 245–50.

Wiedmann, T., and Minx, J.A. (2007) Definition of Carbon Footprint, Research Report 07–01, Durham, UK: ISAUK Research & Consulting.

WRI/WBCSD (2004) *The Greenhouse Gas Protocol: A Corporate Accounting and Reporting Standard*, revised edn, Geneva: World Resource Institute & World Business Council for Sustainable Development.

WRI/WBCSD (2011) *Product Life Cycle Accounting and Reporting Standard – Greenhouse Gas Protocol*, Washington, DC: World Resource Institute & World Business Council for Sustainable Development.

WRI/WBCSD (2014) The Greenhouse Gas Protocol for Community Scale Greenhouse Gas Emission Inventories – An Accounting and Reporting Standard for Cities, Geneva: World Resource Institute & World Business Council for Sustainable Development.

12

WATER FOOTPRINT PRINCIPLES

Philippe Loubet and Guido Sonnemann

Introduction

Water is both a resource for humans and an environmental habitat, which explains the many concerns we have about this 'blue gold'. Of course, water is not as scarce as gold; on the contrary, it is a renewable resource. Water moves continually through a cycle on Earth. There are approximately 1,400 million km^3 of water on Earth, but only 3 per cent is fresh water, of which 69 per cent is locked up in glaciers and snow (Oki and Kanae 2006). The remaining water is usable for humans but it is poorly distributed in the world. More than 2.5 billion people face water scarcity during at least one month of the year (Hoekstra et al. 2012), meaning that available water resources sufficient to meet demand are lacking. The future is not bright, as climate change and population growth tend to increase this threat. Human interventions and activities exacerbate the situation. This is principally owing to agriculture, which is responsible for 70 per cent of water withdrawal, whereas domestic users are responsible for 12 per cent and industrial users for 18 per cent (FAO 2012). Degradation of water quality leads to unavailable water resources for certain usage. Fresh water is also an environmental habitat that can be affected by water scarcity and pollution. In terms of biological value, rivers contain a rich and varied biota, that is, at least 100,000 species – almost 6 per cent of all described species (Dudgeon et al. 2006). Ecosystem destruction due to water abstraction, habitat alteration incurred by damming or water transfer, changes in water chemistry because of pollution, and species removal and additions are the main disturbances caused by anthropogenic activities.

In this context, there is a need for tools to account for water use and pollution generated to produce a good or a service within a specific company or a region. Such accounting should enable identification of the main levers to reduce the pressure from human activities on water resources. This is especially important for regional authorities such as river basin managers. Methods have been developed in order to enable such assessment, first with the virtual water and water footprint concepts, and further refined within the life cycle assessment (LCA) framework. In the present chapter, both methodologies will first be presented. Then, key elements and methodologies for water footprint in the context of LCA will be explored, following the methodology adopted in ISO 14046. In a last part, short examples of water footprint applications for regional authorities will be presented.

Water footprint concept and water footprint assessment

At the beginning of the 2000s, the concepts of 'virtual water' and 'water footprint' were developed in order to account for water issues in supply chains (Jensen 1998). The water footprint concept embraces life cycle thinking as it looks at both the direct and indirect water use by a process, product, company, or region, by taking into account water consumption and pollution through all life cycle stages.

Water footprint is the summation of three different components calculated for the different life cycle stages: blue, green, and grey water footprints. The blue water footprint represents the water that comes from surface or groundwater resources and is consumed – that is, either evaporated to the atmosphere, incorporated into a product, or transferred from one river basin to another. The term consumption does not mean that the water disappears, as water remains in a global cycle and will return somewhere else. However, its local availability is not unlimited, and the blue water that is consumed is made unavailable for other users and ecosystems. Green water is the water from precipitation that is either evaporated, transported, or incorporated into plants. This is particularly relevant for agricultural and forestry products, where it refers to the total rainwater evapotranspiration (from fields and plantations) plus the water incorporated into the harvested crop or wood. The grey water footprint is the volume of fresh water that is required to dilute and thus assimilate pollutants to meet specific water quality standards.

Calculations have been made of how much water is needed to produce the various goods or services: for example, 1 kg of beef represents 15,400 l of water, and 1 kg of coffee requires almost 19,000 l (Mekonnen and Hoekstra 2011). These considerable amounts raise awareness in the public. However, the interpretation of this volumetric approach is questionable. For example, 1 m^3 of water transpired in a wet area is not equivalent to 1 m^3 of water consumed in a

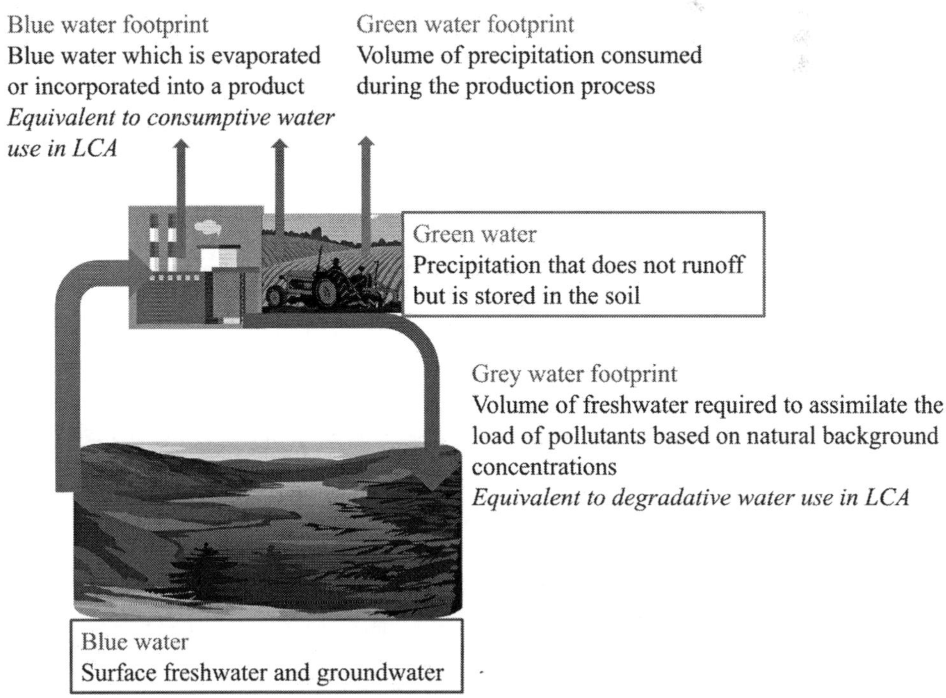

Figure 12.1 Representation of the blue, green, and grey water footprint of an activity in the context of water footprint assessment and similarities with LCA

dry one. In addition, the quantification of pollution in grey water is based on a dilution volume approach that does not consider substance fate, contrary to what life cycle impact assessment models do.

Jointly with the water footprint concept a four-phase process, the water footprint assessment (WFA), has been developed (Hoekstra et al. 2011). It consists of goal and scope definition – for example, to support a specific business on achieving sustainable water management. Then, an accounting phase involves calculation of the water footprint as described in the previous paragraph. It is followed by a sustainability assessment to establish if water use is environmentally sustainable and resources are efficiently and equitably allocated. In the last phase, response formulation, strategies that reduce the water footprint and improve its sustainability can be prioritised for implementation.

LCA-based water footprint

Integration of water use within LCA framework

LCA is extensively described in Chapter 10. It characterizes inventory data (e.g. water consumption, emissions to water) during the different life cycle steps of a product or a service in order to quantify related potential impacts on the environment. In recent decades, many methodological advances have been made in water use assessment. The framework presented in Bayart et al. (2010) for considering water in LCA is now largely accepted. It distinguishes between degradative and consumptive water use:

- Degradative water use is related to pollution of water during its use. It is covered in LCA with impact categories such as fresh-water eutrophication, acidification, and (eco)toxicity. Emissions within the water compartment are characterized into impacts: for example, phosphates emissions in water generate eutrophication and eventually damage ecosystems; emissions of metals in water generate (eco)toxicity and eventually damage ecosystems and human health.
- Consumptive water use refers to water that is not released back to the original river basin and, therefore, not available for downstream users. It includes water that is evaporated, transpired, incorporated into products, released to a different river basin, or directly saline water. The next section describes specifically the impact assessment related to water consumption, because it is at an early development stage, even if new methods are currently being developed and certain ones are operational (Kounina et al. 2012).

Impact assessment of consumptive water use

Methods to assess impacts associated with consumptive water use have focused so far on impacts associated with water consumption from rivers and renewable groundwater (blue water, according to the water footprint concept). Water consumption lowers the quantity of fresh water within a river and, more generally, within a river basin, as water that is consumed is lost from the local watershed to the global water cycle.

Therefore, water consumption leads to water deprivation for downstream water users and ecosystems. Methodologies have first assessed water deprivation midpoint indicators based on water scarcity (Pfister et al. 2009). Water scarcity is defined at the watershed scale as a function of the ratio between total water use by human activities and water availability. It enables a spatial differentiation of water deprivation impacts.

At the end-point level, water use from fresh-water resources disrupts ecosystems by temporarily reducing water availability and damages the 'ecosystem quality' area of protection. Several methodologies have already been developed to assess the damage to ecosystems within rivers, wetlands, or groundwater. Also, damage to human health can be caused by water consumption, but the cause–effect relationship is difficult to assess owing to socio-economic mechanisms. Water deprivation can lead to malnutrition stemming from a lack of water for agriculture. However, if there is sufficient economic wealth in the area, human users adapt by compensating with backup technology (Boulay et al. 2011).

Water consumption from non-renewable or fossil groundwater stocks generates resource depletion. Such depletion leads to reduced availability for future generations. In this case, the 'resources' area of protection is damaged. Green water consumption (precipitation) has already been covered as part of land-use impacts, but Núñez et al. (2013) have built a framework to specifically address this issue. Further methodological advances are currently under development within the group Water Use in LCA (WULCA), founded under the auspices of the UNEP/SETAC Life Cycle Initiative. It aims to provide practitioners with a coherent framework for water use impact assessment.

Water footprint standard ISO 14046 based on LCA

In this context, the water footprint standard (ISO 2013) has recently been developed. It first states that a water footprint should follow the LCA methodology. Second, it states that it should measure the potential impact on environment and not only account for water use volumes. Water pollution-related impacts should also be considered. Finally, a water footprint profile should be presented as a compilation of impact assessment results related to water use: water deprivation, eutrophication, ecotoxicity, etc.

Complementarity of WFA and LCA-based water footprint

It is clear that LCA and WFA have the indirect goal of helping practitioners to preserve water resources. They are also based on four similar phases: (1) setting goal and scope, (2) accounting, (3) impact assessment, and (4) interpretation and solutions (Boulay et al. 2013). Also, both methods use quantitative indicators, but in different phases of the assessment. Whereas WFA particularly relies on indicators in the inventory phase, LCA focuses on impact indicators. The main difference, and the source of the complementarity of the two approaches, is the product focus of LCA and the water management focus of WFA.

The LCA methodology studies the sustainability of products in a holistic approach where water is one impact category among others (e.g. carbon footprint, ozone layer depletion), which enables the identification of burden shifting. On the other hand, WFA uses mostly volumetric indicators and aims to provide information with a wider scope than LCA, as it can be used at various levels: river basin, global, product, or corporate.

Applications for regional sustainable development

This section takes examples from LCA-based water footprint. A case study of regional water footprint from the Water Footprint Network is provided in

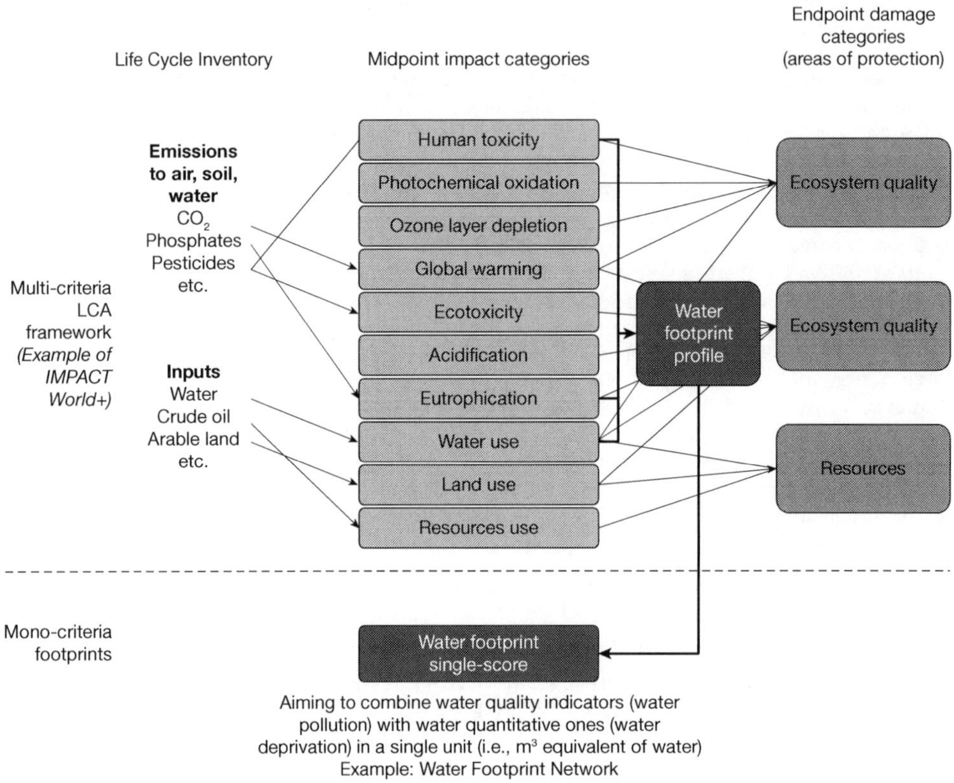

Figure 12.2
Main impact pathways in LCA and presentation of the water footprint profile and single score

Source: Adapted from Impact World+

Chapter 31. Water use impact assessment has already been applied in LCA related to regional development, mainly focusing on (1) the provision of domestic and industrial water and (2) agricultural products.

For example, Muñoz et al. (2010) compare, among other indicators, the freshwater ecosystem impact due to water consumption of two options for supplying water in Mediterranean Spain. The first assessed solution (ERWT) consists of an interbasin water transfer from the Ebro river, and the second solution (AGUA) consists of desalination, as well as improvement of the water management of existing resources. LCA results have shown that the AGUA programme leads to lower water deprivation impacts than the EWRT programme, as it uses large amounts of seawater, which does not lead to water deprivation impact. In order to avoid burden shifting, this study also assesses other impact categories (such as climate change).

Agriculture is the main water consumptive activity, and the incorporation of water use indicators in LCA of agricultural products is therefore highly relevant. For example, Payen et al. (2015) compare the LCA of local French tomatoes and tomatoes imported from Morocco, including water deprivation impacts. They show a trade-off between usual impact categories, mostly energy-related, and water deprivation impacts. Indeed, production of tomatoes in Morocco requires less energy than that of off-season French tomatoes, but it generates a much larger water deprivation as water scarcity is higher in Morocco.

Conclusions

Two main approaches to evaluate water use and related impacts at different life cycle stages of a product or service, or within an activity or a region, have been developed in the last decade: WFA and LCA-based water footprint. They provide indicators to help decision-makers preserve water resources for both quantitative and qualitative issues. They are complementary in that WFA focuses on water management, whereas LCA focuses on the assessment of products. Because of the localized impacts induced by water use from agriculture industry, and cities, it is important to deploy such tools at the regional level as a complement to other life cycle sustainability indicators. They can help regional sustainable development by identifying where the pressures are (both geographically and in value chains), as well as determining levers to better manage water resources at different scales.

References

Bayart, J.B., Bulle, C., Deschênes, L., Margni, M., Pfister, S., Vince, F., and Koehler, A. (2010) 'A framework for assessing off-stream freshwater use in LCA', *International Journal of Life Cycle Assessment*, 15: 439–53.

Boulay, A.M., Bare, J., Benini, L., Berger, M., Klemmayer, I., Lathuilliere, M., Loubet, P., Manzardo, A., Margni, M., and Ridoutt, B. (2014) 'Building consensus on a generic water scarcity indicator for LCA-based water footprint: Preliminary results from WULCA', in *Proceedings of the 9th International Conference on Life Cycle Assessment in the Agri-Food Sector*, San Francisco, CA.

Boulay, A.M., Bulle, C., Bayart, J.B., Deschênes, L., and Margni, M. (2011) 'Regional characterization of freshwater use in LCA: Modeling direct impacts on human health', *Environmental Science & Technology*, 45: 8948–57.

Boulay, A.M., Hoekstra, A.Y., and Vionnet, S. (2013) 'Complementarities of water-focused life cycle assessment and water footprint assessment', *Environmental Science & Technology*, 47(21): 11926–7.

Dudgeon, D., Arthington, A.H., Gessner, M.O., Kawabata, Z.I., Knowler, D.J., Lévêque, C., Naiman, R.J., Prieur-Richard, A.H., Soto, D., Stiassny, M.L.J., and Sullivan, C. (2006) 'Freshwater biodiversity: Importance, threats, status and conservation challenges', *Biological Reviews of the Cambridge Philosophical Society*, 81: 163–82.

FAO (2012) AQUASTAT database, Rome: Food and Agriculture Organization of the United Nations. Available at: www.fao.org/nr/water/aquastat/data/ (accessed 29 July 2016).

Hoekstra, A.Y., Chapagain, A.K., Aldaya, M.M., and Mekonnen, M.M. (2011) *The Water Footprint Assessment Manual. Setting the Global Standard*. London: Earthscan.

Hoekstra, A.Y., Mekonnen, M.M., Chapagain, A.K., Mathews, R.E., and Richter, B.D. (2012) 'Global monthly water scarcity: Blue water footprints versus blue water availability', *PLoS One*, 7(2). Available at: http://journals.plos.org/plosone/article?id=10.1371/journal.pone.0032688 (accessed 29 July 2016).

ISO (2013) *ISO/DIS 14046.2: Environmental Management – Water Footprint – Principles, Requirements and Guidelines*, Geneva: ISO.

Jensen, J.A. (1998) 'Virtual water: A strategic resource global solutions to regional deficits', *Ground Water*, 36: 545–6.

Kounina, A., Margni, M., Bayart, J.B., Boulay, A.M., Berger, M., Bulle, C., Frischknecht, R., Koehler, A., Milà i Canals, L., Motoshita, M., Núñez, M., Peters, G., Pfister, S., Ridoutt, B., Zelm, R., Verones, F., and Humbert, S. (2012) 'Review of methods addressing freshwater use in life cycle inventory and impact assessment', *International Journal of Life Cycle Assessment*, 18: 707–21.

Mekonnen, M.M., and Hoekstra, Y. (2011) 'The green, blue and grey water footprint of crops and derived crop products', *Hydrology & Earth System Sciences*, 15: 1577–600.

Muñoz, I., Milà-i-Canals, L., and Fernández-Alba, A.R. (2010) 'Life cycle assessment of water supply plans in Mediterranean Spain', *Journal of Industrial Ecology*, 14: 902–18.

Núñez, M., Pfister, S., Roux, P., and Antón, A. (2013) 'Estimating water consumption of potential natural vegetation on global dry lands: Building an LCA framework for green water flows', *Environmental Science & Technology*, 47: 12258–65.

Oki, T., and Kanae, S. (2006) 'Global hydrological cycles and world water resources', *Science*, 313: 1068–72.

Payen, S., Basset-Mens, C., and Perret, S. (2015) 'LCA of local and imported tomato: An energy and water trade-off', *Journal of Cleaner Production*, 87: 139–48.

Pfister, S., Koehler, A., and Hellweg, S. (2009) 'Assessing the environmental impacts of freshwater consumption in LCA', *Environmental Science & Technology*, 43: 4098–104.

13

SYSTEM DYNAMIC AND SPATIAL MODELLING IN LCA

Bertrand Laratte and Guido Sonnemann

Introduction

The consideration of the dynamic aspect in LCA (ISO 2006a, 2006b) is quite new. This approach is considered today owing to the challenges generated by the results of conventional LCA, as described in Chapter 10. One can observe the evolution of the environmental impact over time and also the cumulative aspect due to, for instance, the substance lifetime in the environmental impact factor calculation.

A model is a representation of a system (e.g. LCA represents a product system). So what is system dynamic modelling? System dynamics is an approach to understanding the nonlinear behaviour of complex systems over time using stocks and flows, internal feedback loops and time delays.

Up to now, the assessment of environmental impacts has considered a fixed time and fixed (regional) area characterisation factor; this consideration limits the assessment. Indeed, the contribution of a given substance to an impact evolves over time and in a certain geographical area, and there is no particular reason why the calculation of impacts that are (artificially) integrated should coincide with the physical reality of a time series of snapshots. With the usual LCA methodology, the impact result of a punctual emission is given for a fixed time horizon and a global or a given geographical area. For example, the methodology proposed by the Intergovernmental Panel on Climate Change (IPCC; IPCC 2007) to assess the global warming potential (GWP) is usually evaluated with 100 years as a time horizon.

In the context of sustainable regional development, and in particular in eco-innovation, the use of LCA is important to assess and support the environmental evaluation of products and processes, in complementarity with other life cycle approaches. The key advantage is that LCA allows the identification of which processes or which steps of the life cycle are more or less important for the environmental impact calculated. Even if the LCA seems an appropriate tool, one of the main things lacking in the environmental assessment is that temporal and geographical aspects are not considered.

Context and background

In this circumstance, the vision of prospective scenarios incorporating real-time, geographically explicit environmental indicators is essential to further improve the assessment of the environmental benefits of one technology over another. To continue with the example of global warming, one knows that different greenhouse gases contribute to global warming differently, especially in terms of their lifetime. Some fluorinated gases, for example, can have lifetimes longer than 1,000 years! Moreover, climate change will have different consequences, depending on the geographic situation. Therefore, it is important to understand the relevance of potential changes to the actual methodology (with a time horizon at 100 years and with a global approach).

The challenge of including time and regional aspects in life cycle assessments is important but not a new reflection: it has been properly identified for more than 20 years: Owens (1997a, 1997b) identifies the lack of temporal and geographical resolutions (especially in the development of the life cycle inventory (LCI)) as one of the main limits of the LCA, with the effect of limiting the ability to evaluate some environmental issues with relevance. One can easily imagine that, in some cases, it is difficult to estimate the impacts and risks for humans or the environment with due scientific rigour. Owens does not hesitate to conclude that, in certain cases, the LCA is too simplified to be able to provide sufficiently robust environmental impact indicators.

De Haes et al. (1999, 2004) note the lack of temporal resolution, both in the calculation of impacts in LCA and in the development of the LCI. Until now, LCA has been a comprehensive method, but at a linear scale. This is owing, in particular, to the difficulty of including the spatial and temporal variables for all processes.

More recently, Levine et al. (2007) identified the time limit as important, especially in the case of substances that may have different lifetimes. The current LCA method aggregates inventory data and calculates the environmental impacts in the same way for all substances (without lifetime differentiation).

However, some studies suggest a new method, called 'dynamic LCA', that focuses on the greenhouse gas effect (Laratte 2013; Laratte and Guillaume 2014; Potting 2000). Previously, a framework was presented by Potting (2000) for the impact indicators of acidification and human exposure (to hazardous substances). Depending on the application, it can, therefore, be necessary to use one of these site- and/or time-dependent methodologies, using environmental impact indicators and inventory data with dynamic and spatial parameters (factors identified as limiting this methodology; Reap et al. 2008a, 2008b; Zamagni et al. 2009; Finnveden et al. 2009).

Methodology and procedure

As shown previously to assess environmental impacts with LCA, Equation 13.1 is used:

$$E = \sum_i M_i \cdot p_i \qquad (13.1)$$

where:

E = the considered environmental impact

M_i = the mass flow of the substance i

P_i = the characterisation factor of the substance i.

In Equation 13.1, the temporal and spatial aspects are not considered. In other words, the dynamics and the localisation of the evaluated system are not considered. The mass flow represented by M_i is evaluated at time t and in space s; it means that the collection of data is done in one shot and one aggregation, without consideration of the process's evolution and its location. Hence, one of the main limitations of the actual inventory is the lack of time and spatial consideration.

The aim of the characterisation factor, represented by P_i, is to provide an equivalence of the environmental impact for a given substance. For example, methane has an impact factor of 28 for the GWP. It means that one substance of methane contributes 28 times more than carbon dioxide to the GWP. Also, this second parameter doesn't consider the lifetime and spatial aspects of the substances. In this case, it is considered that the potential impact is the same in any place on Earth, and a fixed time horizon is shown for all environmental indicators.

Dynamic and spatial inventory

Until now, generic databases have been used, in which inventory data are available that have been collected by different companies or prepared by exploiting literature or other public sources. However, this compilation of data is not necessarily representative of all technologies; some data were collected in 1996 and are still used in LCA. This lack of time consideration can generate underestimated or overestimated results. For example, generic data provided by databases are usually representative for a certain region; in others terms, when one carries out the LCA of an agriculture system, it is really relevant to collect local data, because the technical details – arable land, nutrients in the soil, and so on – are different in each region. Also, agriculture systems change over time, and every year the yield can be different.

Considering Equation 13.1 and the aspects outlined before, it appears important to integrate dynamic and spatial parameters: dynamic considerations, because a system can change over the time, and spatial aspects, because the region can also influence the final result. Therefore, Equation 13.2 is proposed:

$$E(t_m, s_m) = \sum_i M_i(t_m, s_m) \cdot p_i \qquad (13.2)$$

where:

t_m = the time dependence of the mass substance m

s_m = the spatial dependence of the mass substance m.

Dynamic and site-dependent environmental impact assessment

Consideration of temporal and spatial aspects in the development of the characterisation factor also has a huge interest: it is about providing the necessary information related to the potential contribution of a given substance to a certain impact. Once more, in usual LCA, the lack of temporal and spatial considerations in the assessment can give less precise results. The non-integration of the lifetime of the substance can underestimate its real impact. As an example, one can again take the GWP: as said, the usual LCA considers a fixed time horizon of 100 years.

The spatial aspect can have great importance for some impact indicators, too. The influence of the geographical aspect is fundamental for local pollution, such as, for instance, particulate matter and other emissions to air that affect human health (see Chapter 29 for the application of a site-dependent method) and also eutrophication in water. Eutrophication is a very local phenomenon: it depends on the type of soil (clay, sand or other), the type of nutrient (phosphate or nitrate), the type of water (fresh or marine water), and the concentration level.

Combining the dynamic and spatial LCIs and the dynamic and site-dependent characterisation factors, Equation 13.3 is proposed:

$$E(t_m, s_m, t_p, s_p) = \sum_i M_i(t_m, s_m) \cdot p_i(t_p, s_p) \tag{13.3}$$

where:

t_p = the time dependence of the characterisation factor p

s_p = the spatial dependence of the characterisation factor p.

Discussion and conclusion

These arguments indicate the opportunities of modelling the temporal and spatial aspects in LCA. These considerations are important, because more precise results could influence the stakeholders that make decisions at the regional level on different choices with regard to sustainable development. Understanding time- and site-dependent impacts better can also determine the establishment of a regional policy to reduce pollution (e.g. reduction of the eutrophication of a lake). A case study that involves the site-dependent method of integrating life cycle and risk assessment for air pollutants affecting human health is presented in Chapter 29.

However, until now, the LCA software that is used most often does not propose the implementation of these parameters, mainly owing to the higher work load and missing automatism to consider time and space. Even if there exist studies to propose improvement of the inventory (Collinge et al. 2011; Beloin-Saint-Pierre 2012) and also improvement of the environmental impact assessment (Kendall et al. 2009; Levasseur et al. 2010; Sonnemann et al. 2003; Laratte et al. 2014), a comprehensive method (including inventory databases and characterisation factors) that allows dynamic, spatially explicit LCA, in the strict sense of the ISO 14040 (ISO 2006a) and ISO 14044 (ISO 2006b) standards, to be carried out easily is still missing.

Therefore, the next, huge contribution that can be expected from LCA experts in the upcoming years will be to improve the LCA method by providing LCA databases and assessment methodologies that include temporal and spatial aspects. This will also be of paramount interest for calculation of the water footprint in line with ISO 14046 (ISO 2014), as described in Chapter 12.

References

Beloin-Saint-Pierre, D. (2012) *Vers une caractérisation spatiotemporelle pour l'analyse du cycle de vie* (PhD thesis). Paris: Ecole Nationale Supérieure des Mines de Paris.

Collinge, W., Liao, L., Xu, H., Saunders, C., Bilec, M., Landis, A., Jones, A., and Schaefer, L. (2011) 'Enabling dynamic life cycle assessment of buildings with wireless sensor networks', in *Proceedings of IEEE International Symposium on Sustainable Systems and Technology (ISSST)*, pp. 1–6.

De Haes, H.A.U., Heijungs, R., Suh, S., and Huppes, G. (2004) 'Three strategies to overcome the limitations of life-cycle assessment', *Journal of Industrial Ecology*, 8(3): 19–32.

De Haes, H.A.U., Jolliet, O., Finnveden, G., Hauschild, M., Krewitt, W., and Müller-Wenk, R. (1999) 'Best available practice regarding impact categories and category indicators in life cycle impact assessment', *International Journal of Life Cycle Assessment*, 4(2): 66–74.

Finnveden, G., Hauschild, M.Z., Ekvall, T., Guinée, J., Heijungs, R., Hellweg, S., Koehler, A., Pennington, D., and Suh, S. (2009) 'Recent developments in life cycle assessment', *Journal of Environmental Management*, 91(1): 1–21.

IPCC (2007) *Working Group III, Climate Change 2007 – The Physical Science Basis: Working Group I Contribution to the Fourth Assessment Report of the IPCC*, Cambridge, UK: Cambridge University Press.

ISO (2006a) *ISO14040:2006: Environmental Management – Life Cycle Assessment – Principles and Framework*, Geneva: ISO.

ISO (2006b) *ISO14044:2006: Environmental Management – Life Cycle Assessment – Requirements and Guidelines*, Geneva: ISO.

ISO (2014) *ISO14046:2014 Environmental Management – Water Footprint – Principles, Requirements and Guidelines*, Geneva: ISO.

Kendall, A., Chang, B., and Sharpe, B. (2009) 'Accounting for time-dependent effects in biofuel life cycle greenhouse gas emissions calculations', *Environmental Science & Technology*, 43(18): 7142–7.

Laratte, B. (2013) *Evaluation dynamique et cumulative des impacts environnementaux dans le cadre d'une analyse de cycle de vie (Dynamic and cumulative environmental impact assessment in life cycle assessment)*. PhD thesis, Troyes, France: Université de Technologie de Troyes.

Laratte, B., and Guillaume, B. (2014) 'Epistemic and methodological challenges of dynamic environmental assessment: A case-study with energy production from solar cells', *Key Engineering Materials*, 572: 535–8.

Laratte, B., Guillaume, B., Kim, J., and Birregah, B. (2014) 'Modeling cumulative effects in life cycle assessment: The case of fertilizer in wheat production contributing to the global warming potential', *Science of the Total Environment*, 481: 588–95.

Levasseur, A., Lesage, P., Margni, M., Deschênes, L., and Samson, R. (2010) 'Considering time in LCA: Dynamic LCA and its application to global warming impact assessments', *Environmental science & technology*, 44(8): 3169–74.

Levine, S.H., Gloria, T.P., and Romanoff, E. (2007) 'A dynamic model for determining the temporal distribution of environmental burden', *Journal of Industrial Ecology*, 11(4): 39–49.

Owens, J.W. (1997a) 'Life-cycle assessment in relation to risk assessment: An evolving perspective', *Risk Analysis*, 17(3): 359–65.

Owens, J.W. (1997b) 'Life-cycle assessment: Constraints on moving from inventory to impact assessment', *Journal of Industrial Ecology*, 1(1): 37–49.

Potting, J. (2000) *Spatial differentiation in life cycle impact assessment: A framework, and site-dependent factors to assess acidification and human exposure*. PhD thesis, Utrecht, Netherlands: Universiteit Utrecht.

Reap, J., Roman, F., Duncan, S., and Bras, B. (2008a) 'A survey of unresolved problems in life cycle assessment', *International Journal of Life Cycle Assessment*, 13: 290–300.

Reap, J., Roman, F., Duncan, S., and Bras, B. (2008b) 'A survey of unresolved problems in life cycle assessment', *International Journal of Life Cycle Assessment*, 13: 374–88.

Sonnemann, G., Castells, F., and Schuhmacher, M. (2003) *Integrated Life-Cycle and Risk Assessment for Industrial Processes*, Boca Raton, FL: Lewish.

Zamagni, A., Buttol, P., Buonamici, R., Masoni, P., Guinée, J., Huppes, G., Heijungs, R., Van Der Voet, E., Ekvall, T., and Rydberg, T. (2009) 'D20 blue paper on life cycle sustainability analysis, deliverable 20 of work package 7 of the calcas project', CALCAS.

14

TOWARDS LIFE CYCLE SUSTAINABILITY ASSESSMENT, INCLUDING SOCIAL LIFE CYCLE ASSESSMENT

Bernard Mazijn

Introduction

Since the publication of the report of the World Commission on Environment and Development, 'Our Common Future' – the so-called Brundtland report – it has been universally agreed that sustainable development should focus on three dimensions: environment, social, and economic (WCED 1987).

However, if students in higher education or participants in training sessions are asked to make an evaluation of a textile factory or of a T-shirt, the list of criteria usually focuses on the environmental dimension only. Nevertheless, the April 2013 human disaster in Bangladesh, at the Rana Plaza textile complex, for example, also underlines, once again, the importance of the social dimension surrounding enterprises, retailers and consumers. It is clear that quite a lot of externalities are not included in the market price of products. Another example that was reported lately concerns the cobble-stones used in the medieval cities of Europe. For a long time, they were produced in the region. Nowadays, this pavement material is imported, mostly from Kandla, India, where the stones are cut by children. Expectations of social responsibility now go far beyond an enterprise's own immediate activity.

Furthermore, analysis of so-called megatrends – such as growth in population, urbanization, purchasing power in emerging economies, and so on, leading to an increase in demand for energy and resources, with huge environmental pressure in terms of climate change, loss of biodiversity, and so on – results in recommendations for moving from a linear, globalized throughput economy towards a circular economy at regional level (e.g. Europe, India, North America) within a context of sustainable development (Mazijn 2013).

In this chapter, it will be argued that life cycle sustainability assessment (LCSA) can contribute to the identification of so-called 'hot spots' in the life cycle of products that are problematic from a sustainable development perspective. This information subsequently becomes the basis for a life cycle approach to supply-chain management.

Social responsibility of organizations

In the context of today's globalization, the title of this book, *Life Cycle Approaches to Sustainable Regional Development*, begs some further questions. Should policy-making at the regional level focus on local production processes or also on products? Is there still a benefit for an enterprise to have the life cycle of its products at the centre of its strategy development, or should it focus mainly on the production process within the enterprise?

Although some interaction with suppliers can be observed, for decades it has been the production process that has been at the centre of industrial policies or strategies. The attention paid to what is important stops at the gate of the enterprise, works with 'management systems', and connects with the common socio-economic lines of thinking. Only nearby stakeholders are involved. Public authorities mostly still focus their policy-making only on production processes of enterprises, taking into account the appropriate national legislation.

If taken a step further, policy-making or strategy development focused on the product would not stop at the gate of the enterprise, but would take into account the whole value chain, upstream and downstream alike. The mapping of stakeholders and the identification of the actors along the life cycle of the product would consider the societal concerns within a context of sustainable development (cf. the three dimensions). Chain management would help to address the 'hot spots' in the value chain.

These different approaches are often considered under the title of 'social responsibility'. For organizations, social responsibility has been defined at the international level through a number of global instruments, including formal standards developed with and supported by stakeholders. Through the use of ISO 26000 (ISO 2010) – guidance on social responsibility in an organization (including enterprises) – the foundation is laid, and the guidelines of the Global Reporting Initiative (GRI; GRI 2011, 2014) allow communication with the world at large through the mechanism of a sustainability report.

ISO 26000 and GRI are compatible instruments. They address seven key themes: management, human rights, labour practices, environment, fair operating practices, consumer issues, and community involvement. 'Environment' is only one of the concerns of the international community that is integrated in these international standards. ISO 26000 describes the:

> responsibility of an organization for the impacts of its decisions and activities on society and the environment, through transparent and ethical behaviour that
>
> - contributes to sustainable development, including health and the welfare of society;
> - takes into account the expectations of stakeholders;
> - is in compliance with applicable law and consistent with international norms of behaviour; and
> - is integrated throughout the organization and practised in its relationships.

The latter part refers to another notion, that is, 'sphere of influence', defined as: 'the range/extent of political, contractual, economic or other relationships through which an organization has the ability to affect the decisions or activities of individuals or organizations'. It is expected that an enterprise will exercise 'due diligence':

> a comprehensive, proactive process to identify the actual and potential negative social, environmental and economic impacts of an organization's decisions and activities over the entire life cycle of a project or organizational activity, with the aim of avoiding and mitigating negative impacts.

If an organization 'stays silent about or benefits from such wrongful acts', it is regarded as being complicit. We can see that the requirements for 'social responsibility' are closely related to the purpose of LCSA as described below.

There have been parallel initiatives within business itself. The three dimensions of sustainable development (environment, social, and economic) are often referred to as the 'triple bottom line', a concept that can be seen as similar to the inclusive 3Ps approach: people, planet, and profit. Reference to triple bottom line objectives is increasingly found in public statements by major companies, implemented through a variety of mechanisms, including, prominently, that of corporate social responsibility. The concept transcends the sphere of business. As people and planet imply a collective interest, profit can be interpreted as private interest. Therefore, it is not surprising that the World Summit on Sustainable Development referred instead to 'people, planet and prosperity'.

Life cycle sustainability assessment

The previous points argued that striving towards sustainable development involves more than taking care of environmental considerations. The question 'What is a sustainable product?' should be based on criteria related to the 3Ps: people, planet and profit/prosperity. A laptop with a European eco-label, for example, cannot be called 'sustainable' if it is produced in social and economic conditions outside 'international norms of behaviour' (cf. definition of social responsibility in ISO 26000).

LCSA aims to assess simultaneously the environmental, social, and economic impacts of products throughout their life cycle. A full assessment of a product takes into account private costs and benefits, as well as externalities. LCSA makes use of three techniques: environmental LCA (E-LCA), social LCA (S-LCA), and life cycle costing (LCC). This allows trade-offs to be clarified and sustainability along the value chain to be enhanced. Weaknesses in performance along the life cycle can be identified, actions with chances of positive impacts can be prioritized, and so on. In short, the results of LCSA can be used as a basis for life cycle management/chain management. A more detailed focus on components such as E-LCA and LCC appears in other chapters of this book.

Social life cycle assessment

Research on integrating social criteria into LCA started in the 1990s. The first attempts to include social considerations into E-LCA started by expanding the considerations to health and safety issues. Subsequently, other social concerns were addressed, and soon it became clear that social and socio-economic considerations have their roots in the organization itself, are site specific, and are often qualitative in nature.

S-LCA, as a part of the overarching LCSA, is an impact assessment technique that deals with the social and socio-economic aspects of products, specifically their potential positive and negative social impacts throughout their life cycle, from 'cradle to grave'. The different aspects assessed in an S-LCA are those that may directly affect stakeholders during the life cycle of a product. They may be linked to the behaviour of enterprises, to socio-economic processes, or to impacts on social capital. Indirect impacts on stakeholders may also be considered. Five main stakeholder categories have been identified: workers/employees, local community, society, consumers, and value chain actors. Each can be regarded as a cluster that has shared interests owing to its relation-ship with the product systems being investigated. Additional categories of stakeholders or subgroups as defined in international conventions may be added when an assessment is undertaken. It is within this context that a comprehensive set of subcategories has been presented:

- for the stakeholder 'worker' – for example, 'child labour' and 'fair salary';
- for the stakeholder 'consumer' – for example, 'health and safety';
- for the stakeholder 'local community' – for example, 'cultural heritage' and 'local employment';
- for the stakeholder 'society' – for example, 'corruption';
- for the stakeholder 'value chain actors' – for example, 'supplier relationships'.

Each is measured using different inventory indicators, for which methodological sheets have been developed.

Subcategories are fundamental to an SLCA requiring justification for inclusion or exclusion. They are classified according to stakeholder and impact categories and are assessed by the use of inventory indicators. Several indicators and units of measurement/reporting types may be used to assess each subcategory. Inventory indicators and units of measurement depend on the context of the study. The ultimate objective is to translate this result into major impact categories of human well-being.

Different contexts require varying levels of assessment. For example, legislation may already cover many human and worker rights indicators, but many countries still lack such laws. International standards tend to define 'floors rather than ceilings'. However, it is important to emphasize that this should not be taken for granted. A hot spot analysis is a useful first screening. The Social Hotspots Database publishes country-specific sector risk results and thus directs attention to social issues in need of enhanced engagement (Social Hotspot Database 2015).

Examples of LCSA and S-LCA

The publication in 2009 of guidelines for S-LCA of products (Benoît and Mazijn 2009) gave an impulse to further articles in international journals and case studies. Publications on LCSA now cover examples from the energy sector (alternators, biofuels, etc.), building sector (wood frames, sewer pipes, etc.), waste sector (electronic, municipal and other solid waste, etc.), agricultural sector (olive oil, roses, etc.), transport sector (vehicles, etc.). The Prosuite project (Prosuite 2015) contributed to progress in the methodological development of LCSA. Various technology case studies (biorefineries, nanotechnology, information technologies, and carbon storage and sequestration) allowed further insight into the methodology.

The consolidated expertise built up is sufficient to now move LCSA and S-LCA to the next level. In this light, the Social LC Alliance (Social LC Alliance 2015) is already building a collaborative process to complement the former publications.

As well, various universities and research centres in Belgium, Canada, Peru, and Spain, among others, are interested in applying LCSA and S-LCA to the building and infrastructure sectors.

Linking LCSA to sustainable regional development

Global challenges have featured in numerous international scientific and business reports over the past two decades. However, only in recent years has there been an effort by different public and private institutions to look for linkages between the so-called 'megatrends' that reach deeply into societies and their economies (EY 2010; KPMG 2012; UNEP, etc.). Such reports focus on numerous challenges, including, but not limited to, energy and fuels, climate change, material resource scarcity, water scarcity, population growth, urbanization, wealth, food security, ecosystem decline and deforestation.

These analyses fuel a growing consensus that the answer to our global sustainability challenges is to create a circular economy (instead of a linear throughput economy) on a (sub)continental level, such as Europe, India, China, and North America. Life cycle approaches, therefore, need to focus more and more on cradle-to-cradle material flows and dynamics. The future of economic development decision-making is considered to be at regional rather than global level. Some businesses are already addressing this reality by proposing product–service systems in lieu of selling physical products to consumers. The example of renting rather than buying floor coverings is now well known (Desso 2015; Interface 2015). Dutch entrepreneurs are combining LCSA thinking with considerations of a circular economy: the first generation of Fairphone takes into account social and environmental criteria during the life cycle of the smartphone. The second generation includes elements of a circular economy by being modular, so that a consumer can replace – if necessary – one part, for example, the camera, rather than have to buy a whole new device.

Conclusion

Sustainable regional development will require a re-evaluation of the indicators used in LCSAs. The results of LCSA will be balanced during the interpretation phase against regional as well as global standards. When it comes to S-LCA international conventions may still be a minimum standard, but the regional considerations related to social protection can become more important.

References

Benoît, C., and Mazijn, B. (eds) (2009) *Guidelines for Social Life Cycle Assessment of Products – A Social and Socio-economic LCA Code of Practice Complementing Environmental LCA and Life Cycle Costing, Contributing to the Full Assessment of Goods and Services within the Context of Sustainable Development*, Paris: UNEP-DTIE.

Ciroth, A., and Franze, J. (2011) *LCA of an Ecolabeled Notebook – Consideration of Social and Environmental Impacts along the Entire Life Cycle*, Berlin: Greendelta.

Desso (2015) Desso Group. Available at: www.desso.com/about/desso-group/ (accessed 15 December 2015).

Interface (2015) Home. Available at: www.interfaceglobal.com (accessed 15 December 2015).

EY (2010) *Business Redefined – A Look at the Global Trends that are Changing the World of Business*, London: EY.

Global Reporting Initiative (2011) *GRI and ISO 26000: How to Use the GRI Guidelines in Conjunction with ISO 26000*, Amsterdam: Global Reporting Initiative.

Global Reporting Initiative (2014) *Sustainability Reporting Guidelines*. Available at: www.globalreporting.org/reporting/g4/Pages/default.aspx (accessed 15 December 2015).

ISO (2010) *ISO 26000. Guidance on Social Responsibility*, Geneva: ISO.

KPMG (2012) *Expect the Unexpected: Building Business Value in a Changing World. Parts 1, 2, and 3*, Amstelveen: KPMG.

Mazijn, B. (2013) *Vers une 'nouvelle industrialisation' du et pour le secteur du métal. Une économie circulaire dans le contexte du développement durable*, Brussels: Herwig Jorissen.

Petti, L., Ugaya, C., and Di Cesare, S. (2014) 'Systematic review of Social-Life Cycle Assessment (S-LCA) case studies', in *Social LCA in Progress. Pre-Proceedings of the 4th International Seminar in Social LCA*, eds C. Macombe and D. Loeillet, Montpellier.

Prosuite (2015) Home. Available at: www.prosuite.org/ (accessed 15 December 2015)

Social LC Alliance (2015) Home. Available at: www.social-LCA.org/ (accessed 15 December 2015).

Social Hotspot Database (2015) Home. Available at: http://socialhotspot.org/ (accessed 15 December 2015).

Valdivia, S., Ugaya, C., Sonnemann, G., and Hildebrand, J. (eds) (2011) *Towards a Life Cycle Sustainability Assessment – Making Informed Choices on Products*, Paris: UNEP-DTIE.

WCED (1987) *Our Common Future*, Oxford, UK: Oxford University Press.

15

ORGANIZATIONAL LIFE CYCLE ASSESSMENT FOR REGIONAL AND LOCAL ORGANIZATIONS

Julia Martínez-Blanco, Atsushi Inaba, and Matthias Finkbeiner

Introduction

The pursuit of sustainable regional development implicates a variety of organizations and individuals, each of which works towards sustainability goals but also exerts a footprint through its own activities and decisions. Although the use of standardized environmental management systems is already common, the increasing focus everywhere on supply chain impacts points to an increasing need for more comprehensive analytical and decision-making tools for use within and by organizations.

Environmental analysis tools for organizations have emerged during the last 10 years to assess the full value chain of products, often focusing on a single environmental aspect such as carbon or water. Some of these – mostly voluntary – initiatives, promoted by, for example, WRI and WBCSD (2011), have now also paved the way for the use of life cycle assessment (LCA) of the organizations themselves. Life cycle environmental impacts of the supplier network and of the use phase and end of life of the product portfolio can significantly contribute to the total environmental performance of organizations. Depending on its sustainability objectives, an organization may thus need to intervene in the activities of its entire value chain to reduce both direct and indirect impacts and to avoid outsourcing its problems. This should also include a mechanism to prevent trade-offs (or burden shifting) between different environmental aspects.

Organizational LCA differs from previous product-focused assessment techniques in that its object of study is the whole organization. It differs from other organization-oriented methodologies by its approach, that is, including not only the internal facilities of the organization but also upstream and downstream activities. It provides a complete environmental understanding at the level of the organization and supports an improvement in its total environmental performance.

Organizational LCA is universally applicable. Private and public organizations alike have a role in the protection of the environment. Apart from reducing the direct and indirect impacts from their own activities, public organizations have a further responsibility to act as an example and to implement a policy framework for the wider community to change to more sustainable practices. A number of chapters in this book describe sustainability options and actions for public authorities and regional organizations.

In this chapter, we introduce the general concept of organizational LCA, give details about the existing guides and initiatives, and provide some recommendations about its use for the assessment of and by regional organizations.

Organizational LCA: An overview

Companies, corporations, firms, and regional and local institutions have a key responsibility in efforts to improve environmental performance. The first step to an effective reduction in long-term environmental impacts is a comprehensive analysis that sheds light on their nature.

Organizational LCA is a compilation and evaluation of the inputs, outputs, and potential environmental impacts of the activities associated with the organization, adopting a life cycle perspective (ISO 2014). Most of the principles and requirements of the product LCA standard (ISO 14040; ISO 2006) described in Chapter 10 also apply for organizational LCA, after some minor terminology amendments (Finkbeiner and König 2013).

This methodology is capable of serving multiple goals at the same time (e.g. identifying environmental hot spots throughout the value chain, performance tracking, supporting strategic decisions and informing corporate sustainability reporting). It is important to point out that, according to ISO/TS 14072 (ISO 2014), organizational LCA is not suitable for comparisons of different organizations intended to be disclosed to the public.

Guides and initiatives

LCA for entire organizations is becoming of increasing interest as a complement to product-based analyses, but also as a stand-alone methodology. Several guideline documents have been published, with follow-up initiatives to promote the development of organizational LCA in specific sectors and organizations (see Table 15.1), in particular the ISO/TS 14072, the UNEP/SETAC Life Cycle Initiative, and the OEF/PEF initiative from the EC.

Table 15.1 Overview of current initiatives for the LCA of organizations

	ISO/TS 14072	UNEP/SETAC guidance	OEF guide
Developer	International Organization for Standardization, ISO/TC 207	UNEP/SETAC Life Cycle Initiative	Joint Research Centre and Institute for Environment and Sustainability (DG Environment European Comission)
Main document	ISO/TS 14072: Environmental Management – Life Cycle Assessment – Requirements and Guidelines for Organizational Life Cycle Assessment	Guidance on Organizational Life Cycle Assessment	Organisational Environmental Footprint Guide
Type of document	Technical specification	Guidance (UNEP/SETAC report)	Non-legislative Act
Name of the methodology, acronym	Organizational Life Cycle Assessment, OLCA	Organizational Life Cycle Assessment, O-LCA	Organizational Environmental Footprint, OEF
Current status	International Standard published	Document published; until 2016 road testing phase	Document published; until 2016 pilot phase and development of OEF sector rules (OEFSRs)

In 2014, ISO published the technical specification *ISO/TS 14072: Environmental Management – Life Cycle Assessment – Requirements and Guidelines for Organizational Life Cycle Assessment* (ISO 2014). It harmonizes the requirements and guidelines to apply life cycle thinking and identifies the benefits that LCA can bring to organizations.

ISO/TS 14072 serves as the backbone for the partnership between the United Nations Environmental Programme (UNEP) and the Society of Environmental Toxicology and Chemistry (SETAC). Their joint Life Cycle Initiative launched the flagship project, LCA of Organizations, in 2013. The main outcome of this project is the document *Guidance on Organizational Life Cycle Assessment* (UNEP 2015). This document builds on the ISO specification and goes into greater detail on the capabilities of organizational LCA (O-LCA) and its methodological framework. Eleven case study experiences of 'first movers' were included in the guidance document to illustrate some methodological aspects as well as the benefits that the methodology could bring to organizations. Eight sectors and four regions are represented in the case studies. During the end of 2015 and 2016, the guidance document is being road-tested in more than ten organizations representing different regions and sectors, and with different levels of experience in the use of environmental tools. The road tests are intended to build a foundation on which a larger group of stakeholders can engage to use O-LCA.

In a parallel exercise, the European Commission started a project in 2011 to develop a reference method for organization and product environmental footprinting. In 2013, it launched the *Organisation Environmental Footprint Guide*, along with an equivalent guide for product footprinting (i.e. the PEF Guide). These documents aim to increase reproducibility and comparability by emphasizing prescriptiveness over flexibility to ensure that the methodology is applied consistently (European Commission 2013). At the European level, the PEF and OEF are supposed to achieve an important goal, the implementation of LCA in European environmental policy. It has been challenged, as it does not align with some life cycle standard principles, and some of the introduced requirements have not been extensively tested or used (Finkbeiner 2013). Until 2016, pilot studies were being conducted on selected sectors (copper products and retail) in order to test further the methodology and prepare OEFSRs (OEF sector rules) for the corresponding sectors.

Technical framework

The organizational LCA methodology follows the four-phase methodology that is the core of the product LCA standards (i.e. goal and scope definition, inventory, impact assessment, and interpretation), as introduced in Chapter 10 of this book. The major differences between product and organizational LCA appear during scope definition and inventory assessment, and that is the reason to focus here on the two first steps of LCA (Martínez-Blanco et al. 2015).

The scope defines the breadth, depth, and detail of the study in accordance with the stated goals (Figure 15.1). The elements that are specific for organizational LCA are reporting organization, reporting flow, and system boundary.

The primary purpose of the reporting organization is to define the unit of analysis, and this requires the identification of the subject of study, the consolidation method, and the reference period. The subject of study should

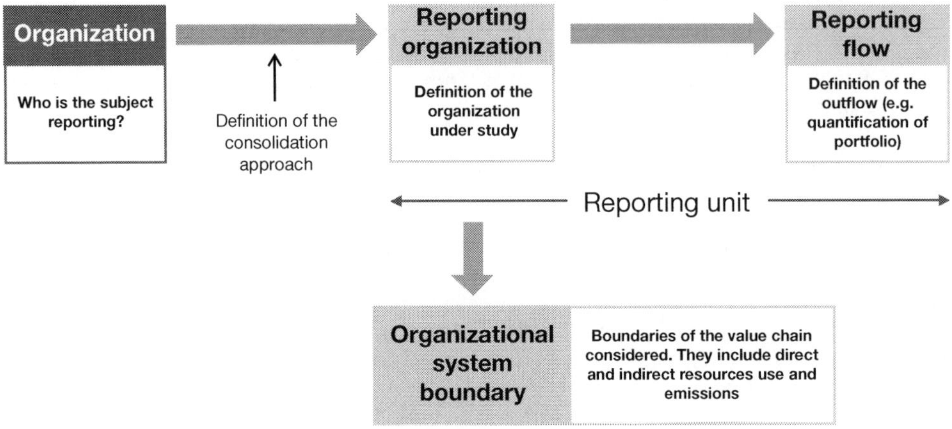

Figure 15.1
Main elements of the scope phase in organizational LCA according to UNEP (2015)

represent a clear unit of operation (e.g. divisions, regions, or facilities). If properly justified, the application of organizational LCA to segments or selected parts of an organization is possible. The consolidation method assists in representing the structure of the organization and its relationships with other organizations, using one of three methods – financial control, operational control, or equity share approaches – to consolidate its units (e.g. facilities).

The reporting flow is a measure of the outputs from the reporting organization during the reference period, on a per physical basis or another, such as economic revenue.

Finally, system boundary determines the direct (i.e. from sources that are owned or controlled by the reporting organization) and indirect (i.e. from sources owned or controlled by another organization, a supplier, or the user/consumer, but a consequence of the activities of the reporting organization) resource uses and emissions. Downstream activities (use phase, end of life, etc.) should be included if products directly consume energy or generate emissions during this phase, or indirectly consume energy or cause emissions during use.

The inventory is the phase of an organizational LCA where data are collected, the systems are modelled, and the life cycle inventory results are obtained based on the previous definition of the goal and scope of the study. The inventory should consist of all resource use and emissions necessary for the activities involved in the provision of the reporting flow and considering the system boundaries definition. Data collection can be performed with a bottom–up approach (at the product level) or a top–down approach (at the organization or facility level). In general, better quality, more specific data can be expected for activities inside the reporting organization.

The impact assessment and interpretation of organizational LCA follow the same recommendations and requirements as product LCA, and the same impact categories and methods can be applied. The selection of impact categories should reflect a comprehensive set of environmental issues related to the system being studied. Inventory-level indicators may be evaluated within the context of organizational LCA, but they should be presented separately from the impact categories. Finally, interpretation should indicate the consistency of the results according to all the aspects defined during the goal definition and scope phase.

Applying organizational LCA in a regional or local context

Many organizations at regional and local level are already engaged in some degree of sustainability assessment. Numerous companies and cities report the carbon footprint of their private and public activities, such as the city of San Francisco[1] or the Norwegian city of Haninge.[2] At least 100 cities across the globe have used the Global Protocol for Community-Scale Greenhouse Gas Emission Inventories to measure their greenhouse gas emissions (WRI, C40, and ICLEI 2014).

Life cycle approaches can support cities in the assessment of their impacts from a multi-impact perspective. Organizational LCA offers the framework for regional and local organizations to account for the resources, emissions and environmental impacts linked to the functions provided, including all suppliers. This information helps organizations on their way to achieving better environmental conditions, by directly reducing their impacts and by promoting better management within citizens and communities.

The above is relevant, not only to companies, but also to many other types of organization, such as municipalities, city councils, regional development offices, regional research institutions, as well as local authorities for education, health, or sport. Most are public organizations, and thus they do not have profit as a primary target, but focus instead on development and providing services to the local/regional community.

The following list states some of the benefits that this approach can bring to regional organizations (adapted from Martínez-Blanco):

- It can reduce pressure on the environment and avoid future negative effects. Some of these impacts may act over the area where the regional organization operates, with a direct effect on the regional communities, but external areas along supply chains may also benefit.
- It can give insight about relationships, main actors, and impacts involved in internal operations and the list of suppliers.
- It can reveal environmental hot spots along the various value chains where the organization should intervene.
- It can track environmental performance of materials inventories and impacts of the organization over time.
- It can provide a solid basis to identify priority actions and targets at different levels, and get estimations of potential future scenarios according to them.
- It can improve organizational procedures, for instance in the gathering and managing of environmental data.
- It can provide a solid foundation for regional sustainability reporting.
- It can show environmental awareness and encourage suppliers in the value chain, citizens, and other regions to adopt environmentally friendly practices. For example, the organization may have the influence to set a list of specifications with which suppliers should comply while operating as a link in the product value chain.

Some of the specificities of regional and local organizations when organizational LCA is being applied are described below. Regarding the goal definition, targets for local communities and for suppliers are expected to have a wider relevance in the study. For the reporting organization definition, it is particularly important to set the limits of the study, as many regional organizations may be part of bigger authorities or institutions and collaborate with other regional and local institutions to provide their functions.

For organizations that are providers of services or social functions, the identification and quantification of the reporting flow is not straightforward. Answering the questions 'What?', 'How much?', 'How well?', 'How long?' in those cases can be a challenging task. A good example is an urban planning office, for which it is difficult to define and quantify the services provided, as they will be project-specific. In such a case, the reporting flow can be quantified, for instance, in terms of the number of hours invested, or per number of people who receive the service provided by the institution. The challenge of defining the functional unit for public services is also addressed in Chapter 28, on territorial LCA.

With regard to data collection, depending on the scale of the assessment, environmentally extended input–output (EEIO) analysis and hybrid IO LCA methods can be used, at global and national level, or at subnational scale, for organizations and industrial sectors, respectively (see Chapter 8 of the book for more information on EEIO analysis).

Conclusions

Although LCA was originally considered for products, its benefits and potential can now be extended to organizations as well. Organizational LCA is a promising new member of the LCA family, not only for companies, but also for regional and local organizations that are mostly in the public arena. Several initiatives and guides contribute to the development of the methodology. The implementation of organizational LCA for organizations at the regional and local level, if possible supported by the regional government, may help them to achieve better environmental performance and drive sustainable innovation within communities and suppliers.

Notes

1 www.sfenvironment.org/sites/default/files/fliers/files/ccar_grp_3-1_january2009_sfe-web.pdf (accessed 31 July 2016).
2 https://www.kth.se/polopoly_fs/1.193966!/Menu/general/column-content/attachment/Carbon%20Footprint%20-%20A%20Case%20Study%20on%20the%20Municipality%20of%20Haninge.pdf

Further reading

Martínez-Blanco, J., Inaba, A., and Finkbeiner, M. 'Life cycle assessment of organizations', in *Encyclopedia Life Cycle Assessment, Vol. 7, Special Forms of LCA*, ed M. Finkbeiner, Dodrecht, Netherlands: Springer.

Pelletier, N., Allacker, K., Pant, R., and Manfredi, S. (2013) 'The European Commission Organisation Environmental Footprint method: Comparison with other methods, and rationales for key requirements', *International Journal of Life Cycle Assessment*, 19: 387–404.

References

European Commission (2013) *Organisation Environmental Footprint Guide*, European Commission, Joint Research Centre, Institute for Environment and Sustainability. Available at: http://eur-lex.europa.eu/JOHtml.do?uri=OJ:L:2013:124:SOM:EN:HTML (accessed 31 July 2016).

Finkbeiner, M. (2013) 'Product environmental footprint – Breakthrough or breakdown for policy implementation of life cycle assessment?', *International Journal of Life Cycle Assessment*, 19(2): 266–71.

Finkbeiner, M., and König, P. (2013) 'Carbon footprint and life cycle assessment of organizations', *Journal of Environmental Accounting & Management*, 1(1): 55–63.

ISO (2006) *ISO 14040: Environmental Management – Life Cycle Assessment – Principles and Framework*, Geneva: ISO.

ISO (2014) *ISO/TS 14072: Environmental management – Life Cycle Assessment – Requirements and Guidelines for Organizational Life Cycle Assessment*, Geneva: ISO.

Martínez-Blanco, J., Inaba, A., and Finkbeiner, M. (in press) 'Life cycle assessment of organizations', in *Encyclopedia Life Cycle Assessment, Vol. 7, Special Forms of LCA*, ed M. Finkbeiner, Dodrecht, the Netherlands: Springer.

Martínez-Blanco, J., Inaba, A., and Finkbeiner, M. (2015) 'Scoping organizational LCA – Challenges and solutions', *International Journal of Life Cycle Assessment*, 20: 829–41.

UNEP (2015) *Guidance on Organizational Life Cycle Assessment*, Paris: Life-Cycle Initiative, United Nations Environment Programme and Society for Environmental Toxicology and Chemistry. Available at: www.lifecycleinitiative.org/wp-content/uploads/2015/04/o-lca_24.4.15-web.pdf (accessed 31 July 2016).

WRI, C40, and ICLEI (2014) *Global Protocol for Community-Scale Greenhouse Gas Emission Inventories (GPC)*, US: World Resources Institute, C40 Cities Climate Leadership Group, and ICLEI – Local Governments for Sustainability. Available at: http://ghgprotocol.org/files/ghgp/GHGP_GPC.pdf (accessed 31 July 2016).

WRI and WBCSD (2011) *Corporate Value Chain (Scope 3) Accounting and Reporting Standard – Supplement to the GHG Protocol Corporate Accounting and Reporting Standard*, World Resources Institute and World Business Council for Sustainable Development. Available at: www.ghgprotocol.org/files/ghgp/public/Corporate-Value-Chain-Accounting-Reporting-Standard_041613.pdf (accessed 16 July 2016).

Part III

CONVERTING ANALYSIS INTO LIFE CYCLE TOOLS

16

CERTIFICATION, STANDARDIZATION, AND INSTRUMENTS FOR REGIONAL SUSTAINABLE DEVELOPMENT – A LIFE CYCLE VIEW

Cristiana Reho and Pia Wiche

Introduction

Environmental sustainability is a topic that provokes growing interest and concern among citizens and consumers. It is, therefore, gaining a higher position on the agendas of public administrations and private organizations. However, in an ever more globalized and competitive market, it is sometimes difficult to distinguish how to make sustainable choices.

Competitiveness is at the core of many regional policies around the world and has been linked to well-being, although with a number of externalities that hinder sustainable development (Thompson and Huggins 2012). Hence, regional policies should consider incorporating instruments to promote sustainable consumption and production by all actors in a territory.

Standards and certification are currently used as trademarks of sustainable production at a process level. Certification can support environmental claims by proving compliance with a standard or regulation, thus easing decision-making, as shown in Chapter 18. Furthermore, regional operators need to be able to recognize the application of standards and certification of compliance in other organizations – especially in the private sector, consumer goods suppliers, or services – that want to operate in the region. Regions can thus choose the 'most sustainable' operators and suppliers.

That being the case, what would be the impacts of such an approach? Which indicators and standards should be used, how would quality assurance be practised, and how would businesses and the community be affected? For instance, following regional standards may be a challenge for multiregional companies, which would have to adapt their operations from site to site.

For many life cycle tools, the matter has been solved by agreement on indicators (such as with a materiality matrix) and standardizing methodologies and then the offer of certification upon the adoption and application of those standards.

An example would be carbon footprint. In general, carbon footprint on a territorial scale would come from international agreements (sometimes also from community interest, such as the notable case of the State of California, in the US (California Environmental Protection Agency n.d.). The indicator would have to follow a standardized methodology, in this case provided by the Intergovernmental Panel on Climate Change, with a particular instrument (input–output calculation). The report would then be verified, and its results could be published for accountability. For other approaches to carbon footprint, please see Chapter 11.

Standardization

The terms 'standard' and 'certification' are commonly used interchangeably, but they mean very different things.

Standards are 'technical *documents*, established by consensus and approved by a recognized body, that provide rules, guidelines or characteristics for activities or their results, aimed at the achievement of the best practice in a given context' (adapted from ISO/IEC Guide 2:2004, Definition 3.2).

In other words, the term 'standard' is commonly applied to a set of recognized benchmark parameters or conditions established by a group of sector stakeholders. Standards shouldn't be confused with legislation, although they sometimes serve as the technical basis for regulations. Standards, unlike formal laws, are developed by collaborative, non-government bodies, such as standards associations, through a process in which appropriate representation of stakeholders and technical expertise take part.

They are useful because they provide a systematic and consistent approach and compile best practices to achieve a result. Each standard focuses on different topics, and it has to be clear that each of them grants different results, as can be seen from Table 16.1.

Measurement standards provide tools to measure specific environmental performance and can be useful to integrate the management standards in providing, year on year, the status of the environmental impacts.

Performance standards indicate the level of performance of the organization, product, or service. They usually specify a set of best practices or impact levels.

Table 16.1 Several types of standard and their scope explained

Scope of the standard	Goal achieved	Example
Test method/measurement	Provide a consistent way of executing a test or analysis, to obtain consistent and reliable results	ISO 8466–2 water quality calibration
Specification/quality	Facilitate purchasing; create standardization; provide technical data	ISO 559:1991; steel tubes for water and sewage
Performance	Provide a definition of the minimum performance level of an organization or worker	Defined by each organization
Practice/environmental	Provide instructions on how to perform a task that does not include a test result; estimate the environmental impacts of an activity	ISO 14046: Water Footprint
Guidelines/management	Map a course of action or provide a framework; manage a company	ISO 14001: Environmental Management Systems
Labels/declarations	Guarantee that a set of targets have been met; speed decision-making	Singapore Green Label Scheme

Management and methodologies can also be standardized, including, prominently, the ISO 14000 series on environmental management tools. ISO also produces guidelines as an interim step (or an alternative) to the standardization process, such as for social impact assessment. ISO 14001 is the most used environmental management standard, recognized worldwide. By December 2014, 324,148 ISO 14001:2004 certificates had been issued in 171 countries (ISO 2015).

Labels or declarations that are supported by certification provide a guarantee that a set of targets has been met, as a minimum, in the environmental performance of the product or service. The benefit is that the user does not necessarily have to carry out their own comparative analysis, as better described in Chapter 18.

Understanding the differences between environmental standards helps the user to choosing the standard that best suits their needs. These depend on the objectives that the organization has:

- Apply a standard and/or be certified.
- Make greener choices.
- Generate collaboration with suppliers/outsourcers on measurement of specific environmental indicators and performance.
- Finance or promote projects that have a certified measure of environmental impacts.
- Communicate performance; help the client/community understand the claims of the organization.

There are also external factors that influence the choice: for example, what the stakeholders require or are sensitive to; the most significant environmental issues of the economic sector and of the geographical area in which the organization operates; the cost and benefits of applying the standard and, eventually, of the certification process, and so on.

Who can develop standards?

The world's largest developer of voluntary standards is ISO. With a global span and a standard-generating process in more than 170 countries, ISO manages to deliver consistent, coherent standards, resulting from a consensus process in which experts from all over the world can participate. At the national level, countries usually have a national standards organization (usually also a member of ISO) that deals with the generation of standards of national relevance and may participate in the standardization processes led by ISO.

Although international standards under ISO are the best known, individual countries or regions may have standards to address local needs or specific sectors. British PAS 2050, French BPX30–323, and the different EPD schemes (see Chapter 23) are all examples of local standards or guidelines developed to provide more specific guidance for organizations wishing to measure or communicate one or more environmental dimensions. Many certification schemes and standards have been developed taking into account different aspects of the life cycle of products – for example, FESC and PEFC for wood and derived products, Bonsucro for sugar cane and derived products, RSPO for palm oil, or MSC for fish.

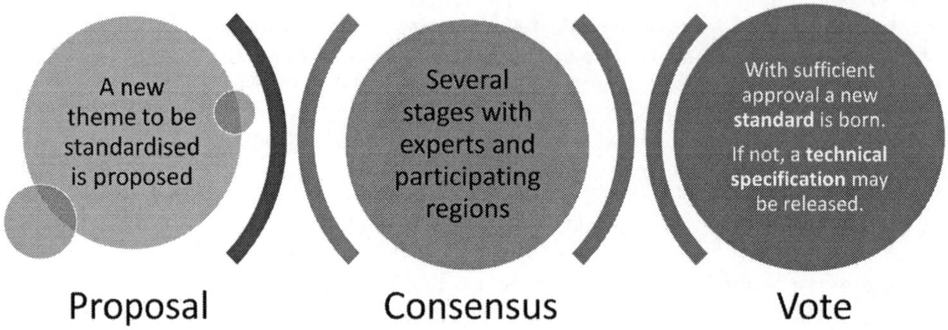

Figure 16.1
Generic process to produce a standard

Apart from the existence of numerous standards and labels, there is currently a gap in the standardization of instruments for regional sustainability, as the focus is mostly on either products or services, not specifically on regions. On the other hand, regions can develop their own standards (such as eco-labels). In general, the process would be similar to that depicted in Figure 16.1.

Not all the proposals become standards. If no consensus is reached, the document will result in a technical specification, awaiting its implementation and feedback to overcome the most pertinent questions and make it a standard.

Certification

Certification can be used to ensure a standard is correctly applied. A certificate is a 'written *statement* provided by a party independent of the company/product/process/person verified that declares and assures that specified *requirements* contained in a standard *have been met*'.

Certification is more than a verification of the fulfilment of requirements, as the results of the assessment are submitted to an independent technical reviewer, and the activity of the certification body is subject to the control of a multi-stakeholder committee safeguarding impartiality, competence, and independence. Many certification bodies are also accredited – that is, regularly verified by a third party – which gives further confidence in the fairness and reliability of their activities.

Certification can be applied to any standard; however, the focus and verification methods applied will depend on the type of standard for which the certification is requested.

Certification on a respected standard puts an organization in a particular position and can sometimes be used as a way to choose suppliers or establish trust with consumers and clients. Many examples of certification can be found on the Standardsmap website (www.standardsmap.org/). Regions can adopt or generate their own certification as proof of sustainable management, in order to regulate activities, reporting, and communication or even to monitor compliance.

Instruments for regional sustainable development

A variety of standards have been created to define sustainability requirements for products and organizations. Certification of conformity to these requirements is increasingly used as a 'recognition tool' to promote and select products, services, or organizations and ensure alignment and consistency, both nationally and internationally. In addition, a number of life cycle tools, such as LCA, have become standardized to ensure consistent application by various partners in different regions (see Chapter 10). For some tools that are not standardized, other guidelines for more consistent application have been prepared.

As commented previously, working with certified instruments is a proof of compliance to the standard and eases selection of service providers. Regions could profit from generating standards or admitting certification in order to select organizations to operate in their territories, or to address performance issues. This would, in turn, allow them to choose the 'most sustainable' operators and suppliers.

There are a few international standards related to sustainable development with a territorial approach.

The greatest and best-known document is surely the *IPCC Guidelines for National Greenhouse Gas Inventories 2006* (IPCC 2006). It contains the guidelines to estimate carbon footprints at national level, sometimes using regions of the country as a subset (bottom–up approach).

Water naturally relates to territory through basins. In *The Water Footprint Assessment Manual* (2011), four out of eight ways of calculating the water footprint were bound to a territory (from a geographical region to a catchment area). For more information on regional water footprint, please refer to Chapter 31.

ISO has developed at least two standards that are specifically for local administrators and are useful with reference to sustainable development of regions: ISO 18091 (ISO 2014b) and ISO 37120 (ISO, 2014c). Both from 2014, they prove that there is a trend to analyse and manage sustainable development from a local perspective. In the same year, ISO 14072 (ISO 2014a)[1] showed a life cycle perspective for an organization, which could be extended to a municipality or administrative region.

Voluntary standards are not the only instruments at hand and should be supported by regional laws and regulations (where possible) to facilitate the management of particular areas. An example is represented by decontamination plans, which can be defined for a specific area in a region and allow for more stringent environmental regulations.

Common gaps in regional sustainable assessments include the definition of impact, the carrying capacity of the environment, and the specific needs of the population in the region. To define them, research on regional natural and social systems, together with consensus on priorities by all actors in the society, is needed.

In order for these gaps to be filled, important players will be universities and research institutions, community organizations, NGOs, as well as the industrial and commercial sectors. The regional government has a critical role in promoting the process and supporting it with regional instruments, that is, through the creation of new laws, or by providing funds for inspection, prosecution, or quality assurance.

The first step for standardizing instruments for regional sustainable development should be the definition of the goal and scope. Put simply:

- What will this standard promote?
  - apply a standard and/or be certified;
  - generate collaboration within the value chain for achievement of specific environmental indicators and performance;
  - finance or promote projects that have a specified range of environmental impacts;
  - provide clear information to the community for informed decision-making;
  - compare performance with other regions;
  - regional priority setting and decision-making.
- What types of instrument can be implemented – voluntary standards, compulsory regulations?
- Which actors will be involved in the process of creation, implementation, and evaluation?
- How will success be measured?
- Under which circumstance will this instrument be revised?

These questions are meant to orient the process; they are open to modification or additions.

Challenges of the application of standards and certification for regional sustainability

A common way of ensuring compliance to a standard is through certification. Certification will commonly require an agent called the 'programme operator', in charge of certifying the implementation of the standard, which poses the challenge of choosing a suitable programme operator. At the same time, certification usually represents a big investment, which may be partially funded by the regional government to support small and medium-sized businesses.

Having a regional standard with a powerful methodology enhances the possibility of comparing the results obtained by different organizations. However, in a globalized world, if the application remains local, it might not allow comparison with imported products.

Standards ensure that life cycle tools are used in a consistent way, allowing comparability across borders, and giving assurance of appropriate procedures along extended supply chains. However, if standards are created for different regions, their comparability might be compromised. For example, the UK, France, Italy, and Switzerland use different schemes to label a product as a 'green product' (European Commission 2015). Although environmental product declarations (EPDs) seem to answer this problem, there are several programme operators, all of which have different requirements for the same types of product, which in turn hinders the main advantage of EPDs: comparability.

Therefore, the scope of the regional standard's application must be well defined to create a tool that will serve the regional decision-making process in the most suitable way.

Conclusions

There is a large pool of experience and knowledge about standards and certification, and there is evidence that they can facilitate the quantification of progress towards sustainable development. Standards and certification can provide timely, relevant information to support sustainable regional development in several ways, such as by certifying low environmental impacts, standardizing communications between organizations and the community, or accrediting fair practices in trading with primary producers, just to name a few.

The setting up of regional standards for sustainable development should respond to the needs of the region and its people, based on the three pillars of sustainable development – environmental, social and economic – and be should based on a life cycle perspective, in order to avoid load transfers between regions.

Regional standards should be built to meet specific goals, as their scope of application will vary according to their characteristics. Furthermore, it is possible that a single standard will not be sufficient to meet all of the region's requirements to advance towards sustainable development.

Therefore, the production of regional standards should include multiple actors: industry, the community, service providers, research institutions, and others, which will in turn provide the knowledge and capacity to find technical specifications, implement the standard, and, later on, certify it.

Finally, there seems to be a trend to create more and better regional standards, thus acknowledging regions' particular needs, ecosystems, and culture.

Note

1 The UNEP Guidance on O-LCA was released in May 2015 and serves as support to the ISO standard and helps clarify some of the methodological changes from a product to an organizational LCA (for more details on O-LCA, see Chapter 15).

References

California Environmental Protection Agency (n.d.) Assembly Bill 32 Overview. Available at www.arb.ca.gov/cc/ab32/ab32.htm (accessed 25 November 2015).

ASTM (2000) Know your types of standards. Available at www.astm.org (accessed 25 November 2015).

European Commission (2015) Single Market for Green Products Initiative. Available at http://ec.europa.eu/environment/eussd/smgp/index.htm (accessed 25 November 2015).

IPCC (2006) *IPCC Guidelines for National Greenhouse Gas Inventories*, eds H.S. Eggleston, L. Buendia, K. Miwa, T. Ngara, and K. Tanabe, Japan: IGES.

ISO (2014a) *ISO 14072:2014: Life Cycle Assessment of Organizations*, Geneva: ISO.

ISO (2014b) *18091: 2014: Quality Management Systems – Guidelines for the Application of ISO 9001:2008 in Local Government*, Geneva: ISO.

ISO (2014c) *ISO 37120: 2014: Sustainable Development of Communities – Indicators for City Services and Quality of Life*, Geneva: ISO.

ISO (2015) ISO Survey. Available at www.iso.org/iso/iso_survey_executive-summary.pdf?v2014 (accessed 25 November 2015).

Thompson, R., and Huggins, P. (2012) 'Well-being and competitiveness: Are the two linked at a place-based level?', *Cambridge Journal of Regions, Economy and Society*, 5(1): 45–60.

17

ECO-DESIGN AND PRODUCT–SERVICE SYSTEMS

Benjamin Walker

Introduction

The history of mankind can be written in terms of the making and trading of goods and services. Local expertise and know-how had economic value, thus creating regional prosperity.

Globalisation has challenged many economies, as the tyranny of distance has been eroded by inexpensive hydrocarbons and instantaneous communication. Coupled with the ever-present quest for profit, these drivers have encouraged identification of inexpensive materials and labour, thereby 'pulling' manufacturing offshore. In developed countries, these factors have generally led to a shift, often a decline, in local economic activity. This need not be the case.

As a strategic tool for sustainable development, eco-design has many attractive qualities. These benefits are not restricted to the reduction in a product's environmental impact, but can potentially trigger new businesses that see the paradigm shift taking place in production and consumption as an opportunity.

It has been claimed that 80 per cent of a product's environmental impact is determined by its design (EU n.d.).

By considering the different stages of a product's life cycle – from material extraction through to its end-of-life scenarios – a designer is led to think more broadly about the resources mobilised and consumed by a product, rather than simply ensuring aesthetic commercial appeal. By taking into consideration all the environmental impacts of a product at the earliest stage of design, the design process should avoid offsetting, or inadvertently redistributing, the negative environmental impacts of a product elsewhere. For example, the elimination of a toxic substance should not lead to higher energy consumption.

This process is known as eco-design and is supported by tools and policy frameworks to help inform design decisions.

Before we examine the relatively new practice of eco-design, it is interesting to take a moment to consider the origins of our consumer culture, and its unintended consequences.

Je t'aime, moi non plus . . .

The nineteenth century saw prolific advancement in technological well-being, notably with the growth of a middle class across the developed world. With the spectacular loss of confidence in financial systems in 1929, the shortcomings of Adam Smith's invisible hand wrought tremendous hardship and misery, notably among the poorer working classes. A young economist by the name of Bernard London suggested a possible solution to the economic stagnation that had immobilised the world (London 1932). Simply put, it was to increase product consumption. Encouraging the consumption of more products would in turn get factories producing consumer goods, providing greater economic activity, and therefore employment. His theory was beguiling: well-made, long-lasting products are an economic nuisance. Short-lived products, with their obsolescence programmed, stimulate consumption and therefore economic activity. But, at what cost?

As the detrimental impacts on the environment of an unchecked consumer society became more apparent, questions were raised as to the role designers played in the egregious, irresponsible use of natural resources. Calls were made for designers to recognise their responsibility for the environmental impacts of product consumption. One of the harshest critics of the design profession was Victor Papanek – a designer himself – who criticised product design as the 'second worst profession in the world' (Papanek 1982) It was trumped only by advertising.

Shortly after Papanek's seminal publication, the OPEC oil crisis triggered a realisation that fossil-fuelled energy was finite and therefore needed to be used responsibly. Coupled with a less-than-optimistic prognosis from the Club of Rome, a great effort was made, and attention was given to how energy was used. The new tools for improving domestic lifestyles were identified as being a source of poor energy use. Initiatives such as the Energy Star® eco-labels were introduced in the United States as a means to identify products that offered a comparable service, but had a lighter energy footprint. Considering energy use in terms, not only of a product's manufacture, but also the energy consumed during its functional life gave birth to the field of eco-design.

Life cycle analysis and the building bricks of the green economy

The 'global village' has allowed millions to be connected with one another, facilitating the exchange of ideas, goods, and services. This sudden promiscuity eventually raised awareness of the damage that a blind, linear economy wreaks. The destruction of the ozone layer and disappearance of Amazonian rainforests became topics of concerned conversation. It became clear that our collective future was under threat. The UN-sponsored Brundtland report of 1987, *Our Common Future*, made the term 'sustainable development' a household term (WCED 1987).

The report laid the framework for an honest analysis of our consumption patterns and production techniques. For the first time, a product had a 'life' represented by clear, determined 'stages'. Each stage had a corresponding environmental load. From material extraction through to fabrication, utilisation, and its end-of-life scenario, a product's impact(s) could be modelled and calculated to determine its overall environmental load. By thorough analysis of each of these life cycle 'stages', strategies to mitigate or reduce a product's environmental load could be identified. This life became a 'cycle', and the new discipline of life cycle analysis (LCA) was born.

By addressing each of these 'stages' in the product life cycle, an eco-design team can ask itself some simple questions, such as:

1 Primary materials: What raw materials go into the creation of this product? How are they harvested, extracted? Under what conditions, and supporting which economies?
2 Manufacturing processes: How is the product manufactured? What energy is used? The 'embodied energy' of a product can differ remarkably, depending on material choice. Are there carcinogens in the product? Where does the waste from the manufacturing process go? Can it be reused?
3 Distribution/transport logistics: How far does the product have to travel? By what means? How is the product transported to the customer?
4 Functional life: How is the product used? What is its energy consumption? What ancillary products does it use (e.g. printer versus cartridge use)? Is it durable? What is its lifespan? What can be done to extend its lifespan? What are the impacts of its maintenance?
5 End-of-life scenario: How is the product disposed of? Can it be reused or remanufactured? Is it recyclable? Does the manufacturer have a take-back programme?

Understanding and taking into consideration these life cycle stages as best one can form the basis for an eco-design approach (See Figure 17.1).

Early eco-designers focused mainly on the materials from which a product was made. This is understandable, as it is the tangible aspect of a product. However, as they were only considering one aspect of the life cycle, sometimes 'benefits' made in one stage of the life cycle, such as using wood instead of plastic for a radio casing, for example, meant that impacts were transferred to another stage – in this case, the increased weight of the radio units increased the carbon footprint during transport.

As LCA tools became more sophisticated, the capacity to undertake granular analysis of a product's environmental impact (based on data sets made available under licence) enabled designers and manufacturers to identify and quantify environmental impacts and build an appropriate design strategy to address these. The increased transparency made possible by such analysis, as well as the capacity to model different manufacturing/distribution/end-of-life scenarios, can provide material for creative discussion within a design team. Local and regional manufacturing scenarios can be modelled to illustrate the environmental impact of a product leaving the factory door. However, the real value of LCA lies

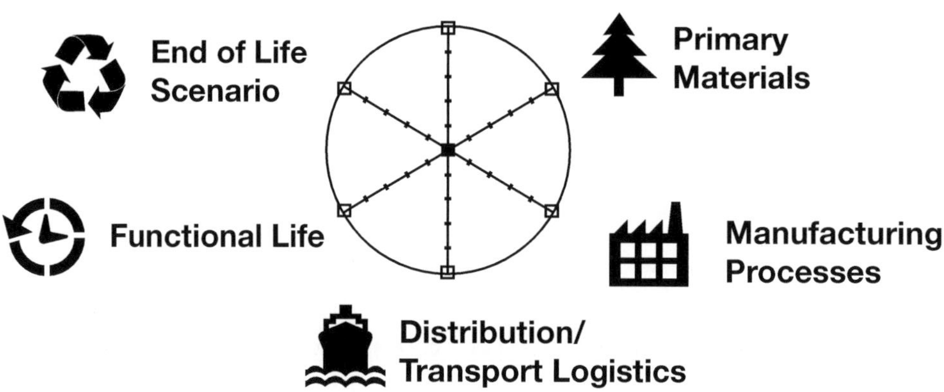

Figure 17.1
Life cycle design process

in comparing, not just manufacturing scenarios, but also user behaviour, to calculate more accurately a product's footprint (see www.pre-sustainability.com/simapro-tutorial). The carbon footprint of the same coffee machine, with equivalent usage, is completely different in Victoria, Australia, from that in Paris, France. Electrical energy is not produced in the same way in these two countries, and this has a significant impact on the carbon emissions provoked by this product over its functional life.

LCA software such as Sima Pro® and Gabi® certainly help put quantifiable metrics behind eco-design choices; however, these tools often require specialised training and a significant financial investment. Efforts to simplify and democratise these tools have been made, with certain success, notably by Open LCA (see www.openlca.org) and e-DEA®, targeted at a broader market.

Driven by converging factors

At both national and international levels, governments are obliging companies to clean up manufacturing process and improve the environmental performance of products. The first government initiative on energy use came from the United States with its Energy Star® programme, launched in the 1970s. The European Union produced the eco-design directive providing a framework for harmonising national legislation on the environmental performance of products (EU n.d.). The International Organization for Standardization (ISO n.d.) has a family of standards – the 14000 series – for environmental management that covers audits, communications, labelling, and LCA, as well as carbon footprints. The ISO 14006 standard provides guidelines for eco-design.

The other important driver for eco-design is the market. Sophisticated, increasingly informed consumers are demanding greater transparency and accountability in their sustainable-product choices. To this end, the practice of eco-labelling has flourished, and it is possible to find labels and certifications for all manner of sustainability agendas (see Chapter 17).

Quantifying environmental impact is only one part of an eco-design approach.

Although it is important to benchmark quantifiable improvements, especially as emissions trading schemes oblige companies and governments to account for their environmental impacts, the true power and potential of an eco-design approach lie in the heuristic value to promote innovation.

Interface, an American carpet manufacturing company, built an entirely new business model on recyclable products while promoting product and material stewardship though an eco-redesign of their products (Interface n.d.). Patagonia has built its business model around eco-design and sustainable development. It makes a significant effort to inform its customers about the environmental, social, and economic impacts of its products and business practices. Its dedicated website, The Footprint Chronicles®, uses transparency about its supply chain to 'reduce adverse social and environmental impacts'. Patagonia demonstrates the benefits of integrating eco-design through celebrating regional specificities and *savoir faire* as a part of its corporate culture (Patagonia n.d.).

As the world moves to a distributed manufacturing system, regional expertise will develop around localised knowledge and skill. This concept is not new. The violin industry of the famed Markneukirchen luthiers in Germany, in the early twentieth century, used the various skills of artisans in adjacent villages in Bohemia to make the components of these popular stringed instruments, which were then exported to the thriving economies of the New World (http://violininformation.webs.com/germanfactories.htm).

In more recent times, the 'makers movement' driven by the democratisation of manufacturing techniques such as 3D printing could hail a new age of prosperous regional development.

Products to product–service systems

LCA enables designers and businesses to rethink a 'product', not just as a seductive package of technology to be acquired, used, and then discarded, (the linear economy) but reconceptualised in terms of the service it renders. Uncoupling physical products from the service they deliver has opened new horizons for the design of services and associated business models – a clear step towards total product stewardship and the circular economy it portends.

There are some experiences my daughters will never know: the pleasure of making a cassette of my favourite music to give to a friend, or the frustration of wrestling with a dysfunctional answering machine. Even our modest DVD collection appears archaic to them, as they explore the world of YouTube and Netflix. They are the new generation of service consumers and live on the cusp of what Thierry Kazazian called 'The age of lightness' (Kazazian 2003). The dematerialisation of products towards product–service systems (PSSs) has been the ambition of environmentally conscious thinking for many years. Intergovernmental organisations such as the United Nations Environment Programme have focused much effort and attention on identifying the potential barriers, as well as drivers, for this new economy of products (UNEP 2011). NGOs and associations such as the Rocky Mountain Institute and the O2 global network have led the debate on how to change consumption patterns and what the impact on product design and stewardship may be, providing examples for designers and industry (von Weizsäcker et al. 1998). Although the conceptual framework for this new approach to product–service design may have been embraced at an early stage by certain practitioners in the design profession, the move towards PSSs has generally remained slow.

Conclusion

Eco-design is a powerful means by which companies can embrace sustainable development. Although it has its origins in reducing the environmental impact of a product, it has now broadened in scope to include the global performance of a product and, increasingly, the service that replaces it.

As we move to an age of greater product transparency, regulatory environmental performance, coupled with increasingly discriminate consumer choices, is driving corporations to new levels of accountability. Products (and PSSs) that demonstrate an understanding of life cycle impacts – and, better still, design decisions to mitigate these impacts – will clearly have a competitive advantage in today's sustainability-conscious market. Green purchasing policies have become more common among OECD governments and can be increasingly observed in developing economies also. As developed countries tackle their environmental degradation, green purchasing policies are seen as a tangible lever to reduce environmental impact, while promoting innovative local businesses.

Eco-design is a federating enterprise that can help build a framework for regional development. Through the identification of regional strengths in terms of resources, know-how, and geography, strategic investment can be made to facilitate circular economies. Following the financial crisis of 2008, the Chinese government embraced the circular economy model as a strategy for reducing the demand made by its economy on imported natural resources, as well as reducing the damage it causes to natural environments. The circular economy concept calls for very high efficiency in resource flows as a way of sustaining improvement in quality of life within natural and economic constraints (see www.indigodev.com/Circular1.html). The eco-design approach is a creative process that structures discussion and action to serve all sustainable development ambitions – especially at a regional level.

References

EU – European Union (n.d.) Available at: http://ec.europa.eu/DocsRoom/documents/5187/attachments/1/translations/en/renditions/native and http://ec.europa.eu/enterprise/policies/sustainable-business/documents/eco-design/legislation/framework-directive/index_en.htm

Interface (n.d.) Available at: www.interface.com/US/en-US/global (accessed 1 August 2016).

ISO (n.d.) International Organization for Standardization. Available at: www.iso.org (accessed 1 August 2016).

Kazazian, T. (2003) *Il y aura l'âge des choses légères*, Paris: Victoires Editions.

London, B. (1932) *Ending the Depression through Planned Obsolescence*, Madison, WI: University of Wisconsin.

Papanek, V.J. (1982) *Design for the Real World*, London: Thames & Hudson.

Patagonia (n.d.) Available at: www.patagonia.com/us/footprint (accessed 1 August 2016).

UNEP (2011) Towards a Green Economy: Pathways to Sustainable Development and Poverty Eradication. Available at: http://www.unep.org/pdf/green_economy_2011/GreenEconomyReport.pdf (accessed 1 August 2016).

Von Weizsäcker, E., Lovins A.B., and Lovins, L.H. (1998) *Factor Four: Doubling Wealth, Halving Resource Use – The New Report to the Club of Rome*, London: Earthscan, Taylor & Francis.

WCED (1987) *Our Common Future*, Oxford, UK: Oxford University Press.

18

ECO-LABELS AND SUSTAINABLE DEVELOPMENT

Benjamin Walker

Introduction

Environmental damage inflicted by resource mismanagement has moved consumers to demand greater discrimination between products based on their environmental performance. An eco-label provides information to help consumers make more informed purchasing choices. Procurement professionals interested in supporting sustainable practices appreciate independent labelling to help identify products that have verifiable environmental or social benefits. But the question remains, how are eco-labels attributed? On what criteria are they based? How can we determine their veracity?

What is an eco-label?

First appearing in the 1970s, eco-labels were intended to help consumers make informed choices about the environmental performance of key products. One initial driver was energy consumption, but labelling has broadened in its scope and now covers an array of other sustainability criteria. By establishing a verifiable link between products and consumer expectations, eco-labels have become an important tool for applying market pressure on industry to minimize environmental and social impacts. Labels now address the entire sustainability agenda – environmental, social, health, economic, fair trade. They are aimed at

Figure 18.1 Eco-labels – greater transparency, better choices

everyone, from home consumers to tourism to big institutions such as banks. Governments and producers have joined civil society groups in developing eco-labels. The choices arising from eco-labels have direct impact on supply and demand in the marketplace. As such, eco-labels are an important link between design and end-users and constitute a vector to guide the market towards greater environmental awareness. In essence, these labels become a tool for decision-making on sustainability issues.

With the rapid increase in eco-labelling programmes across the world and across business sectors, the number of labelling programmes has proliferated. Currently, there are 459 eco-labels in 197 countries, across twenty-five industry sectors (Ecolabel Index n.d.). But, although the idea of a label is simple enough, the criteria and assessment across the life cycle of a product are often complex.

As the market responded positively to eco-labels, questions were raised concerning their validity. Trust in eco-labels is eroded when manufacturers or organizations create their own 'in-house' labels based on unsubstantiated claims. As a result, in 2000, the European Commission published *Guidelines for Making and Assessing Environmental Claims* (EC 2000), which define common environmental declarations – such as 'recycled' or 'biodegradable' – and the criteria of their use. Such guidelines are important in setting a framework for compliance. To further assist this process, ISO created standards for labelling within the ISO 14000 family of standards (ISO n.d.). The ISO 14020 to 14025 series deals with environmental labels and declarations.

ISO defines three types of eco-label.

Type 1 labels – ISO 14020 – also known as 'official eco-labels' – compare products in the same category, awarding labels to those that are environmentally preferable throughout their life cycle. The criteria are set by an independent body and monitored through a certification or auditing process. Many countries now have a national or regional scheme. The German *Blue Angel* was created more than 20 years ago; many others started around 1990. New schemes are emerging all the time.

Type 1 is usually a voluntary, multiple-criteria scheme, managed by an independent third-party (eco-labelling) body that identifies products based on life cycle considerations. The criteria are based on a simplified life cycle inventory of the product.

Examples of national Type 1 eco-labels include:

- Environmental Choice (Canada);
- Eco-mark (Japan);
- European Eco-label (European Commission);
- Blue Angel (Germany);
- Green Seal (US).

Type 2 environmental labels (self-declared claims) are created for their goods by manufacturers, importers, or distributors. They are not required to be independently verified, but companies and organizations understand that, beyond the immediate marketing benefits, consumer groups are increasingly critical of such claims. The criteria are covered in the ISO 14021 standards. Typical examples include:

- ozone friendly;
- 100 per cent recycled paper;
- phosphate-free;
- recyclable;
- totally eco-friendly.

Most countries have laws regarding misleading advertising. These need to be supported by criteria to discriminate between a fair claim and a misleading one, as for example in the Australian Competition and Consumer Commission's guide to green marketing (ACCC n.d.). Criteria that differ between countries create problems for both exporters and consumers.

Although Type 2 labelling can be criticized for its lack of obligatory third-party verification, the potential damage from false claims in an increasingly informed market, supported by social media, encourages a certain level of proper conduct. The EU has published an excellent report on environmental claims (EU 2013), including examples of advertising that have been sanctioned for misleading claims.

Green public procurement remains largely a voluntary instrument in the EU's effort to become a resource-efficient economy (EU n.d.). However, purchasing centres such as UGAP in France use a combination of Type 1 and Type 2 eco-labels to communicate to customers the various sustainability criteria of their suppliers and thus promote responsible purchasing (UGAP n.d.).

Type 3 labels, under ISO 14025, provide a list of the product's environmental impacts throughout its life cycle. These labels are similar to nutrition labels on food products that detail fat, sugar, or vitamin content. The information categories can be set by an industrial sector or by independent bodies. Unlike Type 1 labels, these labels do not judge products; that task is left to consumers. Critics question whether the average consumer has the time and knowledge to determine whether, for example, emissions of sulphur are more hazardous than those of cadmium, but independent experts working for NGOs are often able to interpret such information for their members.

Type 3 declarations provide data on key environmental aspects (e.g. resource depletion, energy use, air pollution, water pollution, solid waste) of products in a format that facilitates comparison of different products by purchasers. A full life cycle analysis is usually a prerequisite for such quantification.

Type 3 declarations have proven to have limited application to the consumer market, as comparisons are difficult. Nevertheless, such quantified life cycle environmental data form the basis for an environmental product declaration (EPD). This is an independently verified document that communicates transparent, comparable information about the life cycle impacts of products (EPD n.d.). Such declarations are treated more thoroughly in Chapter 23.

All environmental labelling claims should adhere to the principles set out in the ISO 14021 standards (see Figure 18.2), these being:

1 Self-declared environmental claims shall be accurate, verifiable, relevant, able to be substantiated and not misleading.
2 Environmental claims shall be based on scientific methodology that is sufficiently thorough and comprehensive to support the claim and that produces accurate and reproducible results.

3 Information concerning the procedure, methodology, and any criteria used to support environmental claims shall be available and provided upon request to all interested parties.
4 The formulation of environmental claims shall take into consideration all relevant aspects of the life cycle of the goods or service, although not necessarily considering a full life cycle analysis.

Although most labels focus on single issues, such as the Energy Star® label in the US, the criteria underlying the attribution of an eco-label often evolve gradually. Initially, the Max Havelaar label focused only on the social and economic criteria behind its 'Fair Trade' label. Recently, the label criteria have grown to also include environmental criteria. The Forest Stewardship Council (FSC; FSC n.d.) label was initially focused on the responsible environmental management of timber. It now embraces the social impacts of forest management, thus leading to a multi-criteria eco-label.

Sustainable procurement policy is an important driver for change (see Chapters 19 and 33). It is not obvious that all relationships between producers and buyers need to be facilitated through the use of a label fixed to the product. This is important in the context of two other trends. First, there has been a gradual shift in the focus of environmental requirements from the physical characteristics of a product or product line (e.g. Blue Angel) to the management of a whole production facility (e.g. the FSC and Marine Stewardship Council (MSC; MSC n.d.), Max Havelaar, the International Federation of Organic Agriculture Movements (IFOAM)) and – in the form of corporate codes of conduct and policy

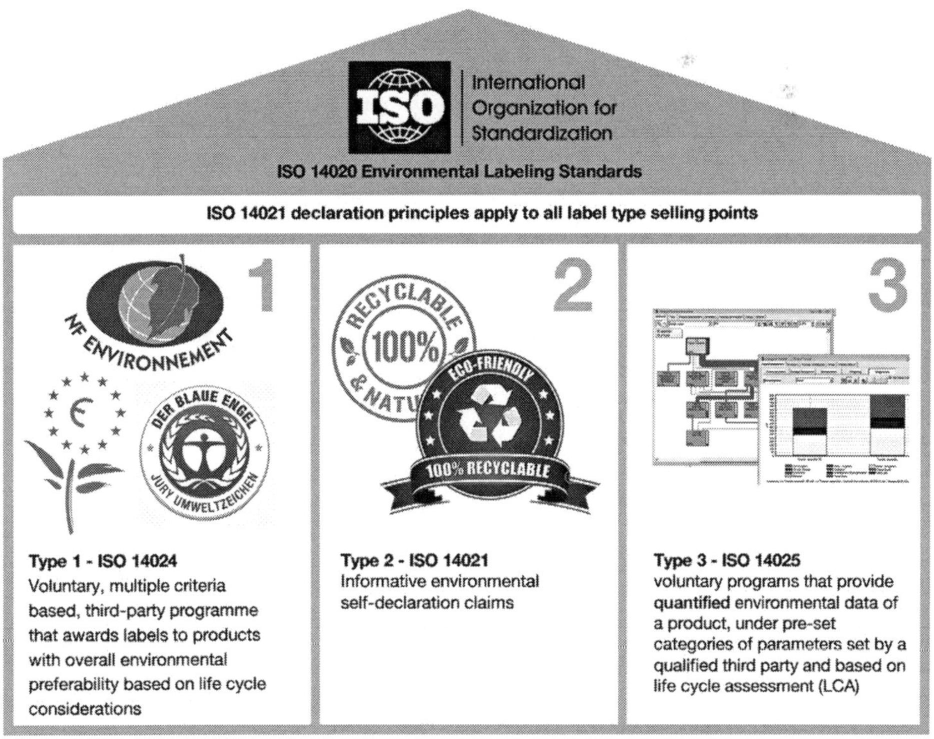

Figure 18.2
ISO environmental labelling standards

statements – to the overall activities and impact of entire companies. No longer are major companies concentrating solely on producing a few green product lines. Their entire corporate policy considers environmental and social responsibility. Increasing attention to supply chain responsibility has accelerated this evolution.

Logos a gogo . . .

As eco-labels proliferated, four certification organizations – FSC, IFOAM, Fairtrade, and MSC – began to work in closer collaboration. Although they covered different sectors, they had similar vulnerabilities in defending their credibility in the face of a growing number of eco-labels. Their alliance to strengthen sustainability standards was quickly joined by other organizations to create the International Social and Environmental Accreditation and Labelling Alliance (ISEAL; ISEAL n.d.). ISEAL now has nineteen full members, two associate members, and several dozen subscribers to offer services and support to their own codes to ensure credible standard-setting procedures. Although the mission of ISEAL is laudable as 'defender of standards', it is difficult to maintain, as labels become more 'granular' in their specificity and domains.

Consumers are now focusing on an increasingly complex range of environmental and social issues. They understand that sustainability (or responsibility) requires a balance between these concerns and are turning to responsible organizations to recommend responsible products and services. In many respects, this is the role originally played by the FSC. Through their association with the NGO movement, such organizations have become the trusted guardians of good environmental practice in the eyes of the general public. Companies that adopt the FSC label are effectively benefiting from the positive associations associated with FSC and its supporters. Because it works outside state regulations, some academics have classified FSC as a non-state market driven form of environmental governance (Cashore 2002).

Perhaps the most significant trend, and one in which it is easy to see the relationship between eco-labels and the larger corporate social responsibility (CSR) agenda, is the increasing demand by market actors for 'proof of compliance' with environmental, social, and ethical norms. A significant characteristic of eco-labelling programmes is that they provide such evidence of compliance. This trend is spreading to governments and intergovernmental bodies and their norms.

Accordingly, the sustainability movement has broadened. Where products were once the 'front line' of scrutiny, increasingly this now covers the entire management and supply chain. CSR (CSR n.d.), covered by ISO 26000, has become an instrument for improving management practices. Although it is not possible to be certified under this family of standards, there are several parallel certifications and tools to help structure CSR indicators to communicate a company's credentials.

Eco-labels are not only for physical products. A suite of management 'eco-labels' has arisen to communicate adherence to corporate best-sustainability practices. The United Nations Global Compact (GC; Global Compact n.d.) encourages 'enlightened global business' by inviting adherence to ten principles, covering human rights, environment, labour, and anti-corruption. Structured as a leadership initiative, the GC sets out a blueprint for corporate sustainability. The emphasis is more on process than on results and sets a framework for continuous improvement through annual reporting. Membership of the GC allows companies and NGOs to participate in thematic working groups, share best-practice examples, and, perhaps most importantly, co-brand their organization with the GC logo.

Regional certification

Long before the invention of eco-labels, the assurance of a certain quality was often associated with a region, for example, 'Swiss-made'. And who could disassociate champagne from its authentic, traditional techniques and methods of producing high-quality sparkling wine – in the Champagne region of France. Although quality may vary between champagne houses, the provenance of the grapes and method of production are assured by the region's wine growers via a common code. Such regional specificity translates directly into quality, and, hence, into a market premium.

The concept has been expanded to other regions. The French AOC (*appellation d'origine contrôlée*) label informs consumers of the authenticity of the regional origins of certain (certified) wines. The success of the AOC label has seen its application extended to other regional quality foods, and it is being copied in other countries also. Although the label is not an eco-label per se, it often does include environmental and sustainability criteria in its ambit.

Many regions have labels to promote themselves, be it the place (tourism) and/or the products it produces (see Chapter 16 also). These labels do not always make direct environmental or sustainability claims; rather, they are inferred from the attribution of such a label by an overseeing body – often regionally based itself. Such a process can even apply to a region itself. In terms of global recognition for cultural uniqueness, UNESCO's World Heritage listing is difficult to surpass.

Figure 18.3
Examples of regional quality labels

Conclusions

Eco-labels can be considered a market-driven tool for environmental governance. Although many levels of engagement exist for compliance, and with differing associated costs, it's important not to lose sight of what these labels represent: an organization's commitment to improve sustainability performance. Recognized third-party labels offer brand prestige and integrity, but transparent self-declarations can be just as important, as they must now withstand critical analysis from an increasingly informed market.

Eco-labels are a potentially powerful tool in regional development, communicating the environmental, economic, and social performance of a product or service in a transparent, verifiable manner. They also allow discrimination of products entering the region or purchased by the administration.

The development of an eco-label is a sophisticated exercise that depends on life cycle insights along the supply chain, often also reaching down to end-consumer level. Such analysis, carried out by independent agents, is necessary to maintain public confidence in the veracity of the labels. But, the incorporation of multiple criteria in a single label will always require further information and interpretation. The recognition of 'the world behind the label' is indispensible for sustainable purchasing decisions within a region, but also for developing its own quality labels based on sustainability criteria.

Further reading

Frankl, P., and Rubik, F. (2005) *The Future of Eco-labelling: Making Environmental Product Information Systems Effective*, Sheffield, UK: Greenleaf.

References

ACCC (n.d.) Green Marketing and the Australian Consumer Law. Australian Competition and Consumer Commission. Available at: www.accc.gov.au/system/files/Green%20 marketing%20and%20the%20ACL.pdf (accessed 1 August 2016).

Cashore, B. (2002) 'Legitimacy and the privatisation of environmental governance: How non-state market-driven (NSMD) governance systems gain rule-making authority', *Governance*, 15: 503–29. Available at: http://onlinelibrary.wiley.com/doi/10.1111/1468-0491.00199/abstract (accessed 1 August 2016).

CSR (n.d.) Corporate Social Responsibility. See: https://en.wikipedia.org/wiki/Corporate_social_responsibility (accessed 1 August 2016).

Ecolabel Index (n.d.) Available at: www.ecolabelindex.com/ (accessed 1 August 2016)

EC (2000) Guidelines for Making and Assessing Environmental Claims. Available at: http://ec.europa.eu/consumers/cons_safe/news/green/guidelines_en.pdf (accessed 1 August 2016).

EPD (n.d.) Available at: www.environdec.com/ (accessed 1 August 2016).

EU (2013) Available at: http://ec.europa.eu/consumers/documents/consumer-summit-2013-mdec-report_en.pdf (accessed 1 August 2016).

EU (n.d.) Available at: http://ec.europa.eu/environment/gpp/index_en.htm (accessed 1 August 2016).

FSC (n.d.) Available at: www.fsc.org (accessed 1 August 2016).

Global Compact (n.d.) Available at: www.unglobalcompact.org/HowToParticipate/Business_Participation/index.html (accessed 1 August 2016).

ISEAL (n.d.) Available at: www.isealalliance.org/ (accessed 1 August 2016)

ISO (n.d.) Available at: www.iso.org (accessed 1 August 2016).

MSC (n.d.) Available at: www.msc.org (accessed 1 August 2016).

UGAP (n.d.) Available at: www.ugap.fr/le-developpement-durable-a-lugap_904737.html (accessed 1 August 2016).

19

SUSTAINABLE PROCUREMENT AND REGIONAL SUSTAINABILITY

Fritz Balkau

Background

Government institutions absorb a significant part of the national flow of goods and services and have a big influence over what comes on to the market. 'Sustainable' purchasing can lower the ecological and social footprint of society's materials flow; however, because the embedded impacts inherent in goods and services are not always immediately visible to purchasers, deciding on what is sustainable can be difficult for individual buyers.

As well as addressing sustainability concerns, a preference for certain goods can enhance local innovation and production. Purchasing in bulk also lowers the cost price of (environmentally) preferred products, with subsequent economic advantages that spread beyond the purchasing agency.

Although a simple idea, sustainable procurement faces major challenges in product identification and life cycle evaluation, as well as in the administrative and regulatory framework needed to implement purchasing criteria.

Introduction

> Sustainable procurement by both public and private sector organisations ensures that their purchasing reflects broader goals linked to resource efficiency, climate change, social responsibility and economic resilience.
>
> (ICLEI n.d.)

Sustainable public procurement (SPP) occurs when governments implement programmes to purchase environmentally and socially preferable goods, services, and infrastructure. Also called sustainable purchasing or green purchasing, SPP is closely related to its business equivalent, sustainable supply-chain management (SSCM), and has been adopted by numerous public institutions, municipalities, and governments around the world. Major international organisations, including the World Bank, UNEP, and OECD (OECD 2015), are promoting and supporting the application of sustainable procurement in the public sector to improve organisational efficiencies and national sustainability performance. NGOs and private institutions such as the International Institute for Sustainable Development (IISD) are assisting in its spread by providing guidance material and case studies. Professional purchasing organisations are also active in this area (Portuguese Public Administration Shared Services Authority (eSPap), Union of Public Purchasing Groups (UGAP)).

In its simplest form, SPP involves choosing a product based on an understanding of its impacts on the environment and on the social conditions at the upstream suppliers. More sophisticated applications are based on detailed knowledge of products' impacts also during their use phase and disposal. IISD defines SPP as, 'governments seeking *value for money across the asset life cycle* rather than simply at the point of purchase'. It is important to recall that 'value for money' includes intangibles such as environmental, social, and cultural values, in addition to financial benefits. Life cycle costing (LCC) is now an indispensable tool to allow such calculations (see, for example, Chapter 33).

Sustainable purchasing can target a range of common products, as shown in Table 19.1, from IISD.

The impact of SPP becomes clear when we realise that purchasing by government institutions can represent up to and more than 45 per cent of national goods and services (see Table 19.2). In addition to the direct reduction in accumulated impacts, the leverage effect in encouraging the production system to manufacture more sustainable products (and to manufacture them more sustainably) can be immense. Product changes at this level may create entire new markets, from which other purchasers then also benefit.

Table 19.1 Areas of frequent public procurement

Products	Services	Works
Office and ICT equipment	Software	New building
Vehicles	IT servers	Renovation of buildings
Indoor lighting	Electricity	Landscaping
Outdoor lighting	Vehicle fleet, other transport	Railways
Paper	Postal	Roads
Office supplies	Waste handling	
Fuel	Catering: food	
Furniture	Catering: beverages	
Apparel	Air conditioning	

Source: IISD n.d.

Table 19.2 Procurement in various countries

Importance of public procurement: Public procurement as a percentage of GDP (approximate)	
European Union	17
US	18
South Africa	32
India	36
Mauritius	36
Brazil	32
Chile	37
Vietnam	43
China	58
UAE	33

Sources: EU, IISD, AusAID and OECD 2012

The outcomes of SPP can be felt well beyond the region where the purchase is made. Through a reduction in pollution and other impacts along the entire supply chain, the beneficiaries include far-away countries where materials are mined, components are produced, or entire products are manufactured.

SPP is thus a powerful change agent for sustainability and efficiency, at little extra effort. Public agencies already buy on a daily basis. No new procedures or additional operations are needed; only some selection criteria need to be amended. Because the outcomes are highly visible, there is increased community awareness on sustainability, and clear signals are sent to the manufacturing or distribution sectors. SPP is thus an effective vehicle to leverage governments' purchasing power to foster green industrial growth.

Underlying factors

For SPP to be fully effective, it is necessary to understand the dynamics of product life chains, with both their upstream and downstream linkages. This 'world behind the product' is often complex. A whole range of operators, suppliers, contractors, and subcontractors may be involved, and various wastes and social impacts occur at every stage.

Also important to consider is that many products actually have greater impact during use than in manufacturing (e.g. the motor car or the light bulb), which means that use patterns (the consumption phase) need to be independently assessed. The disposal of the used product at the end of its life, whether by dumping or recycling, has further environmental implications. It is unlikely that a regional institution will have the necessary data about the life-chain impacts of the products it is considering buying. Short of commissioning an independent life cycle analysis, the institution will therefore rely on third-party information and perhaps published data. Although a huge range of information sources is potentially available, the challenge is to select ones that are reliable for the purpose. Manufacturers' claims are hardly impartial, and so independent eco-labels have been introduced as a way of providing guidance (see Chapter 18). Even then, eco-labels have some shortcomings, including the inability to always provide specific information reflecting the purchaser's sustainability criteria, eco-labels being rather generic. The ISO 14020 series of standards was created to provide greater reliability in eco-labelling, with three categories of labels (ISO n.d.).

Life cycle analysis and derivative eco-labels are only two of the issues to address. Another is to decide which sustainability criteria to include in the consideration in the first place. Many purchasers restrict themselves to one or two topical issues – easier to assess than the totality of environmental, social, and economic qualities embedded in a product. Energy and carbon intensity are popular criteria now, as are water and waste. But some purchasers are also sensitive to social issues, such as worker conditions in far-way places, child labour, or gender issues. Thus, ethical products and fair-trade products have become desirable for many institutions. Locally produced products also have a particular attraction to regional authorities (see Figure 19.1).

A difficulty for the purchaser or consumer – whether industrial, administrative, or consumer – is to recognise the sustainability profile of the materials and components constituting the product. There is no simple general method for

Figure 19.1
Eco-labels and green consumerism

achieving this. Some techniques that can assist in this process are shown below (note that there is some overlap between categories):

- independent life cycle assessment of materials and products;
- compliance with product standards;
- product information from suppliers or manufacturers, for example EPD and PEF (see Chapter 23);
- sustainability or eco-labels and/or environmental awards;
- third-party product certification or endorsement;
- use of corporate indicators – for example EMS, GC, ISO 9000;
- identification of undesirable components, for example lead;
- identification of origins of products or components and their notoriety;
- product preferences as defined by the home organisation.

Incentives and examples

The greater use of SPP is beneficial both from a business perspective and for national or regional development, because it exercises a positive selection pressure on innovation towards clean technologies. By purchasing certain classes of product, public bodies create a volume market that becomes more attractive to suppliers.

Countries are increasingly alert to these benefits (IISD n.d.). The US Federal Green Purchasing Program has published guidelines for purchasing, developed criteria for various products, and engaged in partnership programmes with more than 13,000 companies. A similar programme in the European Union is aimed at reducing environmental impact and as an incentive for innovation. In India, a consortium of government agencies, financing bodies, and industry has developed template specifications on office electronics, furniture, bricks, indoor and outdoor lighting, and textiles. Vietnam has established a centre of expertise on sustainable procurement and strengthened the Central Procurement Platform, reporting directly to the prime minister. Both Abu Dhabi and Chile see green purchasing as a way of supporting their sustainable development programmes. South Africa sees it as a vehicle for black economic empowerment, and Mexico sees it as a driver for sustaining SMEs. The Nova Scotia Sustainable Prosperity Act (2007) has the objective of protecting natural resources, reducing pollution, and fostering renewable energy. It calls for sustainable purchasing and for sustainable meetings for its government agencies.

Although the desire is often there, public bodies' ability to implement such policies is still very uneven. There is a big step between objectives and outcomes. Conceptually, sustainable purchasing merely adds some additional criteria to an already-existing activity, but this is less simple to operationalize than is often believed. The ability to implement depends on factors such as:

- clear responsibility assigned across the entire organisation;
- defined focus areas, targets, and goals;
- administrative rules and trained personnel;
- guidelines for purchasing criteria and assessment techniques;
- outreach to suppliers;
- transparency and accountability;
- monitoring of outcomes;
- regular review of the process and its results.

A further question is whether to focus on only individual specified priority products, or to employ SPP as a generalised procedure for all purchasing. Much depends on where sustainability sits as a strategic policy item with public authorities, and whether there are local reasons for considering only a limited range of products. Some governments have published guidelines for certain types of product, but remain silent about others.

It is interesting to compare the use of SPP by public authorities with the use of SSCM in business (Balkau and Sonnemann 2011). By definition, SSCM is concerned with the upstream part of a product life cycle, that is, the supplier end. Businesses have more ability to exert pressure on their suppliers concerning product specifications than public end-users. Thus, whereas businesses can enter into collaboration with their supply chains, the public purchaser often has little more than a 'buy/not buy' choice. In this case, the message about sustainability criteria is rather slow to work its way up the supply chain.

Barriers to sustainable procurement

Although the benefits are clear, the implementation of SPP faces some serious hurdles, most of them associated with administrations that are resistant to change. Table 19.3 summarises some of the more significant barriers to entry and lists some of the ingredients for a successful application.

Table 19.3 Barriers of entry to SPP

Barriers to sustainable procurement	What is needed for sustainable procurement
Political hierarchies do not see procurement as a driver for change	Visible political leadership and support
Procurement decisions are spread out over multiple positions/hierarchies	Responsibilities clearly assigned and reported on
Focus on value for money at purchase rather than over the asset life cycle	Legal certainty in the process
Absence of multi-year budget frameworks that discourages long-term objectives	Organisational ability to implement
Processes are more extended with additional administrative steps	
Expertise and mindsets – policymakers are not deal-makers	

Source: IISD n.d.

As is the case for SSCM in business, visionary high-level leadership is required to unite the various partners and stakeholders behind an SPP initiative. Once underway, reliable monitoring and reporting of results will usually dissipate further reluctance as the benefits become clear. Transparency in transactions and a collaborative rather than confrontational relationship with suppliers is a further aspect that accelerates the move to a fully operational SPP.

Key players in SPP

The importance of public purchasing in stimulating economic growth and also its interaction with other non-economic sustainability criteria has long been recognised by policymakers in the public sector. Its importance has led to significant studies by major international organisations, such as the World Bank, OECD, UNEP, the EU, and collective organisations such as ICLEI (ICLEI n.d.) and IISD. These have jointly or independently published authoritative reports concerning SPP, as well as its status in countries. Through such studies, the key areas to address have been distilled out to include: legal and policy framework, planning and LCC, use of environmental standards in contracting, professionalism of SPP teams in administration, communication with suppliers and purchasers, and monitoring performance of SPP. Networks such as IGPN (IGPN n.d.) have been formed to assist in promotion and training. The European Commission (EU n.d.) has published guidance documents on the use of LCC for energy calculations to improve the use of SPP as a tool for green energy innovation. ICLEI runs regular workshops under its Procura programme. Tool kits to assist in implementation are available from the UN and ICLEI, for example.

Through the above organisations, a host of case studies from countries, regions, and cities have been collected, some in comparative compilations, as for example the OECD compendium.

Procurement institutions can also be key players within regions. Many governments have established buying centres for their civil service, public institutions, and municipalities. Gradually, these are integrating sustainability criteria into their selection of goods and services. The French Union of Public Purchasing Groups (UGAP n.d.) and the Portuguese Public Administration Shared Services

Authority (eSPap n.d.) are two examples from Europe, and similar services exist in other countries. Societies of procurement professionals are also active in promoting SPP.

Summary: Moving sustainable procurement forward at regional level

Many useful outcomes can flow from an SPP initiative. Substantial guidance material is available, backed by numerous successful case studies.

Nevertheless, a move to SPP will face resistance from certain stakeholders (especially the present suppliers, and perhaps from inside the administration). A careful path needs to be mapped out. The need for high-level leadership was mentioned earlier, as was the existence of a formal policy or strategy that defines the sustainability objectives of a region (or of a purchasing body within it). The purchasing procedures need to be clarified, including the budget implications, as calculated by LCC.

A monitoring and reporting systems needs to be set in place, with information provided to all stakeholders. The cost implications for the regional budget should be part of the reporting system.

Many international organisations are working on procurement frameworks and guidelines. These programmes are unanimous in stressing the substantial economic and social benefits that flow from a rigorous approach. They also provide guidance to public bodies interested in following such a path.

Acknowledgement

The author would like to thank IISD for providing some of the information on which this chapter is based.

Further reading and resources

In addition to the specific references listed below, there is much detailed guidance on sustainable procurement, as shown by the diversity of sources below. In many countries there will also be national guidance material and information. The sources below have been chosen owing to their broad scope and their easy accessibility on Internet.

The International Green Purchasing Network (IGPN; www.igpn.org/about/index.html) has the mission to globally promote the spread of environmentally friendly product and service development and green purchasing activities. It shares information and know-how on green purchasing and environmentally friendly products and services and assists in harmonising the efforts of green purchasing and the development of environmentally friendly products and services from a global viewpoint.

The Procurement and Innovation Platform (PPI) is available at: www.innovation-procurement.org/resources/search?country=256&category=0&topic=17&language=0&product=0&level=0&q=&page=14

SMART SPP consortium, c/o ICLEI, has produced a SMART SPP Guide – Driving Energy Efficient Innovation through Procurement – A Practical Guide for Public Authorities. It is available at: http://s3platform.jrc.ec.europa.eu/documents/20182/84453/SMART_SPP_Guide_2011_EN_FINAL_www.pdf/361edc1c-6b45-4ae3-8f6c-b6e0b7a4e2e8. There is also a SMART SPP Tool and accompanying User Guide for calculating the LCC and CO_2 emissions of different products and services.

The Sustainable Procurement Resource Centre is available at: www.sustainable-procurement.org/

The UNEP Toolkit for sustainable purchasing is available at: www.sustainablepurchasing.org/public/slides/AM2014/SPLC_CSW2_Farid_Yaker_UNEP.pdf. A guidance document for UN agencies is available at: www.unep.org/resourceefficiency/Portals/24147/scp/sun/facility/reduce/procurement/PDFs/BFABW_Final_web.pdf. There is also the UN Global Marketplace, which includes guidelines for a wide range of common products and services. It is available at: www.ungm.org/Public/KnowledgeCentre/Sustainable Procurement/Tools

References

Balkau, F., and Sonnemann, G. (2011) 'Addressing sustainability issues through enhanced supply-chain management', in *Supply Chain Management – New Perspectives*, ed. S. Renko, InTech, pp. 379–400.

CIPS – Chartered Institute of Purchasing and Supply. Available at: https://www.cips.org/Documents/Products/Sustainable_Procurement_Review_%20new_logo.pdf (accessed August 2016).

eSPap (n.d.) Portuguese Public Administration Shared Services Authority. Available at: www.espap.pt (accessed 1 August 2016).

EU (n.d.) European Commission. Available at: http://ec.europa.eu/environment/gpp/pdf/handbook.pdf (accessed 1 August 2016).

ICLEI (n.d.) International Coalition for Local Environmental Initiatives. Available at: www.iclei-europe.org/topics/sustainable-procurement/ (accessed 1 August 2016).

IGPN (n.d.) International Green Purchasing Network. Available at: www.igpn.org/about/index.html (accessed 1 August 2016).

IISD (n.d.) International Institute for Sustainable Development. Available at: www.iisd.org/pdf/2012/procurement_innovation_green_growth_continues.pdf (accessed 1 August 2016).

ISO (n.d.) International Organization for Standardisation. Available at: www.iso.org/iso/environmental-labelling.pdf (accessed 1 August 2016).

OECD (n.d.) Organisation for Economic Cooperation and Development. Available at: www.oecd.org/environment/Aligning-Policies-for-a-Low-carbon-Economy.pdf pp 66,67 (accessed 1 August 2016).

OECD (2015) Smart Procurement Going Green: Best Practices for Green Procurement, OECD (includes case studies from Italy, Portugal, The Netherlands, USA, Canada, and others).

UGAP (n.d.) Union des Groupements d'Achats Publics (French Union of Public Purchasing Groups). Available at: www.ugap.fr (accessed 1 August 2016).

UNEP (n.d.) United Nations Environment Programme. Available at: www.unep.org/resourceefficiency/Consumption/SustainableProcurement/Publications/tabid/101249/Default.aspx (accessed 1 August 2016).

20

INDUSTRIAL ECOLOGY AND SUSTAINABLE PERFORMANCE OF BUSINESS PARKS

Jean-François Vallés

Introduction

The adoption of regional development strategies that are more respectful of the environment is often hampered by the difficulties inherent in implementing effective solutions. By using the potential synergies that already exist between businesses, it is possible to consider new solutions that are sometimes referred to as industrial ecology or circular economy. Such solutions enable the building of new relationships with the areas surrounding the places where they are implemented.

Because of their high concentrations of companies, business parks are conducive to the implementation of collaborative approaches that overcome some of the hurdles faced by individual companies. But, although synergy looks good on paper, spontaneous cooperation between companies is still limited, and it is often useful to call upon a third party to take on the role of 'connector'. A connector can strengthen regional networks between suppliers and clients, help to strengthen infrastructure, and generally assist in creating a more proactive, sustainable industrial dynamic. Local or regional governments are in a perfect position to provide such a service.

The limits of individual approaches

Despite the efforts of companies to reduce their environmental impact, a considerable amount of progress remains to be made. Uninformed attempts at pollution reduction often involve procedures that are costly to establish and result in waste of materials. The common use of 'end-of-pipe' treatments results in further 'secondary' wastes and residues that must again be properly handled. Businesses also generate a number of indirect impacts related to energy and water use, to transport of materials and goods, to impacts occurring in supply chains, and to the ultimate fate of products. Lastly, the location of businesses can accentuate certain problems such as real-estate prices and use of natural resources and of land. Resolving many of these impacts is beyond the ability of most companies when they act alone.

It is, therefore, not surprising that many businesses, especially smaller companies, do not find economically and technically viable solutions for reducing their environmental impact on their own. This is important, because the economic fabric of many regions is made up of a multitude of small businesses, with only a small number of large firms (if any). In France, 97 per cent of firms have fewer than twenty employees, and only 0.2 per cent have more than 250 employees. The responses of small firms to environmental pressures are poorly adapted and very often simply too costly. There is no internal understanding of complex externalities such as environment (let alone of sustainability), and this is often exacerbated by the lack of a sense of 'ownership' of such issues by the heads of the businesses concerned.

From an individual approach to a bilateral one

In this context, industrial ecology and circular economy approaches aim to solve a number of common problems. Proximity to other companies allows access to expertise and sharing or mutualising of scarce resources. Cooperation also engenders dialogue that can improve knowledge and break down barriers to comprehension. The approach is valid for large and small companies, depending for its success on proximity factors and a willingness to engage in managerial cooperation.

Industrial ecology began as a strategy of waste exchange, reusing a production residue from one company (waste, steam, coolant, etc.) within another company. Its best-known example is in Denmark, where multiple synergies were developed among companies in the Kalundborg industrial park (Kalundborg n.d.). In addition to reducing environmental impacts from wastes, this approach generated income for companies from the sale of their residues and savings from lower-cost supply sources in close proximity. Savings from avoided waste treatment were a further economic benefit. Many other examples of industrial ecology are known around the world. Most recently, the Waste and Resource Recovery Initiative, initiated in 2013 by the New South Wales government (Australia) to reinforce industrial symbiosis, has provided savings of AU$75 million for more than 500 businesses (Corder et al. 2014).

These initiatives demonstrate the advantage of using certain life cycle procedures to identify opportunities. Materials flow analysis software (see the chapters in Part II) is useful to evaluate the raw material inputs and the outputs of businesses within an area and to estimate quantity and quality. The NEXT programme, put into place in Wallonia (Belgium) in 2012 by the Walloon minister of industry, identifies potential synergies in the region (NEXT n.d.). This is the transversal, multisector axis of Wallonia's industrial policy, aiming for more efficient management of resources across all sectors. It offers an example of a global, coherent vision of resource management on a regional scale.

On a macro scale, there are numerous theoretical possibilities for using the residues from some companies in the production processes of other companies, but, at the local level, it is not always simple to find complementary companies in close proximity. Even when this is the case, the economic and technical feasibility of a materials exchange may not be viable. Such uncertainties do not encourage business leaders to jump headlong into such ventures, especially as these types of solution do not figure in their traditional business culture. The reasons for successful industrial symbiosis in Kalundborg included regulatory constraints that limited the expansion of businesses, available skills in the development of pipeline networks (presence of a refinery), and a strong mutual understanding between business leaders.

Some regions have tried to overcome the constraints by relocating companies so as to enhance synergies, as was the case with the Port of Moerdijk (MOERDIJK n.d.). Unfortunately, these solutions cannot be applied in every case.

Despite some uncertainties, sufficient examples do exist to confirm that it is possible to develop synergies between businesses. In the Ardennes region of Belgium, the Kaizerbaracke industrial park hosts businesses in the timber industry. The Renogen company has built a cogeneration facility on this site to transform the production residues from three companies and use them to provide the park with electricity and energy. The electricity generated is equivalent to the energy consumption of 16,000 households; heat production is equivalent to 15 million l of fuel oil (see Figure 20.1).

From its origins, industrial ecology well illustrates the advantage of looking to other companies for certain needs. But, with increased political emphasis on reduction, dematerialisation, cleaner production and industrial efficiency, industrial ecology approaches are also gradually evolving to incorporate other criteria beyond waste transfer, including economic and social issues. In particular, industrial ecology needs to focus more strongly on the product cycle, not only on wastes. Figure 20.2 presents an alternative vision of the classical input–output model that is responsible for much of the current level of waste production.

Industrial ecology also needs to look beyond materials flows and contribute to an understanding of social and business dynamics. In furthering the understanding of how companies operate (professions, needs, flow, functions, framework), it is possible to increase the number of local partners and economic exchanges (subcontracting, maintenance, product supply) and to strengthen partnerships that are mutually beneficial. For example, developing closer business-to-business links also leads to a reinforcement of the local economic fabric, while reducing certain environmental impacts, such as transport emissions.

Figure 20.1 Materials flows at Kaizerbaracke

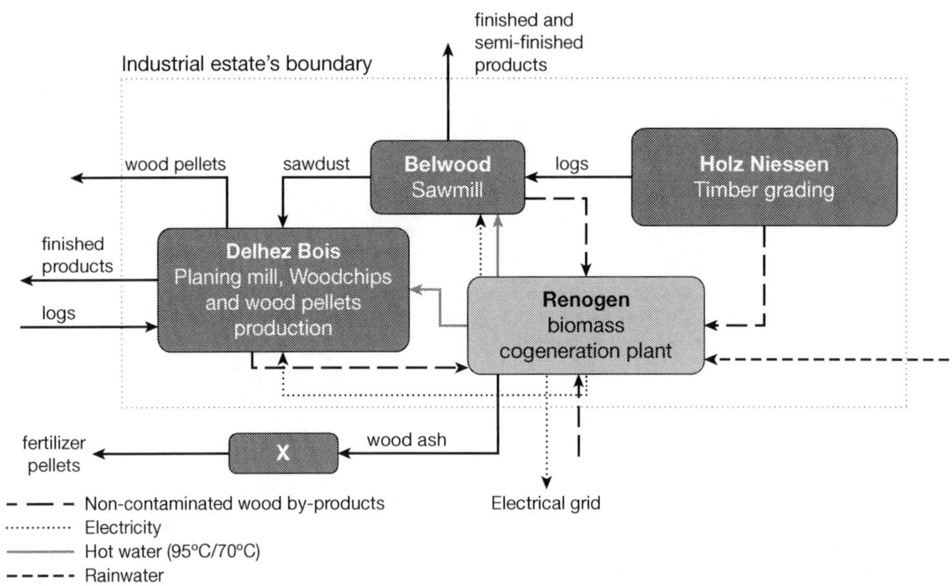

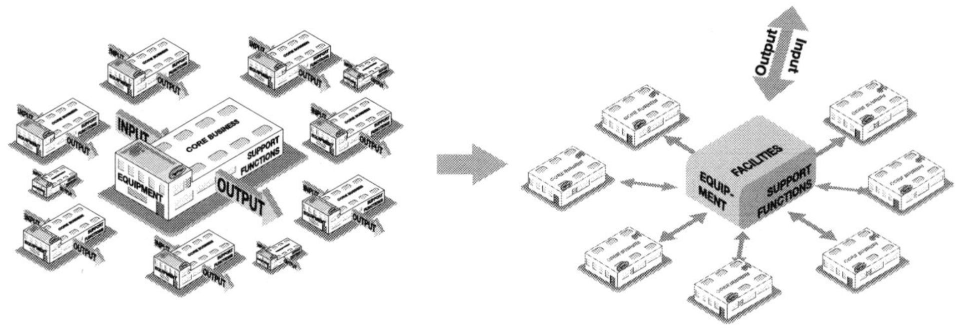

Figure 20.2
Alternative approach to industrial input–output models

From bilateral to multilateral

A number of factors contribute to the creation of synergies. These include:

- existence of pressure to change, either regulatory, commercial, or human;
- understanding of material flows: supply of raw materials, energy and fluids, production and treatment residues, goods and services distribution;
- availability of facilities, buildings, services, and other assets that can be useful to others;
- provision of services geared towards employees, associates, and clients, and so on;
- openness to the idea of sharing and cooperation for mutual benefit.

Where these factors are present, regional authorities, industrial park administrators, or business associations are able to develop solutions that provide a collective response to certain needs, not limited solely to the environmental sector, but also addressing commercial and social aspirations. In the Netherlands, the Rietvelden-Vutter Companies Association has introduced services for its members such as group purchase of energy (gas and electricity), the collection of waste, and pooled resources in managing liquid waste and risk. Moreover, this association has developed a container unloading platform in the industrial zone, with a daily link with the port of Rotterdam. This has helped to greatly reduce vehicle traffic and thus contribute to the optimization of logistics, an important contribution to business viability nowadays.

Also in the Netherlands, the manager of Emmtec (in the polymer sector; EMMTEC n.d.) offers services to companies wishing to set up there: analytical laboratories, logistics and storage, engineering, maintenance, catering, meeting rooms, emergency services, energy supplies, water, nitrogen, compressed air, and so on. However, this experiment has an additional interesting dimension for the local area in the form of partnerships with the region's farmers for the supply of raw materials to the companies. This approach has helped to forge strong bonds that tie businesses to the local area and enable the use of local know-how and natural resources.

Other examples exist in Germany, where many industrial park managers offer advanced services, such as logistics or energy supply management. For example, the chemical company Dow has developed its Valuepark by inviting businesses to 'connect' to the outputs from its petrochemicals business, an example of industrial ecology focused on the product rather than waste (Dow n.d.). Another example is the business park in La Gravelle (France) that has developed a rural concierge service for local businesses (laundry services, tray meals, etc.) and local sustainable farming through an association supporting small farming (AMAP). This concierge company strengthens the links between the employees of the other companies and the region's merchants and farmers. Other examples of synergies and services are discussed in Chapter 34.

Collective approaches by businesses are also useful to local communities looking for solutions to some public-sector objectives, such as economic development, employment, local resource management and preservation of natural resources. This type of innovative reasoning contributes to raising a region's profile and increasing its attractiveness.

Four factors for success

The success of the approaches described above depends on four additional factors:

- The mobilisation of human, financial, and technical resources: Developing synergies and collective approaches between businesses requires the mobilisation of skills to analyse how the synergies work, identify potential new ones and develop sustainable solutions.
- The ease with which quality, sustainable solutions can be implemented as turnkey offers: Businesses do not necessarily have the time or the experience to develop services themselves.
- The degree of confidence businesses have in the service provider: The more the businesses are involved in the structure, the higher their degree of confidence.
- The intensity of needs: Companies that are doing well may not feel that further optimisation or cost reduction is important, especially as cooperative arrangements often reduce their independence to act when conditions change.

Appoint a 'connector' of businesses

The central element of this approach is based on someone's ability to provide the services (material and immaterial) that will enable businesses to operate in a cooperative manner. The connector's role is to have an overall vision of the different flows and facilities in a given region. The connector is positioned to identify possible exchanges between companies, propose pooling of certain equipment, and find outsourcers for services or employment opportunities. The outcome is a set of services and procedures that generate positive impacts for businesses and local areas. As well as using a connector, it is important that the businesses see themselves as stakeholders in such a structure, to increase their levels of confidence and address their specific needs. The interface with regional development policies also has to be considered.

Areas of collaborative activities

With a high concentration of companies, business parks are breeding grounds for the implementation of collaborative approaches. Concentration also allows nearby cities to link with such industrial clusters to create further local synergies. Business parks should also be seen in the light of potential impacts on the environment. Although pollution may be more concentrated, it is also more cost-effective to deal with it on a larger scale.

Despite the numerous advantages outlined above, many current strategies for business parks are not especially creative. Often, they simply locate businesses sides by side, with no real optimisation of potential synergies and complementarities. Fortunately, as explained in Chapter 34, some park and business association managers are now engaging in more progressive approaches. Here, business parks are no longer simply passive sites or real-estate projects. They become the facilitators of business development by offering their services, raw materials, equipment, or skills to enhance synergies and contribute to the further evolution of a regional business dynamic.

The first steps

The success of collective approaches lies in the capacity to create local governance where each party involved in the venture becomes a stakeholder: the businesses, the industrial park administrators, the service users and providers, but also the region or surrounding locality that contributes to, and benefits from, such a project. To allow for the emergence of such approaches, public structures (local communities and chambers of commerce and industry), industrial park administrators, and business associations need to facilitate the following actions:

- encouragement or even organisation of local meetings between businesses, to encourage the growth of a collective culture through simple actions performed together;
- analysis of business operations to identify potential synergies;
- development of model solutions to establish synergies.

In addition, industrial park administrators can contribute to solutions by becoming directly involved in their sustainable performance.

Conclusions

Cooperation between companies has the ability to enhance materials and waste management, but also to provide wider benefits to employees, clients, and other local stakeholders. Cooperation between companies often requires external intervention to ensure commercial neutrality and to bring in the knowledge and expertise required for such a venture. The conditions for such cooperation are most easily found within large, diverse business parks and industrial estates, where the necessary diversity of needs, resources, and materials flows can be found. Collaborative business ecosystems can also become contributors to local development policies, as well as being more competitive in their own right. The establishment and management of business parks must be carefully undertaken to achieve such synergies.

References

Corder, G.D., Golev, A., Fyfe, J., and King, S. (2014) 'The status of industrial ecology in Australia: Barriers and enablers', *Resources*, 3(2): 340–61. Available at: www.mdpi.com/2079-9276/3/2/340 (accessed 2 August 2016).

Dow (n.d.) Valuepark. Available at: www.dow.com/valuepark (accessed 2 August 2016).

EMMTEC (n.d.) Available at: www.emmtec.nl (accessed 2 August 2016).

Kalundborg (n.d.) Available at: www.symbiosis.dk (accessed 2 August 2016).

MOERDIJK (n.d.) Port of Moerdijk. Available at: www.havenschapmoerdijk.nl/en/ (accessed 2 August 2016).

NEXT (n.d.) Programme Next. Available at: http://marcourt.wallonie.be/next/next-299.htm?lng=fr (accessed 2 August 2016).

21

NATURAL RESOURCE USE EFFICIENCY AND ECONOMIC PRODUCTIVITY

Pierluigi Toma, Stefania Massari, and Pier Paolo Miglietta

Introduction – From environmental issues to quantitative analysis

Our society is presently facing a deterioration of the environmental systems and an unsustainable exploitation of natural resources at both global and local levels. These problems have greatly accelerated in recent years, stimulating the development of new environmental policies everywhere.

Environmental issues have assumed such significance now that legislators have imposed regulatory standards and compulsory mechanisms, subject to penalties for non-compliance. However, local authorities should not passively succumb to the restrictions imposed, but rather ensure that environmental legislation becomes a stimulus for innovation. The adoption of more sustainable patterns of production and consumption makes the management of environmental issues more effective (and efficient), ideally transforming an environmental constraint into an opportunity. Subsequent to some recent high-level political meetings, one of the major objectives of global economic development strategies now is to build a low-carbon, resource-efficient economy, as illustrated in Chapter 5. Meeting the new international targets requires countries to significantly increase their efficiency of energy use, decrease CO_2 emissions and other pollutants, and produce energy from renewable resources.

The availability of environmental sustainability assessment methodologies now allows public authorities to develop clearer objectives and build programmes with coherent sustainability targets. Such programmes increasingly contain incentives and economic support for a continuous improvement of environmental performance, rather than simply enforcing legislative constraints. Accordingly, the assessment and improvement of efficiency in the use of natural resources for different economic sectors have major practical implications for a region or territory and have formed an active research field.

Within this framework, quantitative methodologies can increase our comprehension of how effectively we exploit our natural resources (De Leo and Miglietta 2015). The importance of efficiency in natural resource use has been recognized for some time, and several quantitative models have been developed around the issue of environmental complexity (Jebaraj and Iniyan 2006). Natural resource use efficiency analysis can provide decision-makers with quantitative information, allowing policy-making and public communication to become more systematic.

The basis of resource efficiency

The classic paradigm defines productivity as the ratio between an output and the inputs used to achieve it (Vincent 1968). Similarly, according to Lovell (1993), the productivity of an entity is the ratio of its output to its input. If a single input is used to obtain a single output, this ratio is simple to calculate. Where the decision-making unit (DMU) uses multiple inputs to produce multiple outputs, these inputs and outputs must be aggregated so that productivity becomes the ratio of two scalars. When the ratio relates to a single productive factor, we refer to the concept of partial productivity; where every factor of production is considered, we speak of productivity.

Many authors treat both productivity and efficiency as the ratio between output and input, not indicating any difference between the two concepts (Sengupta 1995; Cooper et al. 2000). We define efficiency as the distance measure, that is, the difference between the actual ratio of inputs and outputs and the best possible combination of them. Efficiency and productivity, in any case, remain complementary concepts. Measures of efficiency are more accurate than those of productivity, as they involve a comparison with the most efficient combination, whereas productivity is based solely on the ratio of outputs to inputs (Daraio and Simar, 2007).

A rigorous analytical approach to the measurement of efficiency in production began with the works of Debreu (1951) and Koopmans (1952) and was empirically applied by Farrell (1957). Production efficiency analysis has been employed in environmental studies since the 1980s. International literature contains a large number of surveys and case studies dealing with natural resource use efficiency, a key factor in reaching the global objective of sustainable development (Song et al. 2012). One of the most used methodologies in these environmental surveys is data envelopment analysis (DEA; Zhou et al. 2008).

The use of DEA in studies for natural resource use efficiency

DEA is a non-parametric method to assess the relative effectiveness of DMUs. The exact functional relationship between inputs and outputs need not be considered in DEA technology. Nor does DEA rely on weightings. Weightings can be obtained through a program and no pre-estimated parameters are needed. Therefore, the weights need not be subjective. Output-oriented efficiency is calculated through a model in which efficiency is considered as the ability of a DMU to expand output, given conventional inputs and environmental inputs. Input-oriented efficiency is the ability of a DMU to decrease both conventional and environmental inputs equi-proportionately, maintaining a certain level of output (Reinhard et al. 2000).

It is important to underline that a DMU that is efficient from a technical point of view is not necessarily efficient from an allocative point of view. Allocative efficiency occurs when the marginal rate of substitution among any of the inputs equals the corresponding input price ratio. If this equality is not satisfied, it means that the region or country is not using its inputs optimally. Allocation efficiency, defined as the ability of a DMU to produce a given output at minimum cost, is not taken into account when the DEA method is used, because it involves assumptions about the behavioural rules of the productive units and requires knowledge of the prices of inputs and outputs, which are not always easily available. Conversely, a unit is technically inefficient if it wastes productive resources, does not minimize costs of production, or achieves a final productivity that is lower than the maximum achievable given the inputs used.

From a regional point of view, this input–output model can be useful to investigate the effects of inter-regional trade and to perform *environmentally extended input–output analysis*, illustrated in Chapter 8.

Case studies and representative findings of natural resource use efficiency analysis

A framework for integrated application of LCA (see Chapter 10 for LCA) and DEA has been developed to assess natural resource use efficiency and to benchmark DMUs. A number of case studies below illustrate the role of DEA and the contribution of LCA to improving the consideration of environmental efficiency. This assists in translating environmental benchmarks into economic terms through the monetization of externalities, and *emergy* benchmarks into market-driven indicators for the evaluation of ecosystem services.

Speelman et al. (2008) analysed water efficiency in small-scale irrigation schemes in the North-West Province of South Africa and studied its driving factors. The increasing pressure on water resources, especially in regions where small-scale irrigation systems are important for rural development, generates a need to ensure a more efficient use of water. Through the analysis of technical efficiency and sub-efficiency in the use of water, the authors revealed substantial technical inefficiencies among farmers. Farmers could theoretically reallocate some of the water currently used for irrigation to other uses, without compromising the functioning of small-scale irrigation. Such changes represent valuable opportunities for policymakers in pursuit of greater rationalization and efficiency.

Within a production framework that included intentional and unintentional (undesired) outputs, as well as specific and unspecific inputs, Wang et al. (2013) assessed China's regional energy and environmental efficiency. From the results, the authors found that Eastern China had a higher energy and environmental efficiency and more sustainable development than the Central and Western regions. However, all regions of China had a comparable increase in variation of efficiency during the period 2000–2008.

Assessing the sustainability of a regional agricultural system, Gerdessen and Pascucci (2013) found that, in a variety of decision-making contexts, stakeholders could use different criteria and methodologies, arriving at different, contrasting policies. In order to overcome the complexity of measuring a multidimensional situation, the three dimensions of sustainability need to be considered: economic, social, and environmental. A multidimensional approach to DEA applied to 252 European agricultural regions allowed for the systematic evaluation of their relative performances with respect to these three dimensions (Gerdessen and Pascucci 2013).

Vlontzos et al. (2014) assessed the energy and environmental efficiency of agriculture in EU member states based on a non-radial DEA model. One of the most important findings from the 2001–2008 period is the low environmental and energy efficiency for some countries that otherwise apply high environmental standards, for example Germany, Austria, and Sweden. Low efficiency scores for several Eastern European countries are likely due to a low level of technology in the primary production process.

One of the benefits of applying the LCA+DEA quantitative tool is the improvement in the interpretation of the results. Through the complementary use of DEA and LCA, it is possible to carry out benchmarking and trend analysis within nearly any specific sector, and to integrate the economic aspect with the sustainability assessment (Iribarren et al. 2010).

The LCA+DEA tool was applied by Vázquez-Rowe et al. (2012) to compute the efficiency of producers in forty vineyards in the Galicia region of Spain. Potential natural resource savings of up to 30 per cent were determined via benchmarking with the best-performing units. Environmental benefits ranged from 28 to 39 per cent, and more efficient agricultural practices would also generate significant economic savings.

Fishing was also analysed regionally under the combined approach using a Spanish coastal trawl fishery as an example. These results showed the dependence of environmental impacts on the operational performance of the vessels. Performance improvement targets were subsequently identified for the inefficient vessels. It was recommended that this methodology be used regularly also in other fisheries (Vázquez-Rowe et al. 2010).

Lozano et al. (2009) studied mussel farming in rafts. For each site, an LCA was carried out, and its environmental effects were calculated. Using DEA on the data set, authors defined a DMU that used less input and/or produced more output. The LCA study estimated environmental consequences of technical inefficiencies, helping to convince operators to pursue the double purpose of lowering costs and lowering environmental impact.

Implications and support to public policy and regional sustainable development

Academics in economics and sustainability research have shown increasing interest in assessing natural resource use efficiency, also known as eco-efficiency. Measuring eco-efficiency provides policymakers with valuable information for designing policies for regional sustainable development (Picazo-Tadeo et al. 2011).

The results from these case studies allow policy to better integrate economic activities and the environment. There are clearly several ways to increase a DMU's natural resource use efficiency: for example, through an expansion in environmental expenditure, or strengthening environmental policy to avoid market failures. Figure 21.1 shows that a wider implementation of programmes based on life cycle thinking can be effective in decreasing environmental pressure arising from economic sectors. Assessing natural resource use efficiency through input- and output-oriented DEA approaches helps policymakers to understand if environmental policy should be aimed at minimizing input (natural resources savings) or maximizing output using the current quantity of inputs (increased economic productivity).

Conclusion

As shown in this chapter, the planning of effective environmental policies involves multiple complex considerations. Differences in multi-criteria functions can give rise to unproductive decisions and inefficient policies.

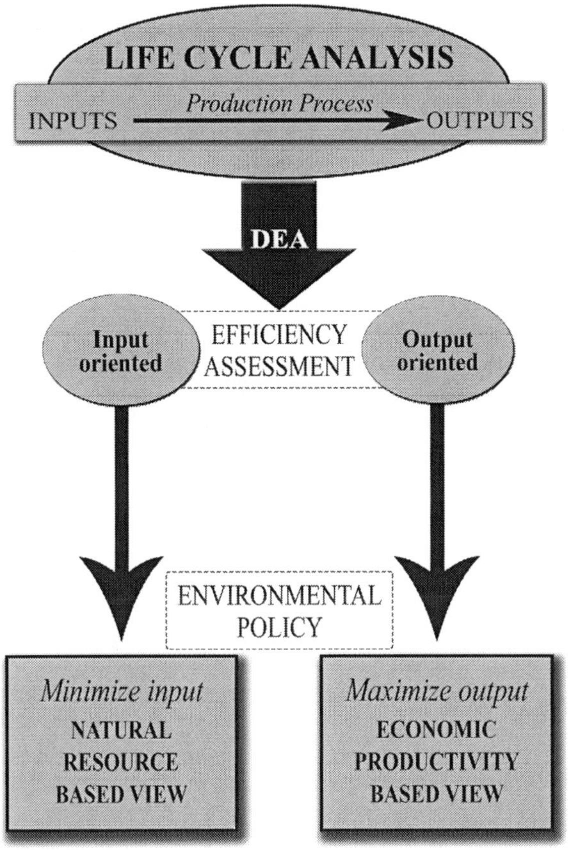

Figure 21.1
From life cycle approach to environmental policy framework through quantitative methods

For this reason, further research is still needed to improve the individual tools that compute the efficiency in the use of natural resources and that can assist in the implementation of environmental and sustainability policies at regional levels.

References

Cooper, W.W., Seiford, L.M., and Tone, K. (2000,) *Data Envelopment Analysis: A Comprehensive Text with Models Applications, References and DEA-Solver Software*, Boston, MA: Kluwer Academic.

Daraio, C., and Simar, L. (2007) *Advanced Robust and Nonparametric Methods in Efficiency Analysis: Methodology and Applications*, Dordrecht, Netherlands: Springer Science & Business Media.

Debreu, G. (1951) 'The coefficient of resource utilization', *Econometrica: Journal of the Econometric Society*, 19(3): 273–92.

De Leo, F., and Miglietta, P.P. (2015) 'The contribution of environmental indicators to the sustainable performance of countries', *International Journal of Sustainability Policy & Practice*, 11(4): 11–26.

Farrell, M.J. (1957) 'The measurement of productive efficiency', *Journal of the Royal Statistical Society Series A (General)*, 120(3): 253–90.

Gerdessen, J.C., and Pascucci, S. (2013) 'Data Envelopment Analysis of sustainability indicators of European agricultural systems at regional level', *Agricultural Systems*, 118: 78–90.

Iribarren, D., Vázquez-Rowe, I., Moreira, M.T., and Feijoo, G. (2010) 'Further potentials in the joint implementation of life cycle assessment and data envelopment analysis', *Science of the Total Environment*, 408(22): 5265–72.

Jebaraj, S., and Iniyan, S. (2006) 'A review of energy models', *Renewable & Sustainable Energy Reviews*, 10(4): 281–311.

Koopmans, T.C. (1952) 'Activity analysis of production and allocation', *Bulletin of the American Mathematical Society*, 58: 395–6.

Lovell, C.A.K. (1993) 'Production frontiers and productive efficiency' in *The Measurement of Productive Efficiency*, eds H.O. Fried, C.A.K. Lovell, and S.S. Schmidt, New York: Oxford University Press.

Lozano, S., Iribarren, D., Moreira, M.T., and Feijoo, G. (2009) 'The link between operational efficiency and environmental impacts: A joint application of Life Cycle Assessment and Data Envelopment Analysis', *Science of the Total Environment*, 407(5): 1744–54.

Picazo-Tadeo, A.J., Gómez-Limón, J.A., and Reig-Martínez, E. (2011) 'Assessing farming eco-efficiency: A data envelopment analysis approach', *Journal of Environmental Management*, 92(4): 1154–64.

Reinhard, S., Lovell, C.K., and Thijssen, G.J. (2000) 'Environmental efficiency with multiple environmentally detrimental variables: Estimated with SFA and DEA', *European Journal of Operational Research*, 121(2): 287–303.

Sengupta, J. (1995) *Dynamics of Data Envelopment Analysis: Theory of Systems Efficiency*, Dordrecht, Netherlands: Springer Science & Business Media.

Song, M., An, Q., Zhang, W., Wang, Z., and Wu, J. (2012) 'Environmental efficiency evaluation based on data envelopment analysis', *Renewable & Sustainable Energy Reviews*, 16(7): 4465–9.

Speelman, S., D'Haese, M., Buysse, J., and D'Haese, L. (2008) 'A measure for the efficiency of water use and its determinants, a case study of small-scale irrigation schemes in North-West Province, South Africa', *Agricultural Systems*, 98(1): 31–9.

Vázquez-Rowe, I., Iribarren, D., Moreira, M.T., and Feijoo, G. (2010) 'Combined application of life cycle assessment and data envelopment analysis as a methodological approach for the assessment of fisheries', *International Journal of Life Cycle Assessment*, 15(3): 272–83.

Vázquez-Rowe, I., Villanueva-Rey, P., Iribarren, D., Moreira, M.T., and Feijoo, G. (2012) 'Joint life cycle assessment and data envelopment analysis of grape production for vinification in the Rías Baixas appellation (NW Spain)', *Journal of Cleaner Production*, 27: 92–102.

Vincent, A. (1968) *La mesure de la productivité*, Paris: Dunod.

Vlontzos, G., Niavis, S., and Manos, B. (2014) 'A DEA approach for estimating the agricultural energy and environmental efficiency of EU countries', *Renewable & Sustainable Energy Reviews*, 40: 91–6.

Wang, K., Yu, S., and Zhang, W. (2013) 'China's regional energy and environmental efficiency: A DEA window analysis based dynamic evaluation', *Mathematical & Computer Modelling*, 58(5): 1117–27.

Zhou, P., Ang, B.W., and Poh, K.L. (2008) 'A survey of data envelopment analysis in energy and environment studies', *European Journal of Operational Research*, 189(1): 1–18.

22

ENVIRONMENTAL AND LANDSCAPE MANAGEMENT SYSTEM

Riccardo Beltramo

Introduction

Although unaffected wilderness areas still exist, much of our landscape is the result of human activity that has shaped our urban and rural surroundings over past centuries. Mountains, moors, coasts, rivers and forests all bear witness to the influence of fire, grazing, cropping and urbanisation. The visual nature of the landscape has a direct link with our social, environmental, and economic fabric, making it an important element of a national or regional sustainable development agenda. We only need to think of the importance of 'beautiful places' as a foundation for much of the tourism on which millions of livelihoods depend to understand the direct economic value of landscape. Yet, careless maintenance and insensitive development often destroy the very resource on which regional prosperity is founded. The recent situation in Dresden (Germany) is an important case in point: the city gained world heritage listing for its scenic river valley and then lost it through inattention to the importance of landscape management.

With some landscapes inherited in their original form and others unintentionally created over extended time periods, we can ask what techniques and procedures have been used to manage such a valuable resource, or whether landscapes should just be left to evolve (and perhaps degrade) in an organic fashion, without any deliberate management at all.

This chapter will examine some methodologies that can answer this important question.

Landscape perceptions

Italy's landscapes form one of the country's most prominent attractions, with a perceived expression of historical interaction between mankind and nature. Likewise, there is a growing interest in landscape issues across Europe, both in terms of conservation and managed improvements. Landscape is simultaneously an expression of environmental quality and a factor for tourist attraction, and thus a generator of economic flows (See Figure 22.1).

Each of us has her/his own conception of landscape and of its evolution, varying with culture, economic situation and other factors. The European Landscape Convention (ELC), signed in Florence in 2000, proposes a definition of land-scape in an effort to establish common policies and share landscape values

Figure 22.1
Cinque Terre: An amazing relationship between man and nature

Figure 22.2
An old cement plant as part of the landscape

and their maintenance: '"Landscape" means a perceived area whose character is the result of the action and interaction of natural and/or human factors'.

This definition includes all areas, not only those appreciated for their beauty. All of the Convention's signatories must consider its recommendations in their landscape management, with actions oriented towards conservation, remediation, and exploitation. This applies to all landscapes, from natural reserves to highly artificial areas, as all have distinctive characteristics that need to be identified, understood, considered, and managed (see Figure 22.2). The ELC definition is intentionally inclusive to ensure a management process that preserves high-quality landscapes and improves weak ones.

Landscape management: Where? When? How?

If landscape issues are sufficiently important to be considered in an international convention, the instruments for their management must be adequate to address the complex relationships between population and environment, as these directly impact on the landscape. Some of these instruments need to be developed; others can be borrowed from related areas of activity. Because landscape includes an environmental component, we have taken inspiration from existing methodologies for the management of interactions between economic entities and environment, in particular the *Environmental Management and Audit Scheme* (EMAS), as approved by the European Commission in 1993.

The value of EMAS certification is attested by the numerous public administrations and private organisations that have used this process to gain recognition for their environmental performance. Such certification is widely regarded by the public as an indicator of environmental quality.

Despite its wide use, EMAS has not so far been visibly used to address landscape management. We need to ask if there is a relationship between a carefully managed environment and a landscape perceived as beautiful. Moreover, is a territorial institution that takes care of environmental quality also able to perform well on the issue of landscape? These are some of the questions that led us to the idea of developing an environmental and landscape management system (ELMS; Beltramo et al. 2011).

The ELMS can be defined as an integrated system, merging the concept of the ELC (an objective) with that of the EMAS (a management process). The ELMS concept was born in 2004, with research undertaken until 2009 under the project for the creation and registration of an ELMS, implemented for the Unione di Comuni Colline di Langa e del Barolo in the north-western area of the Piedmont region, Italy (see Figure 22.3). The ELMS expands the classical EMAS procedure to also evaluate and manage impacts of human activities on the landscape.

As a standardised management system, EMAS is used by both private and public bodies whose activities impact the environment. This study therefore examined environmental declarations of a number of Italian regions, provinces, and municipalities. The most important motivations pushing these organisations towards a rigorous environmental standard are legal compliance and improved environmental quality, with an eye also on what features are valuable to attract tourists. The study attempted to see if elements related to landscape were included in their EMAS documents. It turned out that none of them had considered landscape, presumably because environment and landscape (and their management) are issues dealt with by different administrators and technicians. This finding confirmed the potential for innovation in our proposal.

The similarities and differences between EMAS and ELC were then investigated, with a view to eventual integration. It was found that, although some important differences exist, there are many ways in which the two are complementary:

Figure 22.3
Piedmont landscape

- Although EMAS is aimed at improving the environmental performance of organisations through the implementation of a structured management system, ELC is essentially an agreement by interested parties on landscape issues: rules and regulations are different from country to country, according to the specific details of their legal systems.
- EMAS provides a methodology for an organisation to manage its environmental impact, as 'any change to the environment, whether adverse or beneficial, wholly or partially resulting from activities, products or services of an organisation'. ELC only requires parties to 'introduce instruments aimed at protecting, managing and/or planning the landscape' without, however, providing details of the methodologies or tools that might achieve such specific objectives.

The similarities between EMAS and ELC could, if properly linked, be strong points of connection. For instance, in terms of analysis, EMAS requires that an organisation undertake an environmental analysis (environmental review); ELC states that each country should identify its own landscapes; analyse their characteristics, dynamics, and pressures; and monitor transformations. With regard to goals, EMAS requires the organisation to define and document its objectives and targets. ELC provides that each party should commit itself 'to establish the landscape quality objectives for the landscapes identified and assessed, after public consultation'. This initial review also strongly resembles the inventory phase of life cycle assessment (LCA).

The above ELC definition of landscape has a number of implications. Although there is an extensive literature on landscape quality parameters and indicators, perhaps also connected with landscape ecology, the ELC has chosen only perception as an indicator. 'Perception' is sufficiently subjective to permit landscapes to arouse quite varied and even opposite reactions in different people. The perception of the values of an area's landscape thus needs to be clarified so that distinctive characteristics can be identified that require evaluation, and that can lead to actions to preserve or improve. In any case, ELC needs to include a strong cultural outreach devoted to landscape awareness and management. Given the currently purely programmatic nature of ELC, nothing has so far been planned in terms of instruments for landscape management.

The subjectivity contrasts with the quality of data required by EMAS (and also by LCA) in terms of reliability, accessibility and relevance (Carlson and Pålsson 2001). In recognising perception as an important factor in landscape management, ELC establishes a strong link with sensorial analysis. All three dimensions of data mentioned above are important for allowing an assessment of the quality and applicability of a specific data set. As well as international standards of assessment, training on a combination of methodologies and statistical methods is required to improve data quality, thus improving evaluation and making possible the monitoring of expert panellists' performance over time.

The evaluation of landscapes within environmental and landscape analysis (ELA) can be based on the following:

- UNI 11109:2004 is based on the principles of the ELC and used in studies of environmental impact (UNI n.d.).
- The US Bureau of Land Management Assessment (BLM), used in studies for urban planning, demonstrates its validity in relation to the descriptors of the identified landscape (confirmed by ISO; US Department of Interior, Bureau of Land Management 1980).

A municipality's ELA would aim to analyse the territory in accordance with the EMAS framework, as well as the activities of the municipality that may generate direct and indirect environmental impacts. This would identify significant environmental aspects, leading eventually to improvement objectives and environmental programmes. Under the ELC, each party: identifies its own landscapes in its territory; analyses their characteristics and the pressures likely to modify them; takes note of changes; assesses the landscapes thus identified; and takes account of the specific values assigned to them by stakeholders and the wider public.

In ELA, landscape evaluation follows a series of stages: evaluation of the visual quality of the landscape, evaluation of the quality and effectiveness of the rules of the PRG; and evaluation of relationship between the visual quality of the landscape and the adequacy of the planning rules.

The steps in terms of evaluation of the visual quality of the landscape are:

1. Breakdown of the territory into landscape units: The landscape units are the same conceptual categories as already used for assessing the rules of general urban plans (inner city, new residential areas, industrial or commercial areas, agricultural areas, environmentally valuable areas).
2. Choice of landscape descriptors: From UNI 11109:2004 and BLM assessment, we can identify: morphology, vegetation, hydrography, colour, panoramic views, representativeness, transformations/distortions, to which can also be added built landscape, signs of territorial organisation, symbolic role.
3. Allocation of values: Values are allocated for each unit according to the components considered important, using descriptive criteria on a three-level scale. Evaluation can also be adversely affected by the presence of negative elements.

4 Overall assessment of the municipality from the average value of the landscape units: The rating of all the components is summed to obtain the value of the quality of each landscape unit.

Evaluation of the quality and effectiveness of the town planning scheme rules follows a similar process:

1 breakdown of the territory into landscape units; the choice of units is based on the same conceptual categories used for assessing the rules;
2 identification of a set of indicators for each thematic area (protection of the elements that characterise the building texture, protection of the characteristics related to building typology, protection of formal characteristics, existence and quality of rules encouraging or discouraging building restoration) and, within them, a set of evaluation elements;
3 attribution of scores.

In a final stage, the relationship between a landscape's visual quality and the quality of the planning rules is identified. This is oriented so as to provide useful information to public managers about opportunities and ways to improve the planning rules, where these are not sufficiently effective. The output of ELA is a table of indicators, useful for establishing goals, targets, programmes, and actions.

Time perspective (Beltramo et al. 2011)

The time dimension follows the succession of the plan–do–check–act phases, the iterative process that is the core of EMAS. This dynamic perspective, oriented towards improving performance, resembles the optimisation phase of LCA.

Changes to territory perceived as landscape usually require a long time frame, whereas the possibility and intensity of changes, even if decided on the basis of regulations, reflect the vision of politics with its, normally, rather short-term perspective. There is an overall mismatch between the time perspectives of politicians and those of landscape conservation, protection, and improvement as requested by other stakeholders. Establishing a match between the time horizons of politicians and of stakeholders requires creating a sufficiently powerful public demand for landscape quality to convince politicians to adopt and implement management programmes in the short term. Some interconnected elements within ELMS go in this direction: defined programmes; procedures and operative instructions; awareness, information, and training; monitoring and measurement; audit; and communication.

Developments

A review in 2010 identified data collection and analysis as the most time- and resource-consuming activities in ELA. To overcome this problem, the concept of an eco–land–web–scape management system (Beltramo 2010) has been developed as an evolution that combines an open-source, Internet-based wireless sensors network (WSN) named Scatol8®. The increased use of smart devices is making IoT helpful at various stages: evaluation, monitoring, analysis, training, and knowledge sharing.

Scatol8® is a WSN that detects environmental and landscape variables (i.e. energy consumption, water consumption, air quality, water quality, humidity, temperature, amount of waste, images) and transmits them to a PC, where they are stored, processed, and displayed via an interface, the Crusc8 (dashboard). It can thus provide real-time monitoring of each variable detected, in addition to evaluation of trends, via the display of time series.

The PC is able to merge its data on an Internet server that collects and organises them in a collective database. This allows the data to be published and made accessible on PCs and smart phones, but also in relevant configurations for management purposes (production lines, departments, etc.).

Conclusion

As is already the case for EMAS, ELMS is based on organised information on stocks and flows. The type and quality of information influence operational activities and thus determine the effectiveness and efficiency of outcomes. Some types of information will differ, notably that on the subjectivity of landscape evaluation. In ELMS, this provides information in a format that can lead to improvement actions through a process controlled by the managing organisation. The project has stimulated debate between environment and landscape specialists, providing a multidisciplinary framework that is also suitable for other applications. ELMS could be useful, not only to local authorities, but also to companies that impact the landscape greatly, such as park authorities, conservation bodies, and resource and water management bodies.

Note

1 'Landscape management means action, from a perspective of sustainable development, to ensure the regular upkeep of a landscape, so as to guide and harmonise changes which are brought about by social, economic and environmental processes', Council of Europe, European Landscape Convention, 2000.

References

Beltramo, R. (2010) 'The SCATOL8™: An innovation for shifting from Environmental and Landscape Management System (ELMS) to the Eco-Land-Web-Scape Management System (ELWSMS)', *Romanian Distribution Committee Magazine*, 2(2): 16–23. Available at: www.distribution-magazine.ro/magazine2/ (accessed 2 August 2016).

Beltramo, R., Duglio, S., and Quarta, M. (2011) *Linee guida per un Sistema di Gestione Ambiental-Paesaggistico*, Torino, Italy: Università degli Studi di Torino, Dipartimento di Scienze Merceologiche.

Beltramo, R., Duglio, S., and Quarta, M. (2011) *Sistema di Gestione Ambiental-Paesaggistico – SGAP: Una metodologia per la gestione integrata dell'Ambiente e del Paesaggio*. Rome: Aracne Editrice.

Carlson, R., and Pålsson, A.C. (2001) FAQT – Fundamentals of Data Quality for Industrial Environmental Information Systems. CPM report 2001:1.

UNI (n.d.) 11109:2004: Environmental Impact – Guidelines for Evaluation of Impact on the Landscape for Preparation of Environmental Impact Studies.

US Department of Interior, Bureau of Land Management (1980) *Visual Resource Management Program*. Washington, DC: US Government Printing Office.

23

ENVIRONMENTAL PRODUCT DECLARATIONS TO SUPPORT REGIONAL DEVELOPMENT

Annik Magerholm Fet

Introduction

The increasing pressures of climate change and globalization make sustainable development an essential focus for future progress. As new requirements and regulations are imposed on companies to improve their triple bottom line – environment, society, and economy – pressure builds on them to provide information about their products (Michelsen et al. 2006). The benefits of LCA have been explained and detailed in previous chapters. Extending the results of LCA, this chapter provides an introduction to *environmental product declaration* (EPD), a classification method that provides objective, quantified information about the environmental impact of a product. Translated from LCA results, EPDs are increasingly being used to deliver environmental impact information to professional purchasers, management, government, and consumers and, unlike LCA results, allow easy comparability of information. EPDs maintain the important characteristics of objectivity, comparability, and validity.

EPD context and background

Currently, there are a number of different environmental labels, declarations, and standards used to manage and relay information about products and services, as shown in Chapter 16. ISO the International Organisation for Standardisation, hosts relevant standards in the 14000 series that separate declarations into three types, which all require the consideration of life cycle product information, as described in Chapter 18. Type 1 programmes within the ISO 14000 series are *voluntary* multiple-criteria-based programmes, verified by a third party, that award labels for environmental preferability (ISO 1999a). Type 2 programmes are *self-declared* environmental claims that take life cycle information into account (ISO 1999b). This chapter focuses on Type 3 programmes, which require *third party verification* of LCA in accordance with the predetermined parameters of the ISO 14040 standards series, as well as verification of the resulting EPD document (ISO 2006a, 2006b). By offering third-party certification, Type 3 EPDs provide objective, verified, quantified information about a product in a format that is directly comparable with other similar products. Decisions about product use or purchasing can then be made based on the product's environmental impact compared with that of others.

Product category, a defined group of products that fulfil the same function, is an important term to know in reference to EPDs (Fet et al. 2009). Similarly, *product category rules* (PCRs) establish the criteria for a specific product category and outline the requirements that must be met for EPD achievement. This way, the PCRs standardize the product's functional and performance characteristics, the LCA criteria, and the information to be reported in the EPD, allowing direct comparison of similar products. The development of the PCR document is therefore a critical step, and detailed guidance is provided in ISO 14025 (ISO 2006c).

In addition to a clearly formatted PCR document, EPD programmes must establish a verification system that defines who can be an accredited verifier, the format of the verification, documentation, and meeting of the PCR (Fet et al. 2009). LCA data in the EPD must be verified to conform, at least, to the PCR and to the ISO 14040 series standards. Table 23.1 summarizes the contents of a PCR document and of an EPD, as prescribed in ISO 14025.

There are various Type 3 programmes and resulting product declarations. Some programmes are run by governmental organizations, such as in Sweden and South Korea. Others are directed by private industry organizations, such as the EcoLeaf Program run by Japan's Environmental Management Association for Industry, or research institutes and foundations such as the United States' Earthsure Program run by the Institute for Environmental Research and Education. The Norwegian EPD Program is run by the Confederation of Norwegian Enterprises. Table 23.2 summarizes the location, owner, and scope of existing Type 3 programmes. Because there are so many different programmes, the term Type 3 EPD is used to refer to any Type 3 EPD programme.

Table 23.1 Contents of the PCR document and the EPD

Contents of a PCR document	Contents of an EPD
1 Product category definition and description	1 Identification and description of the organization
2 Goal and scope of definition for the LCA of the product (ISO 14040 series)	2 Description of the product
3 Inventory analysis (incl. data collection, calculation procedures, and allocation of material and energy flows)	3 Product identification
4 Impact category selection and calculation rules (if applicable)	4 Information about the EPD programme
5 Parameters for reporting LCA data	5 PCR identification
6 Requirements for provision of additional environmental information	6 Publication date and period of validity
7 Materials and substances to be declared	7 LCA, LCI, or information module data
8 Instructions for producing the data required to develop the EPD	8 Additional environmental information
9 Instructions on the content and format of the EPD	9 Declaration of content (material and substances)
10 Information on which stages of the life cycle are excluded (if applicable)	10 Excluded life cycle stages (if applicable)
11 Period of validity	11 Statement that EPDs from different programmes may not be comparable
	12 Information about obtaining explanatory materials
	13 Information about verification

Table 23.2 Summary of EPD programmes

Country	Programme name	Scheme owner	Main focus	From	Monitored standards
Denmark	Miljøvare-deklarationen	Fonden Dansk Standard		2009	ISO 14025
Finland	RT Environment Declaration	Building Information Foundation, RTS Confederation of Finnish Construction Industries RT	Building products	1988	ISO 21930
France	INIES	Département Energie, Santé et Environnement	Building products		NF P01–010
Germany	IBU	Institute of Construction & Environment	Building products	1998	ISO 14025, 21930
Japan	EcoLeaf	Japan Environmental Management Association for Industry	Electronics	2002	ISO 14025
Netherlands	MRPI	Industry	Building products	1997	ISO 21930
Norway	Norwegian EPD Program	Confederation of Norwegian Enterprise	Building products and furniture	2002	ISO 14025, 21930
Sweden	International EPD System	Swedish Environment Management Council	Various products	1997	ISO 14025, 21930
South Korea	Type 3 Labelling Program	Ministry of Environment	Electronics	2001	ISO 14025
United States	Earthsure	Institute for Environmental Research & Education	Food & agriculture	2000	ISO 14025
United States	Eco-Profile Program	Scientific Certification Schemes			ANSI LCA standard

Although EPDs are comparable within sectors, different EPD programmes and sectors require the meeting of different standards. Further harmonization between programmes is therefore necessary for universal comparability (Minkov et al. 2015). The European Commission (EC) has set out to address this issue in its Product Environmental Footprint (PEF) scheme. A PEF is defined as, 'a multi-criteria measure of the environmental performance of a good or service throughout its life cycle' (European Commission 2012: 5). Additionally, the PEF is accompanied by PEF category rules, similar to the EPD's PCRs. PEFs are seen as an extension of current EPD programmes, as they solve the problem of cross-programme comparability. The differences between EPDs and PEFs must also be noted, however, as PEFs are more tools for measurement of a product's environmental impacts across its life cycle, rather than a certification of its environmental preferability, as documented in an EPD (Minkov et al. 2015). Because the discussion over the merits of the PEF scheme is still in its early phases (Minkov et al. 2015; Finkbeiner 2014; Galatola and Fant 2014), it is simply noted here as a consideration for the future of life cycle-based methods for product evaluation.

Description

Public procurement, as described in Chapter 19 of this book, has proven to be the largest driver for the development of EPDs. Statistics show that public procurement in Europe represents 19 per cent of the EU's gross domestic product

(GDP), roughly €425 billion (European Commission 2013). This has resulted in the development of sector-specific standards for managing the qualifications of public spending and for incorporating environmental considerations into each.

The construction sector is important within EPDs in its relation to public procurement and meeting of many international standards. It contributes to 10 per cent of EU GDP and is a very resource-, energy-, and waste-intensive industry (European Commission 2015). As a result, there has been an increase in the focus on environmental documentation of building materials. The European Committee for Standardization (CEN) standard EN 15804 provides the core PCR for construction products and services and sets the requirements for the development and verification of construction EPDs. The goal of EN 15804 is to ensure that, 'all environmental product declarations (EPDs) of construction products, construction services and construction processes are derived, verified and presented in a harmonized way' (CEN 2012). EN 15804 is currently supported by ISO 21930, but may soon be replaced by it as core PCR. Considered 'twin' standards, ISO 21930, Sustainability in Buildings and Civil Engineering Works – Core Rules for Environmental Declaration of Construction Products and Services Used in Any Type of Construction Works, may soon pull ahead because of its consideration of a product's reuse, recovery, and recycling potential outside the four traditional life cycle stages that make up EN 15804.

ISO-140250-series standards are therefore not the only international standards that promote EPDs, and EPDs in the construction sector often fulfil the requirements of several standards. For other sectors and product groups, such as food production and the healthcare industries, a similar multi-standard tendency can be observed, which suggests the importance of increased transparency and comparability of product information across sectors and EPD programmes.

Discussion

The benefits of EPDs stem from their comparability and ability to translate complex LCA information into understandable inputs to decision-making. Although LCA does not take economic and social aspects into its assessment, resulting EPDs can provide positive socio-economic effects for regional sustainable development through objectively informing the decision-making and purchasing choices of local consumers and increasing the competitiveness of local producers based on their environmental status.

EPDs are primarily used in business-to-business communication, but can also be used to educate consumers and perform five said functions: (1) management tool, (2) communication tool, (3) evaluation/assessment tool, (4) political tool, and (5) action tool (Hillier et al. 2004). In terms of management, information from the EPD can be used in environmental management and product innovation and design to improve environmental performance (Fet et al. 2014).

When EPDs are related to regional development, the potential of EPDs to communicate environmental information throughout the supply chain and to purchasers and consumers is most relevant. Increased communication of product information increases environmental awareness. When environmental information is available and quantified, and presented in a standardized and comparable format, consumers can recognize and interpret criteria that inform their purchases. Increased consumer and society awareness supports the use of EPDs as a political tool. EPDs offer regulators and governments a tool to increase environmental awareness in the business and public sectors (Hillier et al. 2004). Making environmental product information available and comparable, while also increasing consumer awareness and government requirements, encourages a transition in the way companies produce and present their goods and services. The information identified in an EPD can help a company determine its hot spots – for example, areas of heavy resource or energy use – and restructure production to reduce environmental impacts. Because an EPD is not a fixed document, continuous assessment and updates reflect, and communicate to consumers, a company's improvement. In this way, a company's actions can be shaped by the EPD.

Conclusion

Communicating complex LCA results, EPDs are an objective, comparable way for companies to pass environmental product information throughout their supply chain. Within each EPD programme, a standardized format, verified by a certified third party, easily translates information to consumers and regulators and helps the company to identify areas for potential improvement. Further standardization of EPDs is necessary for universal comparability, however. Although there are a number of international standards that set the requirements for EPDs, different EPD programmes and sectors require the meeting of different sets of standards. This makes international and cross-industry comparison more difficult and should be kept in mind as future EPD programmes, such as the EC's PEF scheme, are developed. Overall, however, the increasing pressure and need to reduce environmental impacts throughout supply chains make EPDs an important tool for the transition to greener industry.

References

CEN (2012) *EN 15804: 2012 Environmental Product Declarations – Core Rules for the Product Category of Construction Products*, Brussels: CEN.

European Commission (2012) *Product Environmental Footprint (PEF) Guide: Consolidated Version*. Available at: http://ec.europa.eu/environment/archives/eussd/pdf/footprint/PEF%20methodology%20final%20draft.pdf (accessed 17 September 2015).

European Commission (2013) Annual Public Procurement and Implementation Review 2013, Commission Staff Working Document, Brussels: European Commission.

European Commission (2015) Growth: Internal Market, Industry, Entrepreneurship and SMEs – Sectors: Construction. Available at: http://ec.europa.eu/growth/sectors/construction/index_en.htm (accessed 23 June 2015).

Fet, A.M., Magerøy, M., Aspen, D.M., and Malnes, D. (2014) 'Harmonization of documentation for EPD-generation: The Norwegian model', in *Pathways to Environmental Sustainability: Methodologies and Experiences*, eds R. Salomone and G. Saija, Dordrecht, Netherlands: Springer International, pp. 225–41.

Fet, A.M., Skaar, C., and Michelsen, O. (2009) 'Product category rules and environmental product declarations as tools to promote sustainable products: Experiences from a case study of furniture production', *Clean Technology & Environmental Policy*, 11: 201–7.

Finkbeiner, M. (2014) 'Product environmental footprint – Breakthrough or breakdown for policy implementation of life cycle assessment?', *International Journal of Life Cycle Assessment*, 19(2): 266–71.

Galatola, M., and Pant, R. (2014) 'Reply to the editorial "Product environmental footprint – Breakthrough or breakdown for policy implementation of life cycle assessment?" written by Prof. Finkbeiner (*Int J Life Cycle Assess*, vol. 19, no. 2, pp. 266–271', *International Journal of Life Cycle Assessment*, 19(6): 1356–60.

Hillier, B., Jonsson, L., and Ryding, S.O. (2004) *International Guide to Environmental Product Declarations*, Stockholm: GEDnet.

ISO (1999a) *European Standard EN ISO 14024: Environmental Labels and Declarations – Environmental Labeling Type I – Guiding Principles and Procedures*, Geneva: ISO.

ISO (1999b) *European Standard EN ISO 14021: Environmental Labels and Declarations – Self-Declared Environmental Claims*, Geneva: ISO.

ISO (2006a) *European Standard EN ISO 14040:2006: Environmental Management Life Cycle Assessment – Principles and Framework*, Geneva: ISO.

ISO (2006b) *European Standard EN ISO 14044:2006: Environmental Management Life Cycle Assessment – Requirement and Guidelines*, Geneva: ISO.

ISO (2006c) *European Standard EN ISO 14025:2006: Environmental Labels and Declarations – Environmental Labeling Type III – Principles and Procedures*, Geneva: ISO.

Michelsen, O., Fet, A.M., and Dahlsrud, A. (2006) 'Eco-efficiency in extended supply chains: A case study of furniture production', *Journal of Environmental Management*, 79: 290–7.

Minkov, N., Schneider, L., Leimann, A., and Finkbeiner, M. (2015) 'Type III Environmental Declaration Programmes and harmonization of product category rules: Status quo and practical challenges', *Journal of Cleaner Production*, 94: 235–46.

Part IV

USING LIFE CYCLE APPROACHES AT REGIONAL LEVEL

24

ECOLOGICAL FOOTPRINT

A local policy perspective

Pier Paolo Miglietta, Federica De Leo, and Pierluigi Toma

Introduction – Ecological footprint as a policymakers support tool

The ecological footprint measures the biologically productive area needed to sustain a certain human community (Rees 1996; Wackernagel and Rees 1998; Wackernagel and Silverstein 2000) or process (Scotti et al. 2009), as already indicated in Chapter 7. Because of its intuitive meaning and ease of computation, it has rapidly taken root as a tool for assessing the human pressure on natural resources and ecosystem services. Accordingly, it is perceived as a tool that helps to set up an agenda for local policies. For the ecological footprint to be of help to local administrations, it must produce a reliable picture of what happens in the territory of their jurisdiction, and this includes impacts due to both citizens and production.

Local policies aiming at sustainability, in the view of life cycle approaches to sustainable regional development, must thus deal with the latter impacts. The ecological footprint, to be effectively used to monitor progress towards sustainability, must account for the land requirement to sustain local activities.

When conducting an ecological footprint analysis, we have to distinguish between the following categories of ecological space (Chambers et al. 2014):

- productive land, including arable land, that is, the portion of land more productive in terms of plant biomass and typically used for major crops; pasture that is less productive than arable land and exploited for cattle ranching; cultivated or natural forested land destined to the maintenance of biodiversity;
- built or degraded land earmarked for the development of human settlements that has no more agricultural capacity;
- coastal areas of production attributable to the areas closest to the coast, typically very productive, in which at least 90 per cent of fishery is carried out (De Leo et al. 2014);
- land for the production of energy, the portion of land needed for the sustainable management of energy demand, that is, the absorption of polluting emissions arising from the model of energy policy adopted;
- land assigned for the preservation of biodiversity.

In the following paragraphs, we review some local applications of the ecological footprint, showing the results for each type of bio-productive area and the interventions of sustainable development policies adopted or planned to achieve a reduction of the index to the level of biocapacity.

A municipal approach to energy land use

Report 11/02 of the University of Oslo, entitled The Ecological Footprint of the City of Oslo, shows that each Oslo citizen has an ecological footprint of 7.77 global hectares per capita (including an area of 12 per cent related to biodiversity protection). Of this area, 3.59 global hectares are attributable to the energy land component, which accounts for 52.6 per cent, representing more than half of the total ecological footprint per capita. The main local activity that influences this footprint component is transport, which is responsible for a 43.7 per cent of the energy land required per inhabitant, followed by the consumption of food, which contributes 31.73 per cent (ProSus 2002).

Calgary's ecological footprint is currently estimated at between 9.5 and 9.9 global hectares per capita; we notice that in this case also the largest component is energy use, which makes up more than 66 per cent of the total footprint, reflecting the contribution of CO_2 emissions from production and consumption activities such as transportation, electricity, residential heat, and hot water (Global Footprint Network 2007).

In Asia, as shown in the report of the WWF, Hong Kong's ecological footprint was 4.0 global hectares per person, having seen its per capita ecological footprint level off and even decline in recent years. Hong Kong is clearly in ecological deficit, most of which is attributable to CO_2 emissions, which represent a demand for global rather than localized biocapacity. In 2007, the ecological footprint component related to energy land use was 60 per cent of the total ecological footprint of consumption. It started to level off at around 1.8 global hectares per capita between 2000 and 2007. Since 2001, the decline in the energy land use has been due to the use of fewer carbon-intense sources of energy, a relatively stable production footprint, and an increase in the CO_2 emissions embodied in exports. After the services sector, the largest energy land-use component of the ecological footprint of Hong Kong's economy is attributable to the construction sector, whose 85 per cent is embodied in imported goods and services. Considering all economic sectors, internal CO_2 emissions account for 26 per cent of the total energy land-use component of the ecological footprint, and the remaining 74 per cent is emitted elsewhere to supply imports to Hong Kong. A significant portion of emissions related to Hong Kong's total consumption is from foreign sources, playing a fundamental role in supporting its consumption pattern, as well as providing intermediate goods for its economic production activities (WWF 2011).

The city of São Paulo's average footprint amounts to 4.38 global hectares per capita, as revealed in *The Ecological Footprints of São Paulo – State and Capital and the Footprint Family* (WWF 2012). It is 49 per cent larger than the Brazilian average and 25 per cent larger than that of the São Paulo state. São Paulo's citizens, contrary to the other cities considered, have a smaller demand for CO_2 absorption areas owing to the lower levels of emissions associated with the energy matrix and the intense use of biofuels in mobile sources of emissions (WWF 2012).

In all case studies and scientific literature, energy land is the most significant part of the ecological footprint. About 83 per cent of reports show percentage values of energy land larger than 40 per cent (Scotti et al. 2009).

For these reasons, some changes in the consumption model of local communities are essential to reduce the energy land component of the ecological footprints revealed by the analyses.

In particular, as Oslo has a much higher consumption of air transport than the national average, an increase in the use of video-conferences and in the choice of holiday destinations that do not involve air transport has been encouraged. Moreover, citizens have been urged to reduce leisure driving and instead to use public transport, all this accompanied by restrictions on private car use (ProSus 2002).

Calgary has started to invest in a sustainable future, acquiring consciousness about the importance of applying ecological limits to decision-making, creating policy choices that reduce the ecological footprint and, in particular, its energy land component. Calgary's wind-powered CTrain project started in 2001 and consists of a rail transit system that consumes wind energy instead of fossil fuel-generated power. This project contributes to a reduction in local air pollution of 23,382 tons of CO_2 emissions per year, a reduction of 6,329 global hectares of the total ecological footprint, the removal of vehicles from roads, and the conservation of fuels (Global Footprint Network 2007).

The measurement of the supply and demand side of management has been introduced to reduce Hong Kong's carbon emissions and to develop a low-carbon city. In 2009, the WWF-Hong Kong launched the 'Low Carbon Living Appliances Guide', the Low-carbon Office Operation Program, and the Low Carbon Manufacturing Program, which, respectively, promoted efficient energy use in the residential sector, the tracking and reduction of GHG emissions associated with sources such as electricity, transportation, and paper consumption, and the equipment of factories with machinery to identify and report areas of cost savings and recognize positive actions by granting labels to manufacturers. The GHG emissions reductions measures identified from the pilot programmes could lead to a potential 74 million ton CO_2e reduction per year (WWF 2011).

Arable land, pasture land, and fishing grounds for sustainable food consumption

For Oslo inhabitants, the arable land component amounts to 0.799 hectares per capita and is influenced almost totally by food consumption, similar to pasture land, which amounts to 0.89 hectares. Forest land amounts instead to 0.752 hectares per capita, 0.541 of which are attributable to housing-related consumption and 0.211 to the consumption of goods excluding food. In summary, the total productive land area of the per capita ecological footprint of Oslo contributes 35.7 per cent to the total, amounting to 2,441 hectares, mainly influenced by the consumption of food, goods, and housing (ProSus 2002).

Consumption and changes in the arable land component are driven by trade. Between the years 2000 and 2007, in Hong Kong, the imports of products such as rapeseed and soybean oils, refined sugar, oranges, grapes, frozen potatoes, and wheat bran decreased. Exports also decreased between these years, although to a lesser extent, and the cropland footprint consequently fell. In 2007, cropland represented 24 per cent of Hong Kong's total ecological footprint. Whereas the cropland footprint has been decreasing, the grazing land and fishing ground footprints have increased since the 1990s, indicating a slow shift in diet that includes more meat and fish products.

Livestock uses 30 per cent equivalent of the Earth's entire land surface, mostly permanent pasture, but also includes 33 per cent of the global arable land used to produce feed for livestock.

The capacity of marine and coastal ecosystems to sustain seafood production and consumption is rarely considered in economic development policies.

De Leo et al. (2014) conducted an in-depth analysis of the marine ecological footprint of Italian Mediterranean fisheries. The positioning of Italian fisheries shows a level of sustainability close to the threshold value, highlighting the importance of absolute indicators in providing rough estimates about human dependence on ecological systems. It also identifies the relevance of those indicators in ensuring a high level of precision and accuracy in quantifying the human activity impact on the environment.

Seafood consumption per person in Hong Kong is 3.6 times that of the global average. Local seafood production in Hong Kong cannot maintain pace with the city's demand. Much of this seafood is produced unsustainably, harming marine ecosystems, and even causing globally significant impacts, such as illegal fishing of threatened species.

In Hong Kong, the WWF has specific regional solutions aimed at supporting consumers who buy sustainable seafood. Individuals, businesses, and government all have a role to play in supporting the consumption of sustainable seafood in Hong Kong. The WWF-Hong Kong Seafood Guide provides practical advice on the environmental impacts of common seafood products and ways to address them through responsible procurement policies.

From the ecological footprint to local politics and administration

As the ecological footprint has been focused for years on the international benchmarking of countries, its use in local politics requires standardization (Lewan and Simmons 2001).

Local ecological footprint accountability quantifies a city's or region's demand on natural capital and compares this demand with the amount of natural capital actually available, giving policymakers the ability to track the local resource metabolisms (Wackernagel et al. 2006).

Although the index of the ecological footprint seems insufficiently consistent for comparing municipalities internationally, it can be of practical use as a tool for local policy-making and can be applied in particular for the following three aims (Figure 24.1):

- ensuring public information, community dialogue, and engagement by means of individual footprints;
- enabling and empowering administrative action by using footprint analysis as part of impact assessments;
- structuring political guidance as a basis for strategic planning and sustainable development.

For its inherent ability to synthesize complex issues in a simple and communicable way, the ecological footprint is unquestionably a powerful tool for public information (Miglietta and Pastore 2010). The accessibility of individual footprint calculators, via the Internet, is a particularly effective means of promotion and public awareness fundamental for the political purpose of environmental information.

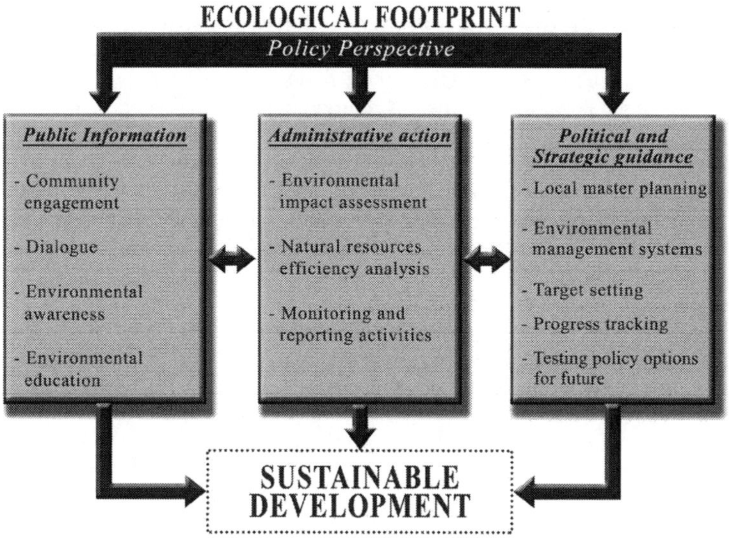

Figure 24.1
From the ecological footprint to sustainable development – A framework for policymakers

A significant example is the interactive lifestyle calculator available for Londoners on the website of the City Limits project, in which citizens can verify if they are on track for a sustainable London in 2050, choose their reduction targets for 2020, and obtain some proposals for concrete activities (i.e. food diet composition, transportation mode, renewable energy installation) to achieve the expected reduction (IWMEB 2002).

Individual ecological footprint calculators could also be introduced in the education system, which would associate this important awareness-raising tool with the specificities of the local social context and monitor the positive effects of these measures, for which we have few documented examples to date (Aall and Thorsen Norland 2005).

The ecological footprint might also be relevant in administrative processes and specific projects, because of its usefulness in evaluating environmental impacts, energy use, land-use management, and natural resource efficiency. For example, it could be useful to employ the data envelopment analysis approach, a non-parametric statistical method used to assess the relative efficiency of decision-making units, already illustrated in Chapter 21.

The ecological footprint assessment is useful in strategic planning processes, such as municipal master-planning or annual reports. It can guide environmental management systems, offering a framework for gathering and organizing data, setting targets and tracking progress, helping environmental reporting activities, and instructing strategic decision-making for regional economic development (Holden 2004; Wackernagel et al. 2006).

Conclusions

The development of environmental policies, as well as other public policies, increasingly involves the participation of various categories of actor, including political, administrative, and economic agents, multi-stakeholders, and experts.

Internal factors, such as ideas, interests, rationality, available resources, strategic considerations and external factors – that is, the political culture and institutions, scientific knowledge, economics and geopolitics, technology, information and expertise available – can all affect the behaviour of these actors and, consequently, the logic underlying the environmental policy-making process.

Many consumers, when making decisions on what to buy and what not to buy, now consider the embedded environmental impact of the particular product. Policymakers recognize the desire of people to include environmental variables in their decisions and create policies that, not only help people to do this, but also lead to economic growth.

The use of footprinting, while introducing new issues into the regional environmental policy agenda, should be accompanied by political goals to obtain results that could affect the political and social context. The first goal for policymakers could be to reduce the ecological footprint over time, but other environmental issues and targets should also be considered.

There is a need to retain flexibility in this process. The drive for international benchmarking may not always give the best results at subnational and local levels, although it remains essential where the export and import of goods and materials are occurring.

References

Aall, C., and Thorsen Norland, I. (2005) 'The use of the ecological footprint in local politics and administration: Results and implications from Norway', *Local Environment*, 10(2): 159–72.

Chambers, N., Simmons, C., and Wackernagel, M. (2014) *Sharing Nature's Interest: Ecological Footprints as an Indicator of Sustainability*, London: Routledge.

De Leo, F., Miglietta, P.P., and Pavlinović, S. (2014) 'Marine ecological footprint of Italian Mediterranean fisheries', *Sustainability*, 6(11): 7482–95.

Global Footprint Network (2007) Toward a Preferred Future: Understanding Calgary's Ecological Footprint 2007. Available at: www.footprintnetwork.org/images/article_uploads/City_of_Calgary_Footprint.pdf (accessed 22 September 2015).

Holden, E. (2004) 'Ecological footprints and sustainable urban form', *Journal of Housing & the Built Environment*, 19(1): 91–109.

IWMEB (2002) City Limits: A Resource Flow and Ecological Footprint Analysis of Greater London. Available at: www.citylimitslondon.com/ (accessed 22 September 2015).

Lewan, L., and Simmons, C. (2001) The Use of Ecological Footprint and Biocapacity Analyses as Sustainability Indicators for Sub-national Geographical Areas: A Recommended Way Forward. Milan, Italy: European Common Indicators Project.

Miglietta, P.P., and Pastore, S. (2010) 'The ecological footprint: Methodologies and applications', *Quality – Access the Success*, 11(116): 1150–6.

ProSus (2002) The Ecological Footprint of the City of Oslo – Results and Proposals for the Use of the Ecological Footprint in Local Environmental Policy. Available at: www.prosus.org/prosusFTP/prosusrep/publications/prosusrep2002_11.pdf (accessed 22 September 2015).

Rees, W.E. (1996) 'Revisiting carrying capacity: Area-based indicators of sustainability', *Population & Environment*, 17(3): 195–215.

Scotti, M., Bondavalli, C., and Bodini, A. (2009) 'Ecological footprint as a tool for local sustainability: The municipality of Piacenza (Italy) as a case study', *Environmental Impact Assessment Review*, 29(1): 39–50.

Wackernagel, M., and Rees, W. (1998) *Our Ecological Footprint: Reducing Human Impact on the Earth* (No 9), Gabriola Island, BC: New Society.

Wackernagel, M., and Silverstein, J. (2000) 'Big things first: Focusing on the scale imperative with the ecological footprint', *Ecological Economics*, 32(3): 391–4.

Wackernagel, M., Kitzes, J., Moran, D., Goldfinger, S., and Thomas, M. (2006) 'The ecological footprint of cities and regions: Comparing resource availability with resource demand', *Environment & Urbanization*, 18(1): 103–12.

WWF (2011) Hong Kong Ecological Footprint Report 2010. Paths to a Sustainable Future. Available at: http://awsassets.wwfhk.panda.org/downloads/hong_kong_ecological_footprint_report_2010.pdf (accessed 22 September 2015).

WWF (2012) The Ecological Footprints of São Paulo – State and Capital and the Footprint Family. Available at: www.footprintnetwork.org/images/article_uploads/Sao_paulo_ecological_footprint_2012.pdf (accessed 22 September 2015).

25

ECONOMIC ANALYSIS OF REGIONAL SUSTAINABILITY – HUNGARY

Klara Szita Tóthné and Judit Roncz

Introduction

The objectives of sustainable development are now a central part of most regional strategies, and it has become important to find appropriate methods to attain these objectives. The integration of an environmental dimension into planning is based on some form of life cycle assessment (LCA), and similar techniques can also be applied to achieve economic and social sustainability. Today, more and more economic and econometric models use life cycle approaches to evaluate pathways to regional sustainability. Life cycle costing (LCC) is a new element of LCA that is especially useful for making choices between project alternatives that, despite fulfilling desired performance requirements, will have different cost implications over their life cycles. Such techniques can highlight the comparative eco-efficiency of various alternatives, making economic analysis an effective tool for measuring the path to regional sustainability.

Measuring regional sustainable development

Regional sustainable development and its relation to LCA and the sustainable product chain were introduced by Balkau et al.in Chapters 2 of this book, and life cycle cost analysis and social life cycle analysis were discussed by Mazijn in Chapter 14. In this chapter, we examine the options for assessing the economic dimension of regional sustainability, keeping in mind the linkages with social and environmental priorities.

The concept of life cycle in economics is founded on the cyclic character of the micro-economy and innovation and has come into common use through the work of Schumpeter (1939). Originally, it was interpreted in relation to products, but was later expanded to technologies and even to organisations, with an increasing focus on the strategic activities of companies, their investments, and the evolution in mission and long-term objectives. The concept itself appeared in environmental economics about 20 years ago, bringing together three important prerequisites for sustainability:

1. limits on growth and the need to minimise the use of non-renewable resources;
2. minimisation of the use of energy and water;
3. reduction in pollutants and avoidance of the use of hazardous materials in production.

Furthermore, sustainable products need to have a longer useful life, with some form of recycling at the end of it. It is also important that sustainable products be widely available at a competitive price (Karlsson and Luttropp 2006). Environmentally friendly products need a pricing mechanism (skimming, price discrimination, mixed pricing) that depends on a combination of consumers' environmental consciousness and willingness to pay, as well as on products' competitive position vis-à-vis other products. In the case of price discrimination, companies segment markets according to consumers' willingness to pay and position products at different prices. If consumers' environmental consciousness is low, a company will use a mixed calculation and may include part of the additional costs in the price of its other products.

The development of environmentally friendly products is greatly assisted by techniques such as LCA that can evaluate parameters such as embedded energy or toxic materials (see also Chapter 17, on eco-design). Reliable metrics for the evaluation of such parameters will allow optimisation of a product's performance with respect to the social, economic, and environmental dimensions of sustainability. There is also a need to determine how these factors interact internally, to avoid burdens being shifted from one priority area to another.

Below, we introduce both commonly used and newly developed methodologies to identify the best approaches to sustainable development from an economic point of view. They serve as a methodological basis for our model, based on three characteristic features: whether it is an econometric model, built upon the LCA approach, or a more indicator-based statistical analysis.

Overview of economic models of sustainable development

Although econometric models are often applied to sustainable development, most seek to forecast only the evolution of economic growth. The scope of these models is limited to measurement of energy and consumption of materials, making them mainly suitable for linking economic development and environmental impact. The PANTA REI model is an ecologically extended version of the sector econometric simulation and forecasting model. Sustainable Europe is a system dynamics model based on physical accounting of resources, with embodied energy calculated as energy cost (Bockermann et al. 2000). The REEIO model incorporates more environmental impact factors relevant to regional sustainable development (Brettel 2003).

Indicator-based statistical models apply selected structural criteria for a core set of indicators to assess the sustainability of a given path. Numerous papers on regional sustainability have been published in the past decade. The EPSILON project has benchmarked sustainable development in European regions relying on the DPSIR framework (Blanc et al. 2005).

LCA-based approaches for sustainability modelling have appeared in recent decades. IO-LCA considers the dynamic movement of material and energy flows linked to economic data. LCA provides an opportunity to assess different levels of economic performance projected into a complex econometric analysis (Szita Tóthné 2007). This kind of application also appears in other models, such as the EVR (Vogtländer et al. 2009), CALCAS (Klöpffer 2008), and EIO-LCA (Hovráth et al. 1998) and REIO- LCA models (Cicas et al. 2007).

LCA was adopted for the evaluation of sustainable energy systems by exergy cost. Cumulative energy consumption during a product's life cycle is defined as the sum of exergy demand for all resources. It estimates the present sustainability paths with an ecoinvent-based LCA of energy transformation/system routes; the cumulative energy demand is expressed as the full equation of exergy costs (Rodríguez et al. 2010).

The eco-cost/value ratio (EVR) model relates to the value chain of the ecological product. Added value and added cost are determined for each production step. The ecological product chain is expressed in terms of money, the so-called eco-cost. These costs are estimated on the basis of technical measures to prevent pollution and resource depletion. The method is used to evaluate products with the same function but of different quality. The cost is added to LCA – for example, to develop government strategies linking the consumer and producer – as the so-called economic allocation indicator (Vogtländer et al. 2009).

CALCAS is a sustainability model of LCA. According to Klöpffer (2008), life cycle sustainability assessment is the sum of three elements: environmental LCA, LCC, and social life cycle assessment. Here, the technological system is approached from three different standpoints.

Our suggested model to analyse regional sustainability

Our model (see Figure 25.1) is based partly on LCA and partly on mutual sectoral relationships. We have retained the input–output model and combined it with LCA models that involve economic and social elements. In developing the concept, we used several models as starting points, selecting their advantages to build a measurable, simplified, transparent, and coherent model specific to our region (NUT2 level, an underdeveloped area that was formerly a heavy industrial centre). The model applies the LCA in a matrix structure. The environmental impacts are aggregated to an environmental performance index, and the economic and social pillars are coupled with trend and scenario analysis.

The regional environmental impacts are aggregated in an environmental performance index (REI) and are given a matrix structure:

$$REI = I_n S I_k^T \qquad (25.1)$$

where I_n is an n-dimension unit row vector, and I_k^T is a k-dimension unit column vector. The regional environmental performance is the sum of the elements of matrix S, representing the single sector environmental burdens from pollutants. Matrix S can be defined as:

Figure 25.1
Structure of the economic model

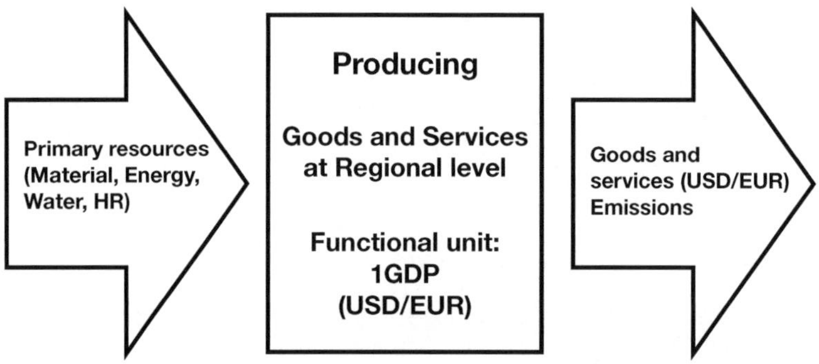

where the rows represent the individual sectors (i = 1, . . ., n) in the region's economy, and the columns represent the measured pollutants (j = 1, . . . , k). The S_{ij} element of Matrix S is the sector i's environmental burden caused by pollutant j. More specifically, a given S_{ij} element is the aggregated environmental burden of the dominant companies from the j pollutant in the environmental burdens of sector i.

The economic and social impacts are measured by scenario and trend analysis. The system boundary is the region's geographical (administrative) border, and the functional unit is the GDP produced in the region. The data for the model are generated from regional material flows and sector-specific production statistics.

Application of LCA to regional level requires (as a standardised use of LCA):

- setting of system boundaries;
- determination of the functional units;
- determination of data quality requirements;
- data collection;
- analysis of the economic background and balance of sector relationships;
- analysis of environmental impacts and identification of the main pollutants;
- effect analysis; and
- reporting.

Our model does not integrate the region's overall economic activity, but focuses on sectors and subsectors, according to:

- where the impact is most significant: that is, sectors that generate the most pollutants (80 per cent of the total emitted; environmental dimension);
- those sectors that produce the highest value added (economic dimension);
- highest employment, layoffs, fluctuation, and number of employees with scientific degrees: these are also considered as factors in our analysis (social dimension).

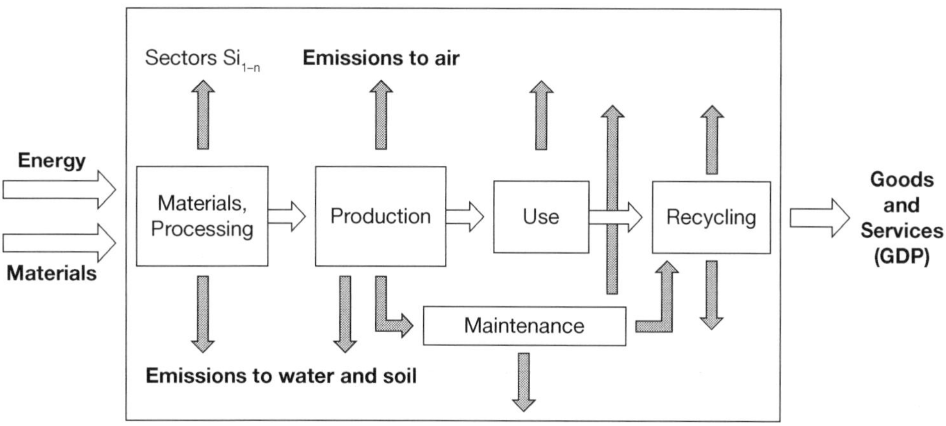

Figure 25.2
Structure of the sectoral assessment

Sector selection is based on analysis of statistical data and material flow analysis. The analysis uses the typical internal structure of the companies, as well as monetary and quantitative indicators of materials and processes. We established economic indicators (ECON) on the basis of the EVR model and the selected regional indicators from the long-term strategic goals of the region in a sustainable development strategy. The same applies to social indicators. The results are shown in a three-dimensional ENV, ECON, SI coordinate system.

To further develop this model, results were converted into a single indicator to gain regional sustainability indices based on the min–max statistical method, sector to sector:

$$I_{ECONS1} = \text{sum } i = (1, \ldots, k) \, [(X_i - X_{imin})/(X_{imax} - X_{imin})] \quad (25.3)$$

where:

I_{ECONS1} = the ECON indicator index of sector S_1

X_i = the observed (i) economic indicator in the sector (i = 1, . . ., n)

X_{imin} = the minimum value of the observed economic indicator or 0

X_{imax} = the maximum value of the observed economic indicator, equating with strategic target value.

When $X_{imax} < X_{imin}$, we use the inverse form $(1 - X_{imin})$.

The regional ECON indicator index is:

$$RSI_{ECON,S} = \text{sum } (i = 1 - n) \, [I_{ECON}/n] \quad (25.4)$$

Figure 25.3
Results of the regional sustainable assessment

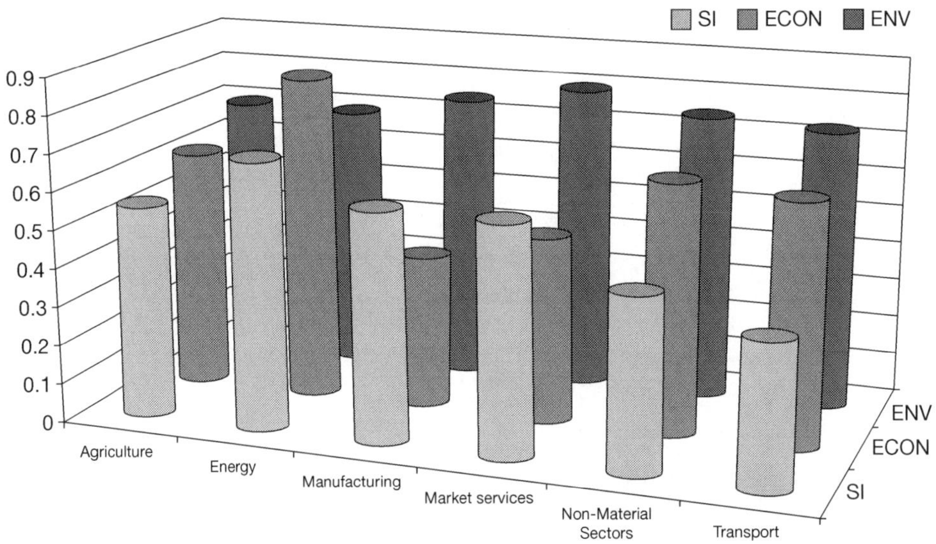

The integrated indicators for measuring the economic analysis are as follows:

- gross value added by sectors and by capita employed in the sector;
- corporate income tax;
- GDP per unit of energy use;
- consumption of electricity from renewable energy sources;
- personal income;
- R&D expenditure.

Major uncertainties of the methodology

Issues related to the approach selection are as follows:

- Interpretation of what we mean by sustainability determines the selected indicator set. How do we simplify the complexity of the analysis?
- Substitutability of pillars: Which weak or strong sustainability concept should we follow, and what conversion rate should we apply to quantify the overall sustainability value?

Issues related to indicator and statistical data selection are as follows:

- Should we use a core set of indicators or new indicators?
- Reliability of statistical data, including the material and energy flows: Data on energy flows are not perfect, but high quality is essential for sustainability evaluation.
- Lack of data: As far as we know, a new material flow accounting process is being developed, but, to date, we have not encountered such time series.
- Territorial borders and their overlap (in case of regions): Environmental agency territories are different from one another.

Issues related to the evaluation criteria are as follows:

- how we define something as sustainable or unsustainable;
- scale and intervals;
- defined limits;
- changes from year to year;
- influence of EU directives/targets;
- translation for policy decision-makers.

Conclusions

All of the models investigated considered the different dilemmas of sustainability measurement. The proposed regional sustainable LCA evaluation methodology determines the data needs and territorial and sector delineation of the research field. It analyses sectors' production, employment, and waste production rates. The energy production, industry (chemical and machine engineering), waste treatment, and mining sectors were prominent in terms of GHG potential/GVA. Given the more limited extent of employment in mining and waste management, the industry, construction, and electricity production sectors should be given priority in the analysis. This led us to a specific subsector analysis and helped us to characterise the further steps needed to improve the model and sustainability assessment.

References

Blanc, I., Friot, D., Margni, M., and Jolliet, O. (2005) 'Assessing regional sustainability with the EPSILON Project', 2nd International Congress with Innovation Fair, *Sustainable Management in Action*, SMIA 05, September, Geneva, p. 9.

Bockermann, A., Meyer, B., Omann, I., and Spangenberg, J.H. (2000) *Modelling Sustainability with PANTA RHEI and SuE*. Project on Modelling Sustainable Europe financed by EU DG XII, implemented by Wuppertal Institute, Wuppertal, and the Universities of Edinburgh, Tamper, and Madrid, pp. 1–31.

Brettell, S. (2003) Econometric Modelling in the Evaluation of Regional Sustainable Development, Cambridge Econometrics, pp. 1–9.

Cicas, G., Hendrickson, C.T., Horvath, A., and Matthews, H. (2007) 'A regional version of a US economic input-output life cycle assessment model', *International Journal of Life Cycle Assessment*, 12(6): 365–72.

Horvath, A., Hendrickson, C., Joshi, S., and Lave, L. (1998) 'Economic input-output models for environmental life-cycle assessment', *Policy Analysis*, 32(7): 184A–191A.

Karlsson, R., and Luttropp, C. (2006) 'EcoDesign: What's happening? An overview of the subject area of ecodesign and of the papers', Special Issue, *Journal of Cleaner Production*, 14: 1291–8.

Klöpffer, W. (2008) 'Life cycle sustainability assessment of products', *International Journal of Life Cycle Assessment* 13 (2): 89–94.

Rodríguez, M.A.R., De Ruyck, J., RoqueDíaz, P., Verma, V.K., and Bram, S. (2010) 'An LCA based indicator for evaluation of alternative energy routes', *Applied Energy*, 88: 630–35.

Schumpeter, J.A. (1939) *Business Cycles. A Theoretical, Historical and Statistical Analysis of the Capitalist Process*, New York, Toronto, London: McGraw-Hill, p. 461.

Szita Tóthné, K. (2007) *Life Cycle Analysis from Theory to Practice* (*Életciklus-elemzés az elméleti alapoktól a gyakorlatig*), Habilitációs tézisek Miskolci Egyetem, p. 136.

Vogtländer, J., Baetens, B., Bijma, A., Brandjes, E., Lindeijer, E., Segers, M., Witte, F., Brezet, J.C., and Hendriks, C.F. (2009) *LCA-based Assessment of Sustainability: The Eco-costs/Value Ratio (EVR)*, Sustainable Design Series, Delft, Netherlands: Delft University of Technology, pp. 1–11. Available at https://www.researchgate.net/publication/258220975_LCA-based_assessment_of_sustainability_the_Eco-costs Value_Ratio_EVR.

26

CIRCULAR ECONOMY AND REGIONAL IMPLICATIONS

Fritz Balkau

Introduction

International programmes on sustainable development regularly invoke circular economy or cradle-to-grave resource flows in their policies and recommendations. The concept of circular economy has many variants, depending strongly on the standpoint of the interested agencies and sectors. It is based on the idea that a managed closed loop of materials is a more appropriate and sustainable way of organising our activities than the current wasteful linear flow that takes natural resources, transforms them into products, then dumps the residues and end-of-life products into the environment. The circular economy concept has particular relevance for a region, as most products brought into the region have to be assimilated there, with many eventually ending up as waste to be disposed of locally. Conversely, secondary raw resources derived from this waste can be a potential source of income.

The idea of a circular economy has gradually evolved and consolidated in the last 20 years. It remains an overarching concept, rather than a methodology, and has strong synergies with other holistic ideas, such as *Cradle to Cradle* (Braungart and McDonough 2002), industrial ecology (see Chapter 20), eco-design (see Chapter 17), and similar approaches, without, however, completely replacing them. Industrial ecology is based on similar foundations of a cyclical materials flow of wastes and raw materials, but without always taking a society-wide approach to closing the loop. For its implementation, circular economy relies on a variety of subsidiary concepts, such as functional economy and extended producer responsibility (EPR n.d.), as well as LCA and other approaches, as introduced in Parts II and III. The Ellen MacArthur Foundation (n.d.) has developed the concept into a sophisticated action programme. Overall, the concept has considerable value; nevertheless, there remains a problem of identity – is circular economy a concept for materials optimisation (i.e. a part of the manufacturing sector), or does it fit better into the culture of waste management? A further issue is its undefined relationship with more holistic policies of resource efficiency, dematerialisation, product–service systems, and eco-design, all of which aim to drastically reduce the waste flows in our community.

Circular economy in action

Despite a plethora of encouraging government statements, success in implementation of circular economy ideas has still only been modest so far. However, the movement is accelerating. One example is the recent initiative of the European Union to launch its Circular Economy Package (see Box 26.1), which includes action both on waste and on products (production and consumption; EU 2015).

BOX 26.1 EU CIRCULAR ECONOMY PACKAGE

Main contents

Introduction

1 Production:
- product design
- production process

2 Consumption

3 Waste management

4 Secondary raw materials

5 Priority areas:
- plastics
- food waste
- critical raw materials
- construction and demolition
- biomass and bio-products

6 Innovation, investment, horizontal measures

7 Monitoring progress.

Conclusion

See: http://ec.europa.eu/environment/circular-economy/index_en.htm

Individual countries are also gradually moving from rhetoric to action. China and France have both put circular economy concepts into national law in different ways (China 2009; Legifrance 2015). Many regions around the world have taken their own independent measures. These range from product standards concerning recyclability, encouraging eco-design, and eco-park efforts (see Chapters 17 and 34), to mandating actual recovery and recycling schemes. For example, in France, it is now a punishable offence to deliberately design a product to have a short service life (see Box 26.2). Meanwhile, Japan is promoting the so-called 3R approach (reduction, reuse, and recycling) at national level and among countries within its nearer geographical region (3R Regional Forum n.d.). Many of these programmes are an offshoot of more integrated approaches to waste management, but selected product controls are gradually becoming more established also. Universal product controls still seem to be some way off.

BOX 26.2 CIRCULAR ECONOMY AND THE FRENCH LAW ON ENERGY TRANSITION

Circular economy is now frequently mentioned in national and regional policy documents and action plans. The French law of August 2015, on 'Energy Transition for Green Growth', outlines a number of measures that have a life-chain *cradle-to-grave* reach. These include:

1 Product level interventions such as:
   - punishable offence to create planned obsolescence in products;
   - bans on certain products such as plastic bags;
   - specifying recycled content in paper purchased by government agencies;
   - offence to discriminate against recycled or secondary raw materials;
   - obligation for mechanics to propose option of recovered/recycled spare parts.
2 Support for system changes towards industrial ecology, product service systems.
3 Waste management measures such as:
   - sorting, separation, and recycling;
   - waste reduction, for example, in surplus unsold food at supermarkets;
   - establishment of waste depots for some sectors.

Measures for implementation include legal obligations, taxes, promotion, co-financing, agreements, and conventions.

See: www.legifrance.gouv.fr/affichTexte.do?cidTexte=JORFTEXT000031044385&categorieLien=id

French territorial regions are obliged to implement this law, but have also independently taken initiatives to move the agenda forwards, as we discuss below.

The Île de France region (i.e. the greater Paris area) has included circular economy in its Local Agenda 21 action plan, under the subheading of 'waste'. This is to be implemented through measures that optimise materials and energy flows, eco-design of products, repair and reuse, recycling, sharing and pooling, relocation of production, among others. The heavily industrialised Rhône-Alpes region has adopted an industrial ecology path to circular economy. It has called for public and industry proposals to use and exchange industrial flows, to further adapt current industrial processes, and to share business services, information, and equipment. Unlike Île de France, there is no mention of intervention at product level. Other regions in France have taken similar approaches, with local variations adapted to their specific needs (Associations des régions de France 2015). The example of Aquitaine has already been mentioned in Chapter 5.

It is not only governmental bodies and waste companies that are picking up on these ideas. The management consultancy firm KPMG includes circular economy as one of its recommended strategies for future cities (KPMG 2015). It proposes a variety of upstream and downstream implementation measures that include taxes on products and wastes, deposit schemes, recycling, urban mining, and waste-to-energy initiatives.

Waste managers and circular economy

With such a strong policy focus on doing something useful with waste, it is not surprising that the waste industry itself is talking more about circular economy and looking harder at the options. As a result, we are seeing numerous private-sector initiatives that translate the concept into practice. But, whereas a holistic approach to circular economy would consider the entire materials cycle, from cradle to cradle, for waste managers, it usually means a variety of recovery, recycling, and reuse actions to bring the materials embedded in discarded products back into use.

Waste managers have traditionally been reluctant to become involved in upstream issues of dematerialisation, cleaner production, and reducing materials consumption, preferring instead to focus on their strength in recovery, recycling, and transformation. But, new thinking in the waste industry is gradually leading to a broader view. The International Solid Waste Association (ISWA) has already raised the issue of holistic waste management in its 2002 strategy paper '10 year perspective', and has recently adopted a more proactive stance to promote this concept among its members (ISWA n.d.). There is more industry talk about, publications on, and projects about, waste management from a circular, cradle-to-cradle perspective.

It remains to be seen how far traditional waste and resource managers can venture into a genuine circular revolution – one that goes beyond recycling towards closed-loop materials re-entry, reuse, and remanufacture as well. Much depends on local circumstances, but also on the policy framework within which the industry operates. In response to various economic and societal factors, some of the major companies are now reconfiguring their service propositions and experimenting with more ambitious material optimisation and asset utilisation strategies.

Although these initiatives are welcome, the mainstreaming of circular economy ideas will nevertheless depend on whether the waste industry at large is ultimately able to make the change from being a waste handler to becoming a materials manufacturer or a resource company as well.

While evolution of government policy and its implementation are still inching along, some private operators are seizing the immediate business opportunities (Perella 2015). Veolia UK is moving from being a service supplier to also become a resource producer to deliver what it calls 'circular customer solutions'. The company is 'mining' wasted resources such as solid waste, effluent, and surplus heat to manufacture green products and green energy. It is working with a leading consumer goods company to reuse its waste detergent in a truck wash facility. Another company, Suez Environnement (n.d.), regards product take-back as part of an integrated circular economy service managed jointly with the customer. Other examples of moving towards a functional economy concern a product–service system where a manufacturer of refuse vehicles leases rather than sells the vehicles. This keeps the vehicles running longer than normal, because the manufacturer provides more regular maintenance than the owner.

The concept is also taking root in developing countries, such as South Africa and India (Banyan n.d.). The Recycling and Economic Development Initiative of South Africa (REDISA n.d.) relies on a government levy on new tyres that is then used to partly finance recovery and conversion to other products.

Regions and the circular economy

Irrespective of immediate operational details, the global evolution to a circular economy has major implications (and opportunities) for regions. The increasing volumes and complexity of wastes that are discarded will remain a major disposal challenge for some time yet, but just as important are the additional burdens on local industries arising from resource shortages and high operational costs from current wasteful manufacturing methods. A local circular economy can allow regions to escape from this double burden, improving their sustainability outcomes at the same time. It is no wonder that many regions have espoused circular economy ambitions in their strategic plans.

All this is no longer just theory. We see many regions now looking beyond waste management towards development advantages coming from circular economy action. The economic opportunities from business innovation in regional materials flows are a strong factor, in some cases even perhaps the main driver for change. The French region of Hauts de France has used circular economy as the foundation of its third industrial revolution, aiming to develop new products and applications from wastes and surplus materials. We recall that the region of Aquitaine has developed a regional circular economy roadmap focusing on recycling, industrial ecology, and functional economy, including eco-design and life cycle management. We also saw that the region of Île de France is implementing circular economy actions, such as optimisation of materials flows, encouraging eco-design of products, establishing repair and recycling facilities, as well as supporting equipment pooling and borrowing (see Chapters 20 and 17).

Conclusions

It has now been demonstrated that circular economy models based on a more holistic approach to materials management can be of significant benefit to regions, resulting in a significantly reduced waste burden and generating local value chains based on a more dynamic, proactive local industry.

However, although the concept is attractive, the actual transition to a circular economy has not always been mapped out in a clear way by its many advocates and promoters. But, in all cases, a new view of waste management is evolving, combining traditional waste handling and recycling with a closer control over materials supply chains and products' downstream user interface. A comprehensive view of the life cycle of raw materials and of the embedded materials in the waste stream can be provided by some of the assessment methods already described in this book. A growing number of regions are now using circular economy concepts as a way of reinvigorating an ageing industrial infrastructure. In the meantime, entrepreneurs in the waste industry are taking their own path, capturing that part of the circular materials cycle to which they have access and from which they can generate an immediate income. Their business model is based on the creation of an interdependent local web of suppliers, collectors, transporters, and remanufacturers, as well as the associated marketing and sales functions. Individually and collectively, they will operate within the new policy framework for wastes and for products (not to mention consumption patterns) being put in place by nations as well as by regions. The methodologies of life cycle assessment, supply-chain management, sustainable purchasing, product–service systems, and other life cycle approaches that increase resource efficiency, as explained elsewhere in this book, will be essential to underpin such initiatives. Ideas and examples will also be borrowed from traditional concepts of industrial ecology, eco-efficiency, and cleaner production.

It also remains to be seen how the circular economy model based on waste recovery can co-exist with, or even incorporate, materials efficiency initiatives of eco-design, dematerialisation, and product–service systems that are intended to lead us to a more sustainable future.

References

3R Regional Forum (n.d.) Available at: www.env.go.jp/recycle/3r/en (accessed 3 August 2016).

Associations des régions de France (2015) French Regions Positive Achievements and Prospects, Committing, Cooperating and Succeeding Together against Climate Change. Available at: www.arf.assoc.fr (accessed 3 August 2016).

Braungart, M., and McDonough, W. (2002) *Cradle to Cradle: Remaking the Way We Make Things*, New York: North Point Press.

Banyan (n.d.) Banyan Sustainable Waste Management Solutions. Available at: www.banyannation.com/ (accessed 3 August 2016).

China (2009) Circular Economy Promotion Law of the People's Republic of China. Available at: www.fdi.gov.cn/1800000121_39_597_0_7.html (accessed 3 August 2016).

Ecological Transition, (n.d.) Available at: http://ecologicaltransition.blogspot.fr/2010/03/what-is-ecological-transition.html and www.neujobs.eu/publications/socio-ecological-transition-and-employment-implications/socio-ecological-transitions-de (accessed 3 August 2016).

Ellen MacArthur Foundation (n.d.) Refuse Vehicles Solutions. Available at: www.ellenmacarthurfoundation.org/case_studies/refuse-vehicle-solutions (accessed 3 August 2016).

EPR (n.d.) Available at: www.oecd.org/env/tools-evaluation/extendedproducerresponsibility.htm (accessed 3 August 2016).

EU (2015) Circular Economy Package. Available at http://ec.europa.eu/environment/circular-economy/index_en.htm (accessed 3 August 2016).

ISWA (n.d.) Available at: www.iswa.org (accessed 3 August 2016).

KPMG (2015) Future Cities: The Long-Term View. Available at: http://kpmg.co.uk/creategraphics/2015/10_2015/CRT049860/Future%20Cities%20-%20The%20long%20term%20view_CRT49860.pdf (accessed 3 August 2016).

Legifrance (2015) Loi no 2015–992 du 17 août 2015 relative à la transition énergétique pour la croissance verte. Available at: www.legifrance.gouv.fr/affichTexte.do?cidTexte=JORFTEXT000031044385&categorieLien=id (accessed 3 August 2016).

Pauli, G. (1998) *Upsizing: The Road to Zero Emissions: More Jobs, More Income, and No Pollution*, Saltaire, UK: Greenleaf Publishing. See also www.zeri.org (accessed 3 August 2016).

Perella, M. (2015) 'Loopy ideas – Closing the circle one project at a time', *Waste Management World*, September–October: 26–7.

REDISA (n.d.) Recycling and Economic Development Initiative of South Africa. Available at: www.redisa.org.za (accessed 3 August 2016).

Suez Environnement (n.d.) Recycling and Waste Recovery. Available at: www.suez-environnement.com/businesses/recycling-and-reclaiming-waste (accessed 3 August 2016).

Veolia (n.d.) Circular Economy. Available at: www.veolia.co.uk/circulareconomy (accessed 3 August 2016).

27

LIFE CYCLE MANAGEMENT FOR REGIONAL DEVELOPMENT IN FRANCE

Example of building sector

Naeem Adibi, Marie Darul, Vanessa Pasquet, Christelle Demaretz, and Christian Traisnel

Introduction and context

Sustainable buildings are the fabric of future lifestyles. This subject is of increasing interest to regions, as building stocks are usually significant, and their age suggests early replacement or refurbishment, but also because a vibrant modern building industry can be an important factor in industrial revitalisation and in the export of materials and expertise.

Buildings also impinge on a region's sustainability vision. Although the building industry and the general public understand the importance of energy efficiency, other environmental issues surrounding buildings and materials remain less well known. The building sector is responsible for more than a third of global resource consumption, including 12 per cent of fresh-water use, and increasingly incorporates a host of chemical products and substrates. Globally, buildings generate 40 per cent by volume of solid waste, most of it in construction and demolition (United Nations Environment Programme 2011).

In order to achieve more sustainable buildings, it is necessary to consider all sustainability issues across the whole building life cycle (from the design to construction to use and finally to the end of life). In this sense, key actors' understanding of the built environment as a complex system that metabolises materials and energy needs to be increased. The notion of building systems and their materials flows is important to eventually being able to optimise the life cycle management (LCM) of sustainability criteria by various combinations of actors in the supply chains, but also by regional planning authorities.

A number of chapters in this book go into more detail about methods of life cycle analysis, materials flow analysis, input–output tables, and systems analysis. LCM is an ideal approach to act on this information from a system perspective and help in making decisions based on scientific facts relating to sustainable buildings and the built environment (Sonnemann and Margni 2015; Forsberg and von Malmborg 2004).

LCM in the building sector

LCM has been defined by the UNEP/SETAC working group as, 'an integrated framework of concepts, techniques and procedures to address environmental, economic, technological and social aspects of products and organisations to achieve continuous environmental improvement from a life cycle perspective' (Hunkeler et al. 2001). LCM is based on the understanding that a collective approach to addressing sustainability issues is more effective than dispersed individual actions, many of which simply shift impacts to another part of the life chain. LCM is closely linked with the notions of supply chains, consumer responsibility, and pollution prevention. When LCM is used to improve sustainability outcomes, two factors need to be considered: first, the extent of the value chain of a product or project that is regarded as important, and, second, a clear identification of the sustainability impacts and outcomes that need to be addressed through LCM efforts.

LCM is an efficient approach to create shared values, reinforcing links between different stakeholders who subsequently act in concert. This concept is the basis of the [avniR] platform in northern France, as shown in Figure 27.1.

The [avniR] platform links professional organisations, business federations, and chambers of commerce to explore and implement ways to help their businesses mainstream LCM. The initiative is supported by the Regional Council of Hauts de France and by the French environment agency, ADEME.

Figure 27.1
Life cycle management

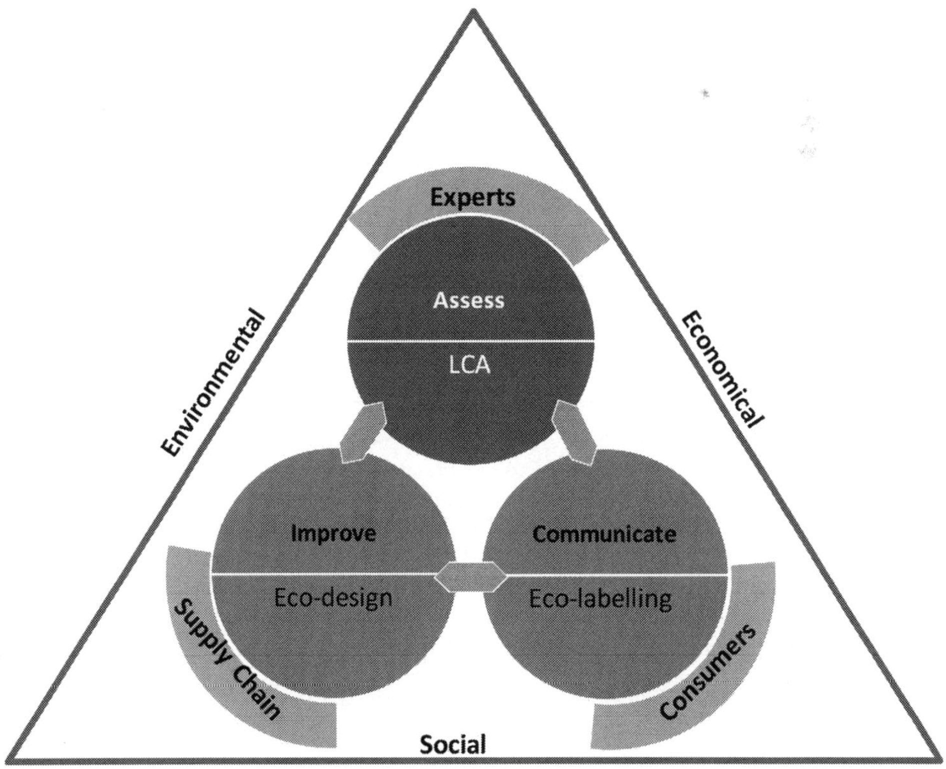

When applied to the building sector, the LCM strategy is based on the following concepts:

1 LCM capacity should be built into existing support organisations, so that businesses receive advice and tools from organisations that they already know and trust;
2 tools and actions to support LCM integration need to be adapted to different sectors to make them as relevant as possible to SMEs; and
3 training and research capacity, also within universities and public research centres, needs to be developed to respond to needs of businesses.

Steps for mainsteaming LCM

Various sectorial approaches launched in 2012 had the ambitious objective of moving beyond applying LCA studies in an isolated fashion, to mainstreaming LCM across entire sectors at the regional level, including the numerous SMEs present. The first four sectors in 2012 (textile, seafood, packaging, and mechanical) enabled the development of a common methodology to establish strategic action plans to mainstream LCM into business, education, and research organisations. A second wave of sectors (food, wood, building, and recycling/end of life) undertook the process in 2013, and the renewable energies sector was developing its strategy in 2015 (Adibi et al. 2015). The success of these initiatives has encouraged its extension beyond northern France through the LCiP project (Pasquet et al. 2015).

The building sector follows the same general approach for developing its LCM action plan as other sectors in the northern France region. Five major steps are followed within the process (Adibi et al. 2015):

1 benchmarking of existing tools and initiatives, panorama of building sector (preliminary study);
2 sector maturity assessment;
3 needs identification;
4 action plan;
5 implementation.

The preliminary study (Steps 1–3) help to identify the following:

- Tools and approaches to underpin LCM within the sector: The building sector seems to be relatively mature in LCA, leading to extensive standards for buildings and building product environmental declaration and use of environmental labels. Although the major manufacturers already use LCA, the use of LCM to co-ordinate with upstream and downstream partners is far from being mainstreamed, especially among SMEs.
- Building hot spots: The planning and design phases, as shown in Table 27.1, are the most decisive steps in which to integrate LCA in building construction and performance. They directly influence the other steps in a building life cycle. Impacts from product manufacturing, the building use phase, and the ease of replacing/dismantling all depend on the design of the building. Figure 27.2 shows the building design/construction steps, environmental hot spots, and the impact cause-and-effect chain.
- Different stakeholders and their influence on the implementation of LCM: As shown in Table 27.1, the project owners, the project manager/design

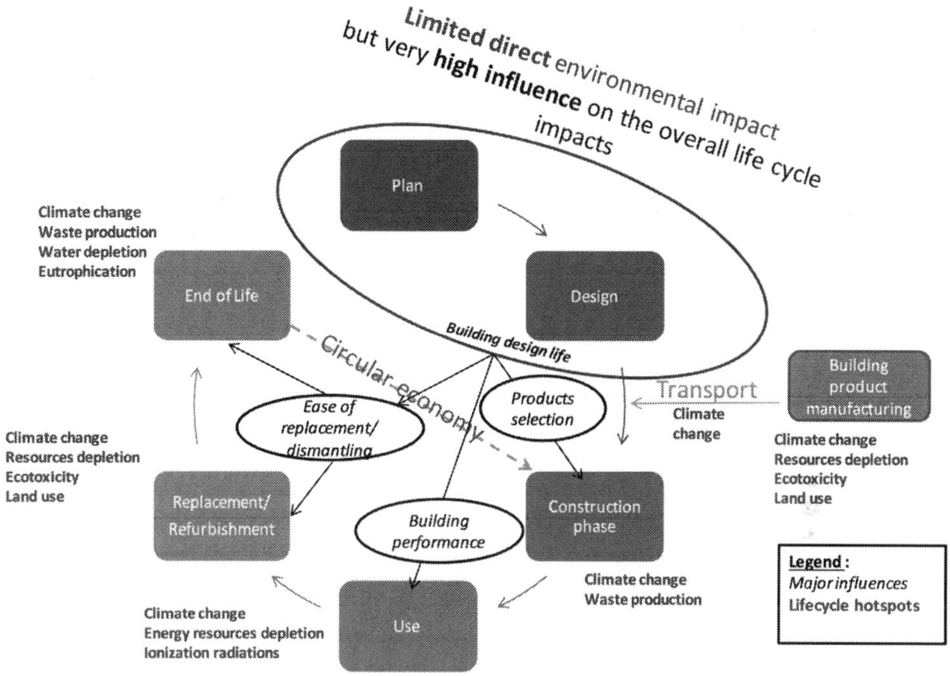

Figure 27.2
Building life cycle, environmental hot spots, and the cause-and-effect chain

Table 27.1 Stakeholder influence in the building value chain

Level of influence for the different stakeholders in the building value chain (from 8, the most influential, to 1, the least influential)	
1 Project owner	8
2 Project manager, engineering, designers	8
3 Building companies	4
4 Building products manufacturers	7
5 Building products suppliers	2
6 Building operator	3
7a End-of-life actors, collectors	7
7b End-of-life actors (recycling/treatment, etc.)	2

team, and the product manufacturers are the most important stakeholders of the building value chain to achieve an effective level of LCM over the building life cycle.

Project owners have a major influence in encouraging the use of eco-friendly products or, indeed, to have an eco-efficient building. The project manager and design team can use LCA methods to choose energy performance and building products, and the manufacturer can limit the environmental impact of his products through eco-design.

The level of maturity of the different stakeholders was established: seventy regional building actors were interviewed in order to identify their ability to integrate LCM in their activity. At the same time, their needs and motivations were identified using the survey. A SWOT matrix was developed as a baseline for the action plan.

Action plan

The action plan is developed through a collaborative approach involving regional stakeholders (regional council, specialised agencies, building federations, architects, etc.), based on their role and their influence.

The plan includes a variety of action types:

- awareness raising (seminars, conference);
- monitoring;
- individual coaching;
- working groups;
- support/development of tools; and
- demonstration projects.

This collective approach links the supply and demand sides of the building cycle and thus of the LCM exercise at the same time. The approach has encouraged the regional building sector to move from its focus on pure assessment to now actively supporting companies in their efforts to improve their sustainability performance across the value chain.

Implementation and results

Demand-side partners (policymakers, project owners, etc.) within the study are increasing their interaction with clients and public authorities to improve awareness about life cycle approaches and the benefits of using an LCM framework for decision-making.

On the supply side, architects, engineering offices, craftsmen and manufacturers are provided with support such as design tools. There is also training of consultants and support organisations to help them promote the LCM approach among value chain partners. The project has a 'knowledge hub' for academics and consultants to share their experience and improve their capacity in LCA, eco-design, and LCM communication.

Several concrete actions under the implementation plan are shown below.

Réhafutur, a showcase rehabilitated building

A large number of houses will require refurbishment in northern France in the coming years. To demonstrate some of the techniques that could be used to improve the sustainability performance of rehabilitated buildings, cd2e and Ekwation cluster undertook the refurbishment of a building called Réhafutur, once an old mining engineer's house and now owned by Maison et Cité. The house was refurbished through a life cycle perspective, using different eco-materials to achieve a very energy-efficient building. Documents developed during the project provide educational materials. The building will be used as a place for training, awareness raising, and education for different regional actors and the general public.

Figure 27.3 Building Réhafutur I, left side

An LCA study is evaluating hot spots and the benefits linked to the major choices inherent in the project (performance and products). The methodology is available to social housing owners interested in using LCM to improve their strategy for the refurbishment of similar buildings.

Simplified communication seminars and flyers on building sector LCM

Activities are underway to assist building actors (especially project owners and product manufacturers) to learn how LCM may help them to be more cost-effective by developing closer links with supply chains. A reduction of the environmental footprint is always well appreciated, but is not strong enough on its own to lead to an eco-design process.

The project organises discussions with professionals to make environmental aspects more relevant for them. As well as explaining the environmental context, it is demonstrated how LCM can reduce environmental impacts and costs at the same time. LCA tools (software, database, etc.), together with available cd2e support, are provided as a way to enhance their eco-design capacities.

More general, awareness-raising seminars are organised on different LCM-related topics, supported by technical documentation and educational flyers for different stakeholders. A simplified flyer called '*Carnets de l'ACV*' informs different users about the advantages of an LCM approach for the sector and motivates them to include additional criteria in their building construction and renovation.

Improving building products and strengthening the link with potential users

The Interreg IVB, CAP'EM project aims to develop the use of 'eco-materials' in north-west Europe. The project promotes alternative materials from abundant, renewable, and recycled sources, thus encouraging local, short supply chains. The materials concerned have at least similar technical performance to conventional products and the lowest health impact achievable. Cellular concrete blocks, lightweight pumice concrete blocks, insulation made from recycled textiles or paper, passive windows made with local wood, unfired clay brick, and acoustic underlayer made with flax fibres are among the materials included in the CAP'EM project. As leading partner in this project, cd2e makes regional industries aware of the options. It also undertakes cradle-to-gate assessments for six regional products as a service to SMEs.

The results are also made available in the CAPEM Compass tool, a free web interface aiming to help building designers to select building products according to technical, but also environmental, criteria.

Training of regional academics and consultants

So that the support available to the sector is broadened, training on LCA and LCM is given to regional academics and consultants. After participation in selected projects, they are able to guide the further implementation of LCM initiatives for various companies and within different projects.

Documentary database on the use of LCA in the building sector

A documentary database of existing LCA/LCM resources is available through the cd2e–[avniR] platform. Users may also contribute to the improvement of this database at the same time.

In the framework of the current sectorial approaches in northern France, synergies are also made with other sectors. These share their experiences in addition to developing transverse links and co-operating on common issues. Thus, the recycling sector collaborates with the building sector to improve the options for buildings' end of life in the region. The wood industry is promoting the integration of sustainable wood products in buildings.

Measurement of LCM integration in the building sector

The [avniR] platform is currently developing a set of indicators to monitor the sectorial LCM implementation, as outlined above.

Among others, the effectiveness of events, training, studies, research, and collaborative projects have been identified, and their efficiency has been evaluated. The leaders of the sectorial studies meet and share their experiences in parallel workshops, three times a year.

This part is still evolving, as the indicators are developed in parallel with actual implementation actions and studies.

Conclusion

Although the programme is still ongoing (in 2015), there are already encouraging outcomes from the [avniR] action plans in the north of France. A variety of actors from the supply and demand sides of the building sector are engaged in the implementation of the action plan. An LCM approach is now systematically adopted in the majority of regional development projects in northern France. As well as the involvement of primary actors in the building value chain, the emphasis on building capacity among associated stakeholders in research, education, consulting, and communication has been vital in implementation, backed by adequate documentation and assessment services.

The work has shown that active participation by regional government and by various authorities and institutions in the region is a key success factor in fostering an LCM approach. Active participation by and support of industry federations and trade associations are also vital, as smaller companies – a strong feature of the building sector – rely heavily on their advice and guidance. Above all, a dedicated project team is essential to start the process off and then to co-ordinate the various actions, until all the partners operate with confidence in such a life cycle regime.

Acknowledgements

The authors would like to express their sincere thanks to the Hauts de France Regional Council and the ADEME for their valuable support and feedback.

References

Adibi, N., Bricout, J., Demaretz, C., Bogaert, C., Duffrène, A.V., Orlans, P., Beutin, C., and Meillier, J. (2012) 'Introducing a sectorial framework to better evaluate life cycle thinking maturity', *Proceedings of the 2nd International [avniR] Annual Conference 2012*, Lille: Avenir.

Adibi, N., Pasquet, V., Roy, A, Salamon, A., Bricout, J., Beutin, C., Renault, Q., Darul, M., Callens, F.X., Haquette, M., Orlans, P., Meillier, J., Boudehenn, J., Reynaud, S., Cabaret, S., Bogaert, C., and Demaretz, C. (2015) 'Mainstreaming the use of life cycle management in small and medium sized enterprises using a sector based and regional approach', in *Life Cycle Management*, Dordrecht, Netherlands: Springer, pp. 79–90.

Forsberg, A., and von Malmborg, F. (2004) 'Tools for environmental assessment of the built environment', *Building & Environment*, 39(2): 223–8.

Hunkeler, D., Rebitzer, G., Jensen, A.A., and Margni, M. (2001) 'Life cycle management: Bridging the gap between science and application', *International Journal of Life Cycle Assessment*, 6(6): 384–90.

Pasquet, V., Roy, A., Adibi, N., Coppee, S., Echard, P., and Rocha, C. (2015) 'Life cycle in practice – Capacity building aiming European SME's', *Conference: 17th European Roundtable on Sustainable Consumption and Production, Conference Proceedings*, Portoro: ERSCP, p. 11.

Sonnemann, G., and Margni, M. (2015) *Life Cycle Management*, Dordrecht: Springer Netherlands.

United Nations Environment Programme (2011) *Buildings Investing in Energy and Resource Efficiency*, Paris: UNEP.

28

FROM PRODUCT LCAS TO TERRITORIAL LCAs

Methodological principles

Eléonore Loiseau and Philippe Roux

Introduction

Life Cycle Assessment (LCA) is widely used all around the world to assess the environmental impacts of products or services (see Chapter 10). Initially designed as a 'product-oriented' approach, recent proposals have been made to broaden its scope of analysis towards meso-level systems, including municipalities (Guinée et al. 2011). Additionally, Hellweg and Milà i Canals (2014) showed the expanding nature of LCA and the development of country LCA in order to guide decisions on national sustainable consumption and production policy setting. The main purpose of LCA is to identify pollution transfers, both in terms of environmental impacts (i.e. multicriteria analysis) and life cycle stages. Applying this LCA characteristic at territorial scale will allow a supply chain perspective to be added to spatial planning and will link micro-scale policies with policies at macro scale (which is essential to identify pollution transfers between territories). Broadening the LCA scope (from a product approach towards assessing all activities at territorial scale) allows an important need to be filled for quantitative environmental indicators to support policy-making at different scales, ranging from municipalities and regions to nations. However, the implementation of LCA in these territorial systems is not straightforward, as a territory, a municipality, or a region is a complex system, encompassing many production and consumption activities, which associates a group of stakeholders and the geographical space that these stakeholders use, develop, and manage (Moine 2006). Consequently, Loiseau et al. (2013) proposed to adapt the standardized LCA framework to assess such systems and developed a new typology of LCA, called 'territorial LCA'. Territorial LCA aims at assessing all consumption and production activities that take place within a given territory in order to support decision-making, both in terms of territorial diagnosis and land planning scenario comparisons. The purpose of this chapter is to describe the main features of the territorial LCA approach and to focus on its operational side. Lastly, some development perspectives will be discussed.

Methodology

Territorial LCA maintains the main principles as 'product-oriented' LCA, that is, this is a life cycle, multicriteria, functional approach. However, the LCA standardized framework developed by ISO (2006a, 2006b) has been completely overhauled in order to be adapted to the study of territories (Loiseau et al. 2013). The territorial LCA framework (see Figure 28.1) is based on four main

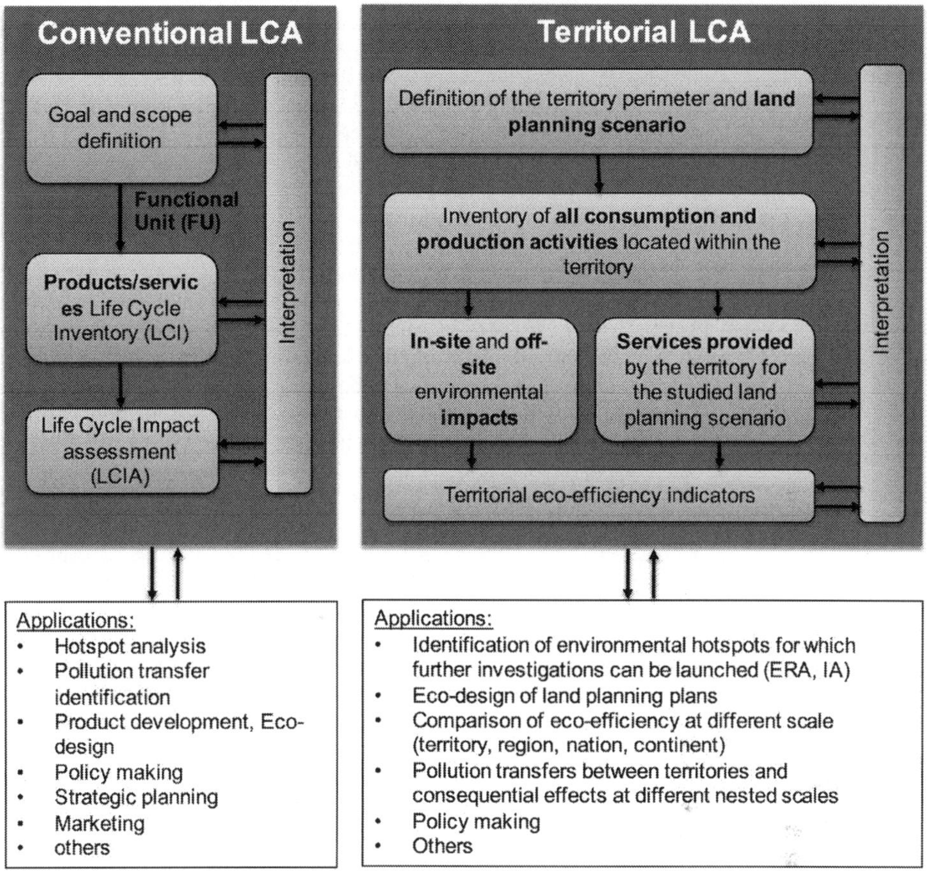

Figure 28.1
Comparison of conventional LCAs with the territorial LCA framework

adaptations: (1) functional unit definition, (2) boundary selection, (3) data collection, and (4) provision of indictors relevant for local decision-making.

In 'conventional LCA', the studied functional unit is the LCA starting point and the output is a vector of environmental impacts. In territorial LCA, the starting point is the definition of a geographical area associated with a land planning scenario (i.e. the baseline scenario or a prospective scenario defined by local stakeholders). As a territory is multifunctional (e.g. economic, environmental, societal land-use functions; Pérez-soba et al. 2008), no one main function can be assigned to this system. In addition, the land-use functions (i.e. services provided such as wealth generation, access to facilities, or hosting a population) are the outputs of a given land planning scenario implemented in a geographic area. Consequently, two main categories of indicator are quantified in territorial LCA: (1) a vector of environmental impacts and (2) a vector of land-use functions. These two outputs can be used to assess the territory eco-efficiency, as defined by Seppälä et al. (2005), by dividing the functions per impacts (e.g. 'wealth generation' or 'inhabitant housed' per kilogram of CO_2 emitted).

The second adaptation is about boundary selection. The issue behind boundary selection is the issue of the territorial responsibility in terms of environmental impacts. The first approach to adopt can be to consider all upstream and downstream economic flows related to consumption activities, as well as all upstream economic flows related to production activities located in the studied territory. Eder and Narodoslawsky (1999) described this perspective as 'total responsibility'. Although, this conservative approach means that there is a double counting at a global scale, it gives a complete picture of the activities that take place in the territory and allows identification of all impacts induced by territorial activities and that occur in the rest of the world.

In order to perform an exhaustive life cycle inventory (LCI) of all consumption and production activities in an operational way, a procedure combining bottom–up and top–down approaches was developed (see Figure 28.2). For all these activities, data collection can be performed in two steps. The first step is to gather data on the types of goods and service consumed or produced and the amounts annually consumed or produced. In order to reduce the workload induced by data collection, the activity inventory is based on public data available in many countries, such as local, regional, or national statistics, agricultural statistics, or administrative documents on companies or local communities (e.g. financial balance sheets). Once this activity inventory has been performed, the second step is to connect the activity inventory data to existing LCIs. Activity inventory data can also be used to quantify certain land-use functions (e.g. the number of community-based facilities or the number of main and secondary residences). Two main approaches can be used for collecting data. The first one, the bottom–up approach, relies on local data for activity inventory. Only data on specific territorial activities are used, such as the annual output of a given plant, or the amount of household waste collected annually. Then, these data can be linked to LCI based on process LCA by using existing databases such as Ecoinvent.[1] Process LCA is based on the modelling of physical flows for a specific process. On the other hand, activity inventory data can be collected using top–down approaches, which are based on national or regional data. This means that these data have to be downscaled to the territory studied. Different downscaling factors can be used, such as the number of inhabitants, household expenditure, or the number of workers (Browne et al. 2011). In addition, top–down approach relies on environmental input–output (EIO) LCA to compute the LCI of production and consumption activities (see Chapter 8). EIO tables account for the quantities of pollutants emitted and of natural resources consumed for each monetary unit of production for a given sector (Suh and Huppes 2005). The choice between the two approaches depends on the most appropriate modelling for a given activity (i.e. process LCA for a specific activity such as the modelling of a wastewater treatment plant, or EIO LCA for an entire sector of activity such as the food industry) and on data availability. Moreover, the type of data gathered during the first step will also influence the type of LCI modelling. On one hand, data in the form of physical flows will be connected to process LCA. On the other hand, EIO-LCA is used for monetary flows.

Finally, the last adaptation is about the need to provide relevant indicators for local decision-makers. All impacts could be quantified through the use of existing life cycle impact assessment (LCIA) methods, such as ReCiPe (Goedkoop et al. 2009) or Impact 2002+ (Jolliet et al. 2003). However, it is recommended that a clear distinction is made between on-site and off-site environmental impacts when the results are shown. On-site impacts are incurred by consumption of resources or emissions of pollutants that take place directly in the studied territory. In contrast, off-site impacts are due to environmental flows induced by territorial activities, but those that occur in other territories. This distinction

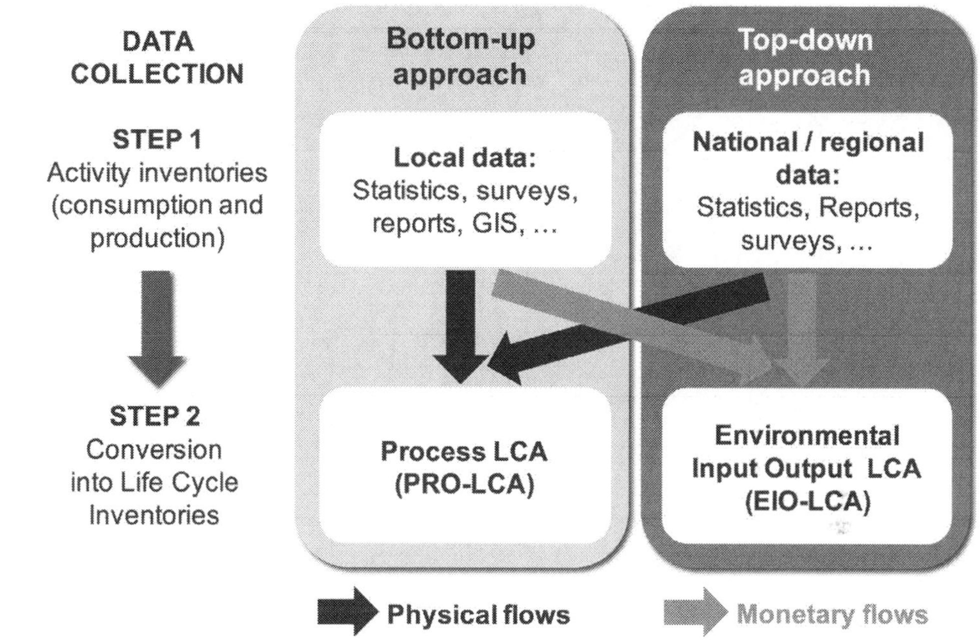

Figure 28.2
A procedure combining bottom–up and top–down approaches to perform the LCI of all consumption and production activities

is essential to emphasize the pollution transfers that take place between territories, usually from developed countries to developing countries (Muradian et al. 2002).

Discussion and perspectives

The territorial LCA approach has been implemented in a real case study in order to perform an environmental diagnosis of a French Mediterranean territory (Loiseau et al. 2014). The time required to perform the whole study was reasonable (2–3 months), and a lot of data were publicly available. In the absence of data, strategies based on proxy were used to fill the gaps. This first implementation highlights the important role that territorial LCA can play to provide useful indicators for decision-making on spatial planning. First, territorial LCA supplies quantifying indicators on both environmental impacts and services provided by different land-use planning scenarios, thus assessing their eco-efficiency. Second, this approach identifies pollution transfers between impact categories and territories. Third, territorial LCA allows the main environmental issues within a territory to be determined, that is, the environmental hot spots and the main activities contributing to these issues. From this diagnosis, it is, hence, possible to define the main drivers of the increase in the territory environmental performances and the environmental aspects that need to be monitored. Consequently, territorial LCA appears to be a very comprehensive approach for a first screening of a territory as a whole. However, a more accurate analysis is still required to improve the robustness of the approach, both in terms of LCI and LCIA. The fact that data collection is based on a combination of bottom–up and top–down approaches decreases the consistency of the modelling. Indeed, process LCA and EIO-LCA rely on inherent differences, such as the existence of cut-offs in process LCA that could lead to an underestimation of the overall LCI (Mongelli et al. 2005). Results should, hence, be interpreted with caution, and methodological developments are still required to improve the LCI with, for instance, the use of hybrid approaches (Suh et al. 2004). Then, until now, existing LCIA methods have not considered spatial variability (Potting and Hauschild 2006), and, regardless of where the resource withdrawal or the emission occurs, the impacts on the environment will be the same. Spatial variability should at least be considered for environmental hot spots, in order to decrease their uncertainties (see Chapter 29). This improvement should be made easier with the upcoming release of regional-based LCIA approaches such as Impact World+.

All these potential improvements could enhance the robustness of territorial LCA approaches. Yet, they will be pointless without simultaneous work on the best practices for disseminating model results for decision-makers and the way to involve stakeholders at an early stage of the assessment.

Note

1 Ecoinvent is the world's leading supplier of consistent and transparent life cycle inventory (LCI): see www.ecoinvent.org/

References

Browne, D., O'Regan, B., and Moles, R. (2011) 'Material flow accounting in an Irish city-region 1992–2002', *Journal of Cleaner Production*, 19(9–10): 967–76.

Eder, P., and Narodoslawsky, M. (1999) 'What environmental pressures are a region's industries responsible for? A method of analysis with descriptive indices and input–output models', *Ecological Economics*, 29: 359–74.

Goedkoop, M., Heijungs, R., Huijbregts, M., Schryver, A.D., Struijs, J., and van Zelm, R. (2009) ReCiPe 2008 Report I: Characterisation, The Hague, Netherlands: Ministry of Housing, Spatial Planning and the Environment.

Guinée, J.B., Heijungs, R., Huppes, G., Zamagni, A., Masoni, P., Buonamici, R., Ekvall, T., and Rydberg, T. (2011) 'Life cycle assessment: Past, present, and future', *Environmental Science & Technology*, 45(1): 90–6.

Hellweg, S., and Milà i Canals, L. (2014) 'Emerging approaches, challenges and opportunities in life cycle assessment', *Science*, 344(6188): 1109–13.

ISO (2006a) *ISO 14040: Environmental Management – Life Cycle Assessment – Principles and Framework*, Geneva: ISO.

ISO (2006b) *ISO 14044: Environmental Management – Life Cycle Assessment – Requirements and Guidelines*, Geneva: ISO.

Jolliet, O., Margni, M., Charles, R., Humbert, S., Payet, J., Rebitzer, G., and Rosenbaum, R. (2003) 'IMPACT 2002 +: A new life cycle impact assessment methodology', *The International Journal of Life Cycle Assessment*, 8(6): 324–30.

Loiseau, E., Roux, P., Junqua, G., Maurel, P., and Bellon-Maurel, V. (2013) 'Adapting the LCA framework to environmental assessment in land planning', *The International Journal of Life Cycle Assessment*, 18(8): 1533–48.

Loiseau, E., Roux, P., Junqua, G., Maurel, P., and Bellon-Maurel, V. (2014) 'Implementation of an adapted LCA framework to environmental assessment of a territory: Important learning points from a French Mediterranean case study', *Journal of Cleaner Production*, 80(1): 17–29.

Moine, A. (2006) 'Le territoire comme un système complexe: Un concept opératoire pour l'aménagement et la géographie', *L'Espace géographique*, 35(2): 115–32.

Mongelli, I., Suh, S., and Huppes, G. (2005) 'A structure comparison of two approaches to LCA inventory data, based on the MIET and ETH databases', *The International Journal of Life Cycle Assessment*, 10(5): 317–24.

Muradian, R., O'Connor, M., and Martinez-Alier, J. (2002) 'Embodied pollution in trade: Estimating the "environmental load displacement" of industrialised countries', *Ecological Economics*, 41: 51–67.

Pérez-soba, M., Petit, S., Jones, L., Bertrand, N., Briquel, V., Omodei-Zorini, L., Contini, C., Helming, K., Farrington, J.H., Tinacci Mossello, M., Wascher, D., Kienast, F., and de Groot, R. (2008) 'Land use functions – A multifunctionality approach to assess the impact of land use changes on land use sustainability', in *Sustainability Impact Assessment of Land Use Changes*, eds K. Helming, M. Pérez-Soba, and P. Tabbush, Berlin: Springer, pp. 375–404.

Potting, J., and Hauschild, M.Z. (2006) 'Spatial differentiation in life cycle impact assessment: A decade of method development to increase the environmental realism of LCIA', *Journal of Cleaner Production*, 1(1): 11–13.

Seppälä, J., Melanen, M., Mäenpää, I., Koskela, S., Tenhunen, J., and Hiltune, M.R. (2005) 'How can the eco-efficiency of a region be measured and monitored?', *Journal of Industrial Ecology*, 9(4): 117–30.

Suh, S., and Huppes, G. (2005) 'Methods for Life Cycle Inventory of a product', *Journal of Cleaner Production*, 13(7): 687–97.

Suh, S., Lenzen, M., Treloar, G.J., Hondo, H., Horvath, A., Huppes, G., Jolliet, O., Klann, U., Krewitt, W., Moriguchi, Y., Munksgaard, J., and Norris, G. (2004) 'System boundary selection in life-cycle inventories using hybrid approaches', *Environmental Science & Technology*, 38(3): 657–64.

29

INTEGRATED LIFE CYCLE AND RISK ASSESSMENT OF HUMAN HEALTH IMPACTS IN CATALONIA

Guido Sonnemann, Marta Schuhmacher, and Francesc Castells

Introduction

Public authorities in a region need to make decisions on the way waste is managed. This includes choices among various treatment options, such as incineration with different filter systems, and among several transports options, such as routes through more- or less-populated areas. To identify the best options with regard to human health impacts, the use of a life cycle approach that integrates life cycle and risk assessment was used in Catalonia, Spain; this involved the generation of an important set of data needed for the regional waste management policy context.

The integrated life cycle and risk assessment method applied in this chapter, in line with Sonnemann et al. (2004), is based on the life cycle assessment (LCA) methodology described in Chapter 10. It provides a case study for the spatial modelling in LCA, as introduced in Chapter 13, but does not look at the cost implications, as done in life cycle costing (Chapter 33). LCA application in regional waste management is also the topic of Chapter 30, where, however, the focus is on the life cycle evaluation of landfill gas recovery and biogas generation in a region of Chile.

An integrated method is applied to assess the human health impacts of two situations and one scenario concerning a municipal solid waste incinerator (MSWI) process chain in Catalonia:

- Situation 1 is the basis for all comparisons. It describes the SIRUSA incinerator in Tarrgaona in its state in 1996. It includes the treatment of scrap in Madrid, ash treatment at the company TRISA in Constantí, and the disposal of the treated ash and the slag in Pierola and Reus, respectively. In the ash treatment plant, cement is mixed with the ash. The cement is transported from Sta Margarida i els Monjos.
- Situation 2 is the situation after the installation of an advanced acid gas removal system in 1997. Besides the features in Situation 1, there is additional transport of CaO for acid gas treatment. CaO is transported from Sta Margarida to the waste incinerator in Tarragona. The data for Situations 1 and 2 are taken from the existing LCA study (Servei de Tecnologia Química 1998).

- As a third alternative, a scenario has been created based on the modular model for the waste incineration process (Hagelüken 2001). Besides the features in Situation 2, a fictitious DeNox system is installed to eliminate NOx emissions. Ammonia for this purpose is transported from the industrial area of Tarragona to the incinerator. It has to be stressed that this is a scenario with an unvalidated MSWI model. However, it is introduced into this case study in order to discuss possible results of a further improvement of the flue gas treatment. This alternative is called Scenario 3.

The function of the MSWI is its service of treating the municipal solid waste of Tarragona. In this context, the functional units presented here are 1 year of treated waste (which equals 153,467 t). However, 1 t treated waste or 1 kWh of produced electricity could also be chosen.

Applying dominance analysis in the LCA-based methodology of environmental damage estimations for industrial process chains developed by Sonnemann (2002), the focus in this study was restricted to the transport and the incineration process. For comparison of the integrated life cycle and risk assessment method results with midpoint indicators, two life cycle impact assessment methods with the human toxicity potential (HTP) are used: CML (Heijungs et al. 1992) and EDIP (Hauschild and Wenzel 1998). For the weighting of the human health impacts, the physical impact parameter 'population exposure' (PE) and the 'disability-adjusted life year' (DALY) concept are applied. As only the area of protection 'human health' is taken into account here, no further aggregation is carried out.

Life cycle inventory analysis

In order to get the life cycle inventory (LCI) data for this abbreviated waste incinerator process chain, the volume flow (Nm^3/a) of the waste incineration plant is multiplied by the unit emission files for the waste incinerator in the different cases, and the overall transport (tkm) is multiplied by the unit emissions files for the transport. These unit emission files provide the emissions of different air pollutants per Nm^3 and per tkm, respectively. The above-mentioned data and the unit emission files for SIRUSA and the transport can be found in Sonnemann (2002).

The most obvious result is that the overall tkm increases significantly from Situation 1, through Situation 2, to Scenario 3. By this, the additional transport processes somehow compensate for the decrease in the emissions due to the improvement of flue gas treatment. How far the transport compensates for the damages avoided through more powerful gas treatment has to be shown by applying the integrated life cycle and risk assessment method.

Life cycle impact assessment with spatial differentiation

The HTP according to the CML method (Heijungs et al. 1992) and the EDIP method (Hauschild and Wenzel, 1998) is calculated (the nomenclature of the latter is changed). In addition, site-dependent impact factors have been determined for the pollutants that are most relevant for the impacts on human health.

In addition to the assignation of the corresponding region to each process, in this application, the transport process has to be differentiated with respect to the districts crossed. The size of the region, here district level, determines the number of regions, by which a mobile process – that is, transport – is represented in the eco-technology matrix developed by Sonnemann (2002).

For the study, the transport related to each process taking place in each class of population density and meteorology class, as well as the distance in the provinces outside Catalonia, is calculated. The emission height for the municipal waste incinerator is 50 m, and that for the transport processes it is 5 m. That means that the product of mass and distance of transport (tkm) is differentiated with respect to the districts crossed. Using the transport distances in each class and the available tkm, it can be calculated how many tkm refer to which class of population density and meteorological conditions. The multiplication by the unit emission file of transport leads to the spatially differentiated LCI, the eco-technology matrix. An example of such a spatially differentiated eco-technology matrix is presented in Table 29.1 for the former Situation 1. As can be seen, the transport is divided into eleven classes, and one class presents the MSWI, as its site is well known.

A damage-assigning matrix for DALY (Egalitarian) is calculated according to Sonnemann (2002), based on the results for the site-dependent impact indicator, expressed as PE per mass of emitted pollutant. Another damage-assigning matrix has also been established with the values of the impact indicator itself, expressed with the unit [persons $\mu g/m^3$ * a/kg]. It has to be mentioned that the impact indicators for the heavy metals arsenic (As), cadmium (Cd), and nickel (Ni), as well as for benzo[a]pyrene (BaP), are supposed to be the same as for PM2.5 (particulate matter). It is assumed that these substances are adsorbed on particles PM2.5 and, therefore, behave with the same terms of fate and exposure. However, the DALY value is specific for each substance, as the dose-response and exposure-response functions are substance-specific as well.

Table 29.1 Eco-technology matrix for the former Situation 1 (kg/a)

	MSWI	Transport (h_{source} = 5 m)										
Pollutant	IIIA[e]	IB	IIA	IIB	IIIA	IVA	IVB	HU	ZZ	SO	GU	M
As	15	0	0	0	0	0	0	0	0	0	0	0
BaP	0	1E-5	2E-5	7E-6	2E-3	4E-5	1E-4	1E-4	5E-4	8E-5	2E-4	9E-5
Cd	15	7E-5	1E-4	4E-5	9E-3	2E-4	6E-4	6E-4	2E-3	4E-4	1E-3	5E-4
PM10	20,411	6	10	3	877	19	58	54	226	39	106	44
Ni	22	0	0	0	0	0	0	0	0	0	0	0
NOx	142,333	101	158	53	13,677	296	906	838	3,522	616	1,658	684
SO_2	602,643	7	11	4	912	20	60	56	235	41	111	46
O_3[a]	0	14	22	7	1,929	42	128	118	497	87	234	96
O_3[b]	142,333	101	158	53	13,677	296	906	838	3,522	616	1,658	684
Nitrate[c]	142,333	101	158	53	13,677	296	906	838	3,522	616	1,658	684
Sulphate[d]	602,643	7	11	4	912	20	60	56	235	41	111	46

Notes: (a) as NMVOC, (b) as Nox, (c) as NOx, (d) as SO_2, (e) h_{stack} = 50 m.
IB, IIA, IB, IIIA, IVA, IVB: combined wind speed and population density classes in Catalonia, with district population densities I: agglomerated – urban, II: agglomerated – rural, III: rural – central city, and IV: rural – rural, and wind speeds A: 0–2 m/s, B: 2–3 m/s, and C: 3–4 m/s.
HU = Huesca, ZZ = Zaragoza, SO = Soria, GU = Guadalajara, M = Madrid.

Source: Sonnemann et al. 2003

The values for nitrate and sulphate do not differ either. This is owing to the fact that these are secondary pollutants for which country averages have been calculated. Only the value for the sulphate in the first row differs slightly. This is owing to the stack height of the waste incinerator (50 m stack height), which differs from that of the transport processes (5 m stack height).

For the primary pollutants, it can be said that the highest values appear in Madrid and the lowest in Huesca. These values are calculated for the transfer of impact factors to other regions. As Madrid is densely populated and Huesca is scarcely populated, in comparison with Catalonia, these results are obvious.

Results and discussion

Table 29.2 shows the damage matrix for the former Situation 1 resulting from the multiplication of the eco-technology matrix and the damage-assigning matrix. If the first column is considered, the first box (the first element of the diagonal) represents the damage for the waste incinerator. The second element in the second column represents the damage for the transport in district class IB, and so on. If one leaves the diagonal, the impact of the processes at other (fictitious) locations is shown. For example, the last element of the first column shows the damage from the waste incinerator if it were located in Madrid. Of course, it has to be admitted, however, that the impact indicator applied to this cell refers to a stack height of 5 m (transport) rather than 50 m (waste incinerator). The damage would be twice as high as it is currently. Nevertheless, this shows impressively the importance of spatial differentiation.

Next, the diagonal elements of the damage matrix are added to produce the damage profile, and then the part corresponding to transport is compared with the value of the waste incinerator. In this way, the damages due to the waste incinerator and transport can be compared.

The most obvious result of this environmental damage estimation study for industrial process chains is that the contribution of transport to the overall

Table 29.2 Damage matrix with DALY for former Situation 1 (a*a/kg)

	MSWI	Transport (h_{source} = 5 m)										
IIIA*	*36.12*	0.01	0.01	0.00	0.83	0.02	0.06	0.05	0.21	0.04	0.10	0.04
IB	55.17	*0.01*	0.02	0.01	1.34	0.03	0.09	0.08	0.35	0.06	0.16	0.07
IIA	50.97	0.01	*0.01*	0.00	1.22	0.03	0.08	0.07	0.31	0.05	0.15	0.06
IIB	42.73	0.01	0.01	*0.00*	1.01	0.02	0.07	0.06	0.26	0.05	0.12	0.05
IIIA	40.54	0.01	0.01	0.00	*0.96*	0.02	0.06	0.06	0.25	0.04	0.12	0.05
IVA	37.77	0.01	0.01	0.00	0.88	*0.02*	0.06	0.05	0.23	0.04	0.11	0.04
IVB	34.46	0.01	0.01	0.00	0.80	0.02	*0.05*	0.05	0.21	0.04	0.10	0.04
HU	29.81	0.01	0.01	0.00	0.69	0.01	0.05	*0.04*	0.18	0.03	0.08	0.03
ZZ	32.38	0.01	0.01	0.00	0.75	0.02	0.05	0.05	*0.19*	0.03	0.09	0.04
SO	29.53	0.00	0.01	0.00	0.68	0.01	0.05	0.04	0.18	*0.03*	0.08	0.03
GU	29.81	0.01	0.01	0.00	0.69	0.01	0.05	0.04	0.18	0.03	*0.08*	0.03
M	76.05	0.01	0.02	0.01	1.87	0.04	0.12	0.11	0.48	0.08	0.23	*0.09*

Note: * this row refers to h_{stack} = 50 m

Source: Sonnemann et al. 2003

damage increases significantly from Situation 1, through Situation 2, to Scenario 3. On the one hand, this is owing to the sharp decrease in the overall damage. On the other hand, the decrease in damage due to the improvement of flue gas treatment is partly compensated by the additional transport processes.

Figure 29.1 shows the results for all three cases, differentiated for waste incinerator and transport, for PE, DALY (Egalitarian), and HTP (EDIP). From the results of all situations, it can be seen that the ratio between the damage due to transport and the overall damage is in the same magnitude for all chosen indicators in this study (PE and DALY). Whether the contribution of transport to the overall result is more significant for PE or DALY in the case of each pollutant depends on the relationship between the toxicity and the mass of pollutants emitted. The more toxic a substance, the higher the increase in DALY.

Conclusions

From the results using the end-point indicators derived in this study on the one hand and the HTP on the other hand, it seems that the HTP concept underestimates the environmental importance of transport. Whereas the share of transport in the first case is 2.1 per cent/1.7 per cent for HTP (CML/EDIP), it is between 3.2 per cent and 4.2 per cent for the end-point indicators derived in this study. Whereas, in Situation 2, the share of transport for the end-point indicators reaches between 17.1 per cent and 19.6 per cent, the share for HTP is still quite small (12.0 per cent/5.8 per cent). The gap widens even more in Scenario 3: 35.5 per cent/20.7 per cent for HTP and 44.1–49.2 per cent for the end-point indicators, respectively. The reason for the differences between the results for the HTP and the end-point indicators are clear. The HTP concept considers the fate of the substances, but does not include exposure information. As the environmental impact of transport is very much dependent on the location where it takes place, the deviation from the HTP results is obvious. The results using the impact indicators of this study, however, show the limits to the ecological benefits of the further technical improvement of flue gas cleaning. Scenario 3 still indicates a clear overall reduction. However, it shows also that nearly half of the overall environmental impact is due to transport. Therefore, further technical improvement at the waste incinerator will only be carried out if there is no significant increase in transport.

The study indicates that quite a quantity of data collection and subsequent modelling is needed to apply the integrated life cycle and risk assessment method. This is the price to pay for moving towards more spatial differentiation when using life cycle approaches for regional development. However, thanks to the ever-increasing availability of data nowadays and the growth in modelling power, the application of such a method becomes more and more attractive to actors for decision-making at the regional level, be it on waste management or other choices that can impact human health and the environment. Evidently, the results provided are much closer to the actual damage level with regard to human health affected or protected by adequate measures. Therefore, the authors could imagine that, in the future, more regional decision-makers will move towards the use of life cycle impact assessment with spatial differentiation for choices in relation to public services such as waste and water management.

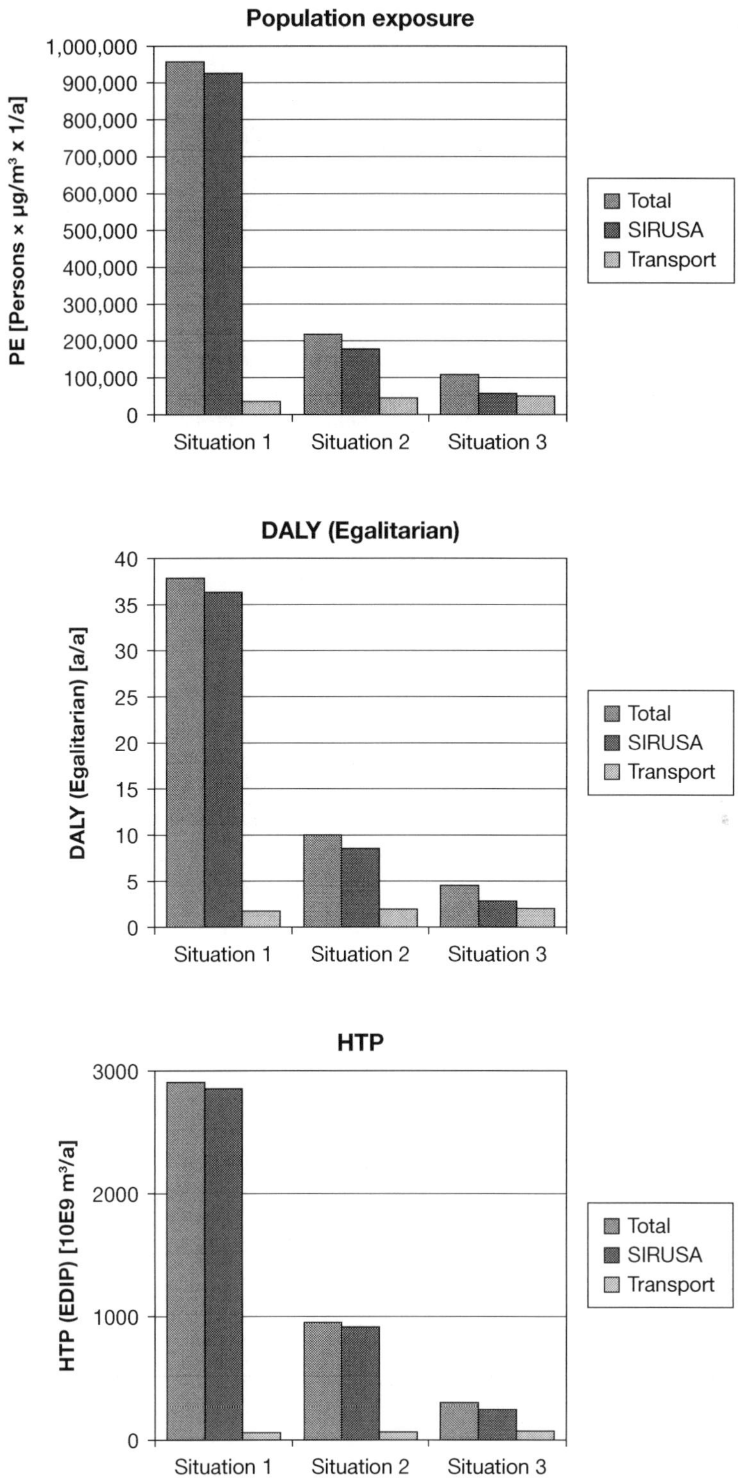

Figure 29.1
Case study results for all three situations/scenarios for PE, DALY, and HTP (EDIP), where SIRUSA stands for the MSWI process

References

Hagelüken, M. (2001) *Effects of Different Models of Municipal Solid Waste Incinerators on the Results of Life-Cycle Assessments*. Student research project, Berlin, Tarragona, Spain: TU Berlin, Universitat Rovira i Virgili.

Hauschild, M., and Wenzel, H. (1998) *Environmental Assessment of Products – Scientific Background*, vol. 2, London: Chapman & Hall.

Heijungs, R., Guinée, J.B., Huppes, G., Lankreijer, R.M., de Haes, H.A.U., and Wegener-Sleeswijk, A. (1992) Environmental Life Cycle Assessment of Products – Guide and Backgrounds. Technical Report, Leiden, Netherlands: CML, University of Leiden.

Servei de Tecnologia Química (1998) *Análisis del ciclo de vida de la electricidad producida por la planta de incineración de residuos urbanos de Tarragona*. Technical report, Tarragona, Spain: Universitat Rovira i Virgili.

Sonnemann, G. (2002) *Environmental Damage Estimations in Industrial Process Chains* (PhD thesis). Tarragona, Spain: Universitat Rovira i Virgili.

Sonnemann, G., Castells, F., and Schuhmacher, M. (2004) *Integrated Life-Cycle and Risk Assessment of Industrial Processes*, Boca Raton, FL: Lewis.

30

LCA APPLICATION IN REGIONAL WASTE MANAGEMENT IN CHILE

Alberto Bezama

Introduction

Defining the best waste management solution for municipal solid waste (MSW) is not an easy task and is definitively not a one-size-fits-all type of decision process. This challenge is even greater when it comes to decisions regarding isolated regions, such as the case of the city of Coyhaique, capital of the region of Aysén in Chilean Patagonia. At the time of this study, the region lacked an appropriate waste management infrastructure, as no sanitary landfill was operating in the region, and no formal recycling activities took place. It actually consisted of the collection of unsorted waste from households and direct disposal in an uncontrolled dumping site. This led to several social and environmental impacts, among which the most relevant considered is groundwater infiltration of the resulting leachate, emission of generated landfill gas, as well as the presence of uncontrolled sanitary vectors around the disposal site. For this reason, during the late 2000s, the construction of a sanitary landfill for the city of Coyhaique was planned, designed to be the cornerstone for future waste management concepts for the city. The sanitary landfill was conceived with a basic engineering structure to provide the necessary multi-barrier concept, to ensure appropriate waste disposal in Coyhaique and neighbouring cities (IASA 2003).

This project can be seen in the light of a regional energy system with a total installed capacity of 37.7 MW, of which about 54 per cent is generated through diesel-operated power plants, 42 per cent through hydropower, and about 4 per cent through wind-power facilities (Comisión Nacional de Energía 2013). This energy matrix proves to be mainly dependent on imported resources, which is a further problem for an already geographically isolated region.

The goal of this work was to evaluate potentially suitable waste-to-energy technological alternatives for the case study region, based on the proposed sanitary landfill project, that could support both the waste management as well as the energy generation sectors. Two major sub-goals were thus formulated:

1 to define technological alternatives that support both the energy generation and the waste management systems in Coyhaique, considering the design of the sanitary landfill as the basis scenario; and
2 to evaluate these alternatives from a life cycle perspective in order to identify the most suitable alternative for the conditions present in the city of Coyhaique and its region.

This chapter presents the steps taken to fulfil the established working goals, with the aim of discussing how the use of LCA processes can support decisions in the waste management sector, especially in cases as special as the case of Coyhaique. It is not intended to present a detailed analysis of the technical aspects of the LCA process carried out. In case of interest in this regard, please refer to Bezama et al. (2013).

Characteristics of the study region at the time of the project

The region of Aysén in Chilean Patagonia is one of the southernmost regions in Chile, far away from the country's metropolitan areas. This region is characterized by a low level of economic activity in comparison with the central regions in Chile, but with healthy, steady growth. Besides tourism, the region is sustained economically by animal husbandry, the forest industry, and by the extraction and industrialization of marine goods. The city of Coyhaique generates around 14,000 t/a MSW, with a total attended population of 45,422 inhabitants (Douglas 2010), and is thus characterized with a per capita generation of 0.84 kg d^{-1}, which is slightly high for a small city in Chile (average: 0.6 kg d^{-1} $person^{-1}$). MSW in Coyhaique is characterized by a high organic content (51 per cent in wet mass), which is lower than the average value in Chile of about 65 per cent. Further major components of the generated MSW in Coyhaique are wood (5.7 per cent), plastics (7.9 per cent), tyres and rubber (0.7 per cent), textiles (3.5 per cent), metal (3.5 per cent), glass (3.4 per cent), paper (2.6 per cent), cardboard (3.4 per cent), hazardous waste (0.6 per cent), and others (17.7 per cent). This characterization was performed during the framework project for the design of the sanitary landfill (IASA 2003). At the time of this study, the projected sanitary landfill had been approved, and thus it was considered, for the purposes of this study, as the basic scenario of the regional MSW management system.

Definition and design of technical alternatives

A conservative approach was taken for the definition of the technical alternatives, meaning that, for this study, the intention was to define the 'next step' to be taken for the enhancement of the MSW management system, rather than the design of an 'optimal waste management system', and, thus the collection and transport system was not considered in the evaluation. In this regard, please refer to the results section, as a small analysis of this decision is given. As presented in Figure 30.1, three different alternatives were modelled to assess the appropriateness of waste-to-energy technologies for the city of Coyhaique. The first alternative, Alternative 0, corresponds to the current situation, that is, it reflects the design of the management system based on the operation of the sanitary landfill designed for the city of Coyhaique. As previously commented, for the modelling of this technological alternative, the environmental impact assessment and the technical design report for the sanitary landfill were considered the basis of information. From these reports, it was possible to build the necessary mass and energy balances related to the construction, operation and closure of the sanitary landfill, including the different inputs for the operation of the sanitary landfill, such as fuel, water, and energy (IASA 2003). As for the potential landfill gas recovery from the operational landfill, the recovery system was considered to have a 60 per cent efficiency, with the remaining 40 per cent of the landfill gas considered as emitted directly (without further treatment) into the atmosphere (IASA 2003).

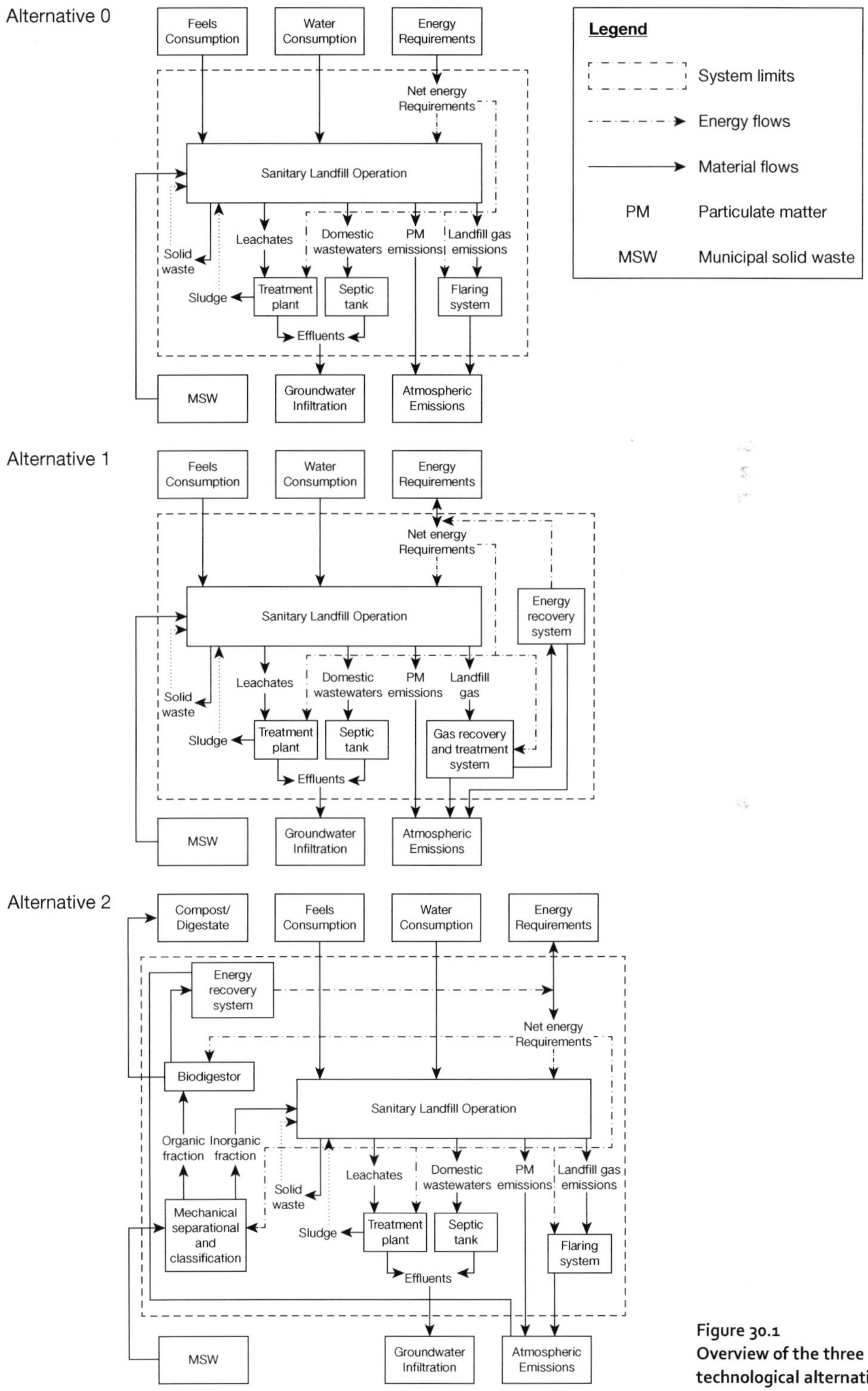

Figure 30.1
Overview of the three technological alternatives

The first technological alternative that includes the waste-to-energy concept is Alternative 1. In this concept, after the recovery of the landfill gas, a further energy generation plant was conceptually designed and modelled, with the state of the art considered when it was scaled and adapted to the case study area (Douglas 2010).

Finally, Alternative 2 contemplates the implementation of an anaerobic digestion system prior to the disposal of the MSW in the sanitary landfill. As observed in Figure 30.1, this conceptual design includes the implementation of a mechanical separation process, followed by direct disposal of the inorganic fraction and digestion of the organic fraction. The anaerobic digestion process was modelled based on similar case studies, with the process adapted to the study area conditions (Douglas 2010; Bezama et al. 2013). The decision to send the inorganic fraction directly to the sanitary landfill, without considering its recycling alternative, was taken based upon two major facts. First, there is neither market nor infrastructure in the region to recover potentially recyclable materials. And second, the remoteness of the region in terms of the distance to the country's big markets, as well as the lack of an appropriate transport infrastructure to connect the region to these markets, makes it economically unattractive to design a waste recycling system.

Life cycle assessment: Considerations

The LCA process was carried out based on the ISO 14040:2006 and ISO 14044:2006 standards. The objective of the study was to compare the three alternatives advanced for MSW management in the city of Coyhaique, with the aim to identify the one most environmentally suitable in the context of existing regional conditions. For this study, the functional unit was defined as 1 metric ton of managed MSW (in a wet basis). As this is a comparative study, the simplification suggested by Finnveden (1999) and by Muñoz and Navia (2011) was utilized in order to define the limits of the system, and thus the common stages of the three alternatives were excluded from evaluation.

As for the outputs, leachates and biogas generation were examined in the analysis. For the estimation of leachate generation, actual data and the technical design of the sanitary landfill environmental impact study (IASA 2003) were utilized. For the estimation of landfill gas generation, the prediction model LandGEM (EPA 2005) was used, which applies to the generation and recovery of landfill gas in municipal landfills. This model was selected in order to provide a conservative value of gas generation, and also to allow for variables such as rainfall in the area, deposited MSW amounts, MSW composition and the number of years of operation of the sanitary landfill. This model provided detailed information relating to conditions in the city of Coyhaique and promoted the development of a more effective assessment of biogas generation.

Through the established mass and energy balances of the three analysed alternatives, the comparative life cycle assessment evaluation was carried out, taking into consideration global warming potential (GWP), abiotic depletion potential (ADP), and ozone layer depletion potential (ODP) as impact categories, as well as required raw energy and produced energy as comparative regional-specific indicators; see Figure 30.2. For a detailed description of the LCA steps, please refer to Douglas (2010) or Bezama et al. (2013).

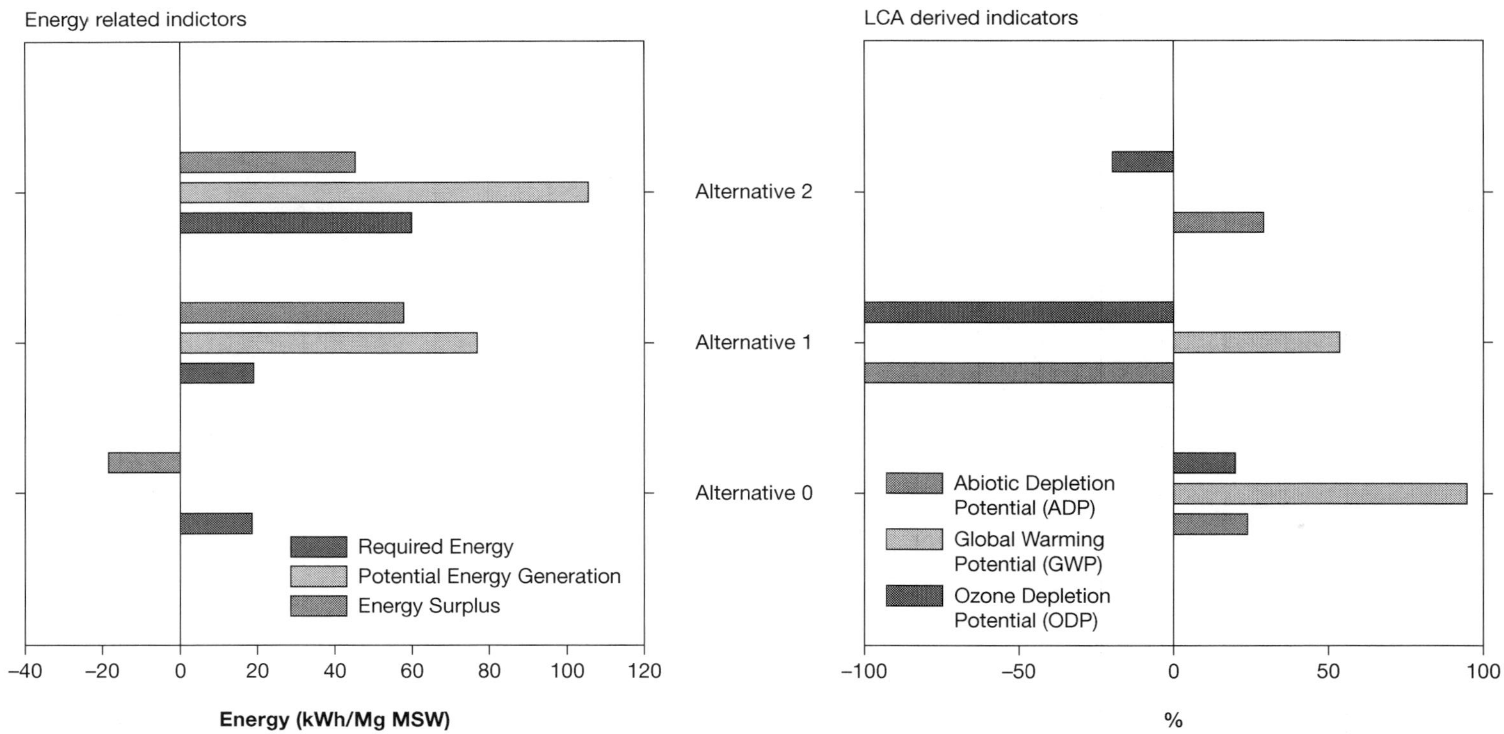

Figure 30.2
Summary of results for the three scenarios

Results and discussion

The results, summarized in Figure 30.2, allow recognition of the contribution of the scenarios associated with Alternatives 1 and 2 in terms of the enhancement of the studied impact categories and indicators. The energy generated in each case is not only enough to fully supply the operation of the infrastructure associated with the technological alternative, but can also generate a surplus of energy. This excess of energy can be used as electric power and injected into the electrical grid of the region of Aysén. Potentially, this energy could provide a clean electricity supply to the city of Coyhaique, in a sufficient amount to cover the needs of about 2,300 inhabitants. This would open up the possibility of creating a market for electricity from landfill to service part of the city's electricity requirements. However, such development should be analysed in a subsequent financial feasibility study, as this was not part of this particular work.

It could be concluded that Alternative 1 – that is, the energetic utilization of landfill gas recovered from the sanitary landfill – implies greater positive environmental impacts in terms of the indicators studied. This was actually a surprising result, considering that, before the study was carried out, it was speculated that the best alternative would be Alternative 2 (i.e. the production of biogas in an anaerobic digester), because the energy production and the surplus generated were greater than those of Alternative 1. However, this result can be explained by the conditions present in the region of Aysén. In particular, a major role can be attributed to the level of the regional industrial and technological development, as the region does not have an implemented infrastructure that could have a demand for the potentially recyclable fraction of the mechanical–biological treatment system envisaged in Alternative 2. In addition, the isolation characteristic of the region and its distance from potential recycling markets do not encourage the design or implementation of a recycling-oriented waste management system, thus reducing the positive impacts of recovering a recyclable material stream from a mechanical–biological treatment such as the one designed in Alternative 2.

Moreover, it is important to take into account that, from an environmental perspective, despite the fact that Alternative 1 was found to achieve higher results within the scope of this evaluation, the implementation of Alternative 2 would still diminish the negative environmental impacts that exist today at the landfill, as it would also help to improve the environmental situation of the city of Coyhaique.

On the other hand, previous experiences (i.e. Bezama et al. 2007) have discussed the advantages of implementing technological alternatives regarding the infrastructural change of the waste management system (e.g. separate collection at source), rather than an 'end-of-pipe' solution approach such as the one presented in this study. Therefore, the question of the infrastructural deficit in the region with regard to providing a mainstream of potentially recovered goods out of a recycling process should be addressed by considering further technological and structural alternatives, in order to determine the most appropriate waste management concept for the city of Coyhaique. Moreover, for appropriate decision-making, the decision should therefore be made on the basis of a combination of studies, such as macro- and micro-economic analyses of region-specific sustainability issues (e.g. socio-economic aspects), which can complement the evaluation of the potential global impacts predicted in classical approaches of life cycle assessment such as the ones presented in this study.

Conclusions

According to the results and discussions, it is possible to show the relevance of introducing life cycle-based tools in the evaluation of waste management concepts, as it may help the identification of hot spots along the life cycle of the analysed technological alternatives. In this study, the results showed that local conditions, especially in the case of isolated regions, play a deciding role in the evaluation, as they might differ from the expected results based on experiences in regions with more advanced infrastructure and technological development.

Decision-makers should be aware, however, that the results obtained in an LCA process are not sufficient information on which to base a decision on the best alternative for waste management. LCA should, therefore, be used as one part of the evaluation's data sources, together with micro- and macro-economic assessment, particularly with the inclusion of regional socio-economic indicators. Nonetheless, it is important to highlight the appropriateness of using the life cycle framework as part of the decision-making process, such as the one presented in this chapter, as it allows the assessment of the case study systems from different levels of aggregation, thus allowing identification the potential impacts of the systems and providing relevant information for the (in this case) regional stakeholders.

Finally, one relevant aspect to consider at this point is that the results presented in this chapter should not be taken as a generalized result, transferable to other communities. It is important to note that all the regional characteristics were taken into account in the evaluation and in the modelling process. Therefore, even though the results can be used as discussion points and/or references for future studies, they should not considered as the basis for the modelling itself in those studies.

References

EPA (United States Environmental Protection Agency) (ed.) (2005) 'Landfill gas emissions model (LandGem) version 3.02. EPA-600/R-05/047'. Available at: http://www.epa.gov/ttn/catc/dir1/landgem-v302-guide.pdf

Bezama, A., Szarka, N., Navia, R., Konrad, O., and Lorber, K.E. (2007) 'Lessons learned for a more efficient technology transfer to South American countries in the fields of solid waste and contaminated sites management', *Waste Management & Research*, 25: 148–61.

Bezama, A., Douglas, C., Méndez, J., Szarka, N., Muñoz, E., Navia, R., Schock, S., Konrad, O., and Ulloa, C. (2013) 'Life cycle comparison of waste-to-energy alternatives for municipal waste treatment in Chilean Patagonia', *Waste Management & Research*, 31(10S): 67–74.

Comisión Nacional de Energía (2013) *Energías, electricidad, sistemas eléctricos, sistema de Aysén*. Available at: www.cne.cl/energias/electricidad/sistemas-electricos/345-sistema-aysen (accessed 10 June 2013).

Douglas, C. (2010) *Evaluation of Waste-to-Energy Alternatives for the City of Coyhaique Based on Life Cycle Analysis* (BEng thesis Environmental Engineering; in Spanish). Concepción, Chile: University of Concepción.

Finnveden, G. (1999) 'Methodological aspects of life cycle assessment of integrated solid waste management systems', *Resources Conservation & Recycling*, 26: 173–87.

IASA (Ingeniería Alemana) (2003) *Estudio de Impacto Ambiental Relleno Sanitario Coyhaique*. Available at www.e-seia.cl (accessed 4 August 2016).

Muñoz, E., and Navia, R. (2011) 'Life cycle assessment of solid waste management strategies in a chlor-alkali production facility', *Waste Management & Research*, 29(6): 634–43.

31

WATER FOOTPRINT ASSESSMENT AT THE REGIONAL LEVEL

Guoping Zhang and Mesfin M. Mekonnen

Introduction

The water footprint (WF), introduced by Hoekstra and Hung (2002), is a volumetric measure of fresh-water use of a product, service, or any human activity. It accounts for both direct and indirect water use, as described in Chapter 12. The WF concept makes itself distinct from the traditional water use indicators that focus only on direct water withdrawals and meeting discharge permits (Zhang et al. 2013). The WF addresses the importance of water use in supply chains, which is key in effectively managing water resources in the global context and building linkages between consumption in one location and production in another.

WF sustainability and response strategies are the critical aspects when we speak about the 'water footprint', as they give a meaningful context to what a WF value entails. Water footprint assessment (WFA) is an emerging interdisciplinary research field that has been applied extensively in industries and geographical regions to support strategy-making or to feed policy dialogues. WFA is the methodological framework guiding the stepwise activities of WF studies.

A wide range of WFA studies at various geographical scales, from global and regional to river-basin scale, have been presented in the literature. The studies of Hoekstra and Mekonnen (2012) and Hoekstra et al. (2012) assessed the WF of humanity, illustrating the global dimension of water consumption and pollution, and revealed that water scarcity affects more than one-third of the global population. Mekonnen et al. (2015) evaluated the WF in Latin America and the Caribbean area and showed that there is a significant space for improving water-use efficiency and reducing pollution from both the production and consumption perspectives. A number of WFA studies at the national and subnational (state/province) levels (e.g. Aldaya et al. 2010; Schyns and Hoekstra 2014; Chouchane et al. 2015) were carried out from various perspectives to support policy discussion aiming to achieve more sustainable development. The body of literature on WFA studies at the river-basin level (e.g. Mekonnen et al. 2012; Dumont et al. 2013; Zhang et al. 2014) has been increasing, and the motivation for these studies ranges from research on methodologies to exploration of paradigm shifts in water policies for sustainable water governance and regional development.

In this chapter, we introduce the methodology for geographical WFA presented in Hoekstra et al. (2011) and use two basin WFA studies (Mekonnen et al. 2012;

Zhang et al. 2014) to showcase the approach to WFA at the regional level and highlight the value of the WFA in sustainable regional development.

Methodology

WFA

Hoekstra et al. (2011) described how the WFA refers to the four phases of study activities (Figure 31.1), including: (1) defining the study goals and scope; (2) quantifying the WF of a process, product, producer, or consumer or the WF in a geographic area in a given time period; (3) assessing the sustainability of the WF from environmental, social and economic aspects; and (4) identifying response strategies for WF sustainability improvement. WFA is instrumental in analysing the complex relations between the indicators within the driver–pressure–state–impact–response (DPSIR) framework that are related to water use and governance, water resources systems, water scarcity and pollution. It can therefore help us understand what can be done to achieve more sustainable use of fresh water in our activities.

WFA at the regional level

The approach to regional WFA can follow the methodology for geographic WFA. It also takes the four phases of the WFA, as described above. We elaborate more on the generic methods of accounting and sustainability assessment hereafter.

WF accounting for a geographic region

WF accounting for a region can be a full WF accounting or the WF within the geographically delineated area of the region. The 'region', depending on the interest and scope of a study, can be a country, a region including a number of countries, a region within a country, or a river basin. The full WF accounting for a region takes into account both the 'WF of consumers in the region' and the 'WF within the region' (Hoekstra et al. 2011).

WF sustainability assessment

WF sustainability assessment for a region is a task to evaluate whether the WF within the region exceeds the 'sustainability' thresholds of three dimensions: environmental, social and economic. The sustainability thresholds are to be established according to the set of sustainability indicators and criteria, and based on the local situations. WF sustainability assessment is a young and growing research field where sustainability indicators and criteria are under development, particularly those related to social and economic sustainability. We therefore provide a brief introduction to the general method for environmental sustainability assessment.

The environmental sustainability of the WF within a region can be assessed using indicators such as 'blue water scarcity' and 'water pollution level'.

- Blue water scarcity (BWS): the BWS of a region – for example, a river basin – is the ratio of the total blue water footprint (BWF) in the basin to the blue water availability (BWA) of the basin. The BWA is the total natural runoff of the basin minus the environmental flow requirement of the basin. When total BWF is larger than BWA, that is, the BWS is larger than 1.0

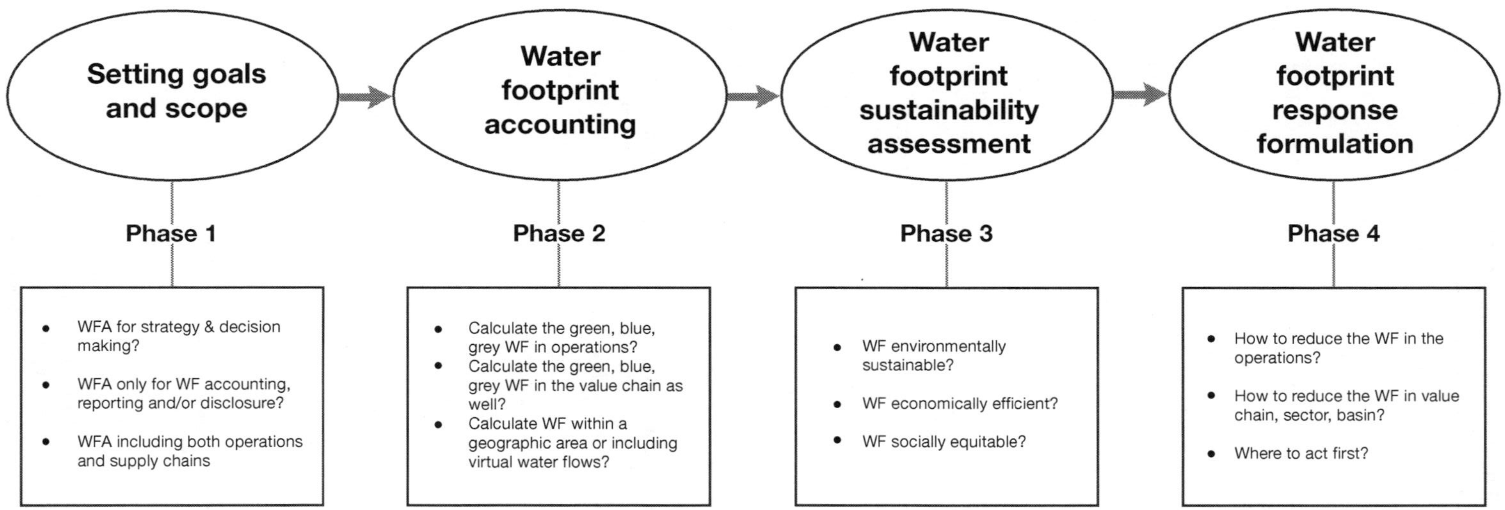

Figure 31.1
The four phases of the WFA framework

(or 100 per cent), the total BWF in the basin is not sustainable, because the available blue water has been fully consumed, and the environmental flow requirement is violated.
- Water pollution level (WPL): the WPL of a region – for example, a river basin – is defined as the ratio of the total grey water footprint (GWF) in the basin to the actual runoff of the basin. As soon as the total GWF in the basin is greater than the actual runoff, meaning that the WPL exceeds 1.0, the pollution assimilation capacity of the basin is fully consumed. A region with a WPL larger than 1.0 is an environmentally unsustainable hot spot, and thus any contribution to the total GWF in the region is unsustainable.

WF response formulation

The WF response formulation is a process in which strategies and actions for improving the regional WF sustainability (e.g. reducing BWS and/or WPLs) are identified. Sustainable water use and management require stakeholder participation, individual contributions, and collective action, although the roles of each stakeholder can be different. However, setting up and observing the basin WF caps based on the maximum sustainable WF, product WF benchmarks, fair WF shares (reasonable WF of communities), and sustainable regional water dependency are the general guidelines for searching for appropriate response strategies (Hoekstra 2013).

Applications of WFA at the regional level

Case 1: WFA for the Hertfordshire and North London area, UK

Background

Changes in climate, land use, population, and lifestyle, and the interactions between them, increase the pressure on water resources and water management in south-east England, UK. Currently, water abstraction in the country is regulated through licensing (since the 1960s), but only considering human demand, without linking that to water availability in space and time. Such a management system is inadequate to cope with the pressure now and in the future (Environment Agency and Ofwat 2011). To support the search for ways to reform the current water abstraction management, a comprehensive WF assessment study for the Hertfordshire and North London area, the northern part of the Thames river basin, was carried out (Zhang et al. 2014).

Study scope and approach

The study area, covering about 3,200 km^2, is densely populated, with more than 6.5 million inhabitants. The total water abstraction, up to about 500 million m^3/year and largely depending on vulnerable groundwater resources, is mainly for public water supply for industries and domestic use, and less than 10 per cent of the abstraction is for agriculture. This WFA covers:

- WF accounting for thirty-five sub-catchments in the area under the baseline (average over 1999–2007) condition, including the blue, green, and grey WFs of agricultural, industrial, and domestic sectors;
- WF sustainability assessment – blue water availability, blue water scarcity, and WPL for surface water and groundwater;

- projection of WF, blue water scarcity, and WPL with climate change and water demand scenarios; and
- recommendations for water abstraction management based on the WFA findings.

Major results and findings

Under the baseline condition, groundwater use accounts for approximately 55 per cent of the total BWF in the area. Ninety-five per cent of the total BWF is due to water transfer through sewerage systems, within and outside the study area. Nearly half of all sub-catchments experience moderate to severe BWS. BWS is strongly related to the transfer of water through sewage treatment works, from one sub-catchment to another. With the future climate changes scenarios, the projected overall BWS will become more severe across the entire study area.

Thirty per cent of the total GWF in the study region is the result of point-source pollution on surface water, 48 per cent is from point-source pollution on groundwater, and 22 per cent is from diffuse sources. The largest GWFs occur in sub-catchments where large-scale sewage treatment works are located and/or with a high pollutant concentration in the effluents, and when large amounts of effluent are discharged into the ground. More than one-third of all sub-catchments have an annual average surface WPL at a significant or severe level.

This study showed that the blue and grey WF, BWS and WPL in the study region are highly affected by the water transfer between sub-catchments through sewage treatment works, that is, influenced by the water management practices. The results for GWF highlight the limitations of the current water quality management system using effluent discharge permits, which is based on pollutant concentration dictated by water quality standards.

Based on the results and findings, BWS and WPL are the criteria recommended to be used to issue and/or re-evaluate water abstraction licences and discharge permits. In such issuing and re-evaluation processes, the licences and permits should be reviewed considering the following aspects:

1 total demand on the consumptive water use from all users in the catchment;
2 total catchment water availability and total catchment pollution assimilation capacity;
3 seasonal/monthly variability of the catchment blue water availability and pollution assimilation capacity;
4 whether the best available technologies and/or best practices have been applied or can be applied with reasonable costs.

Depending on the level of severity of BWS and/or WPL, the abstraction licences would be conditional on time limit and/or seasonal volume limit. Considering the degree of WPL and the type of pollution (i.e. point source or diffuse source), the discharge consent/permit would include conditions such as the requirement for diffuse-source WF reduction strategies.

This study clearly illustrated where and when water use (by whom) creates problems: decrease in water availability and diminishing of waste assimilation capacity (increase in WPL). The study results led to useful recommendations to be considered in future water management regulations.

Case 2: WFA for the Lake Naivasha basin, Kenya

Background and objective of the study

Lake Naivasha is Kenya's second-largest fresh-water lake and was designated as a Ramsar wetland in 1995. The lake's water level has declined significantly in recent times, which coincided with the establishment of horticultural industry in the area in 1982 (Becht and Harper 2002). Lake Naivasha has become the key site for Kenya's horticultural industry (mainly cut flowers), which is the country's third most important foreign exchange earner. To analyse the possibility of mitigating excessive water consumption and pollution of the lake to safeguard more sustainable socio-economic development, the WF assessment study for the Lake Naivasha basin, in relation to cut flowers, was conducted by Mekonnen et al. (2012).

Major results and findings

The study showed that about 98 per cent of the BWF and about 61 per cent of the GWF in the basin can be attributed to the commercial farms around the lake. Cut flowers contributing about 98 per cent to the total BWF related to crop production around Lake Naivasha. The Netherlands, the UK, and Germany together take more than 90 per cent of the virtual water export (VWE) due to the export of cut flowers from the lake basin. The VWE related to the export of cut flowers from the basin has increased significantly, almost doubling within 10 years, from 1996 to 2005.

WF mitigation strategies

The cut-flower industry in the region is economically and socially important, as it contributes a large share of Kenya's export value while providing employment and infrastructure for a large population around the lake. Therefore, the long-term conservation of the lake ecosystem and sustainable regional socio-economic development require sustainable use and management of the water in the basin. Sustainable management and use of the lake water require coordinated, collective actions taken by the stakeholders in the region. The maximum allowable decline of the lake water level due to water abstractions must be decided, which is subject to the maximum sustainable blue and grey WF in the basin.

Actions from the stakeholders along the cut-flower supply chain, involving consumers, retailers and traders, could be effective in lessening the pressure on the water resources in the lake region. One proposal of this study was to establish a 'water-sustainability agreement' between major agents along the supply chain, including a 'water-sustainability premium' on the final product at the retail end. The premium on flowers sold by the retailer will generate funds that can be used for financing watershed management projects and measures to reduce the blue and grey WF within the Lake Naivasha basin. Besides, it can also create awareness among consumers of the value of water. The mechanism of the water-sustainability premium can enhance the sustainability of the regional development, while reducing the risk of Kenya losing its horticulture business in the long term.

Conclusions

WFA is a holistic tool unifying both quantity and quality aspects in integrated water resources assessment, planning, and management. It is powerful when analysing the complex relations between the indicators, within the DPSIR framework. The case studies presented above demonstrated that WFA can help identify where the pressures on water resources are, who contributes to unsustainable water use and management, and what and how water resources can be better managed as a contribution to sustainable development in a region.

References

Aldaya, M.M., Garrido, A., Varela-Ortega, C., Novo, P., and Casado, R.R. (2010) 'Water footprint and virtual water trade in Spain', in *Water Policy in Spain*, eds A. Garrido and M.R. Llamas, Leiden, Netherlands: CRC Press, pp. 49–59.

Becht, R., and Harper, D.M. (2002) 'Towards an understanding of human impact upon the hydrology of Lake Naivasha, Kenya', *Hydrobiologia*, 488: 1–11.

Chouchane, H., Hoekstra, A., Krol, M.S., and Mekonnen, M.M. (2015) 'The water footprint of Tunisia from an economic perspective', *Ecological Indicators*, 52: 311–19.

Dumont, A., Salmoral, G., and Llamas, M.R. (2013) 'The water footprint of a river basin with a special focus on groundwater: The case of Guadalquivir basin (Spain)', *Water Resources & Industry*, 1–2: 60–76.

Environment Agency and Ofwat (2011) The Case for Change – Reforming Water Abstraction Management in England, Report GEHO1111BVEQ-E-E, London: Environment Agency and Ofwat.

Hoekstra, A.Y. (2013) *The Water Footprint of Modern Consumer Society*, London: Earthscan from Routledge.

Hoekstra, A.Y., Chapagain, A.K., Aldaya, M.M., and Mekonnen, M.M. (2011) *The Water Footprint Assessment Manual: Setting the Global Standard*, London: Earthscan from Routledge.

Hoekstra, A.Y., and Hung, P.Q. (2002) Virtual Water Trade: A Quantification of Virtual Water Flows between Nations in Relation to International Crop Trade, Value of Water Research Report Series No 11. Delft, Netherlands: IHE.

Hoekstra, A.Y., and Mekonnen, M.M. (2012) 'The water footprint of humanity', *Proceedings of the National Academy of Sciences*, 109: 3232–7.

Hoekstra, A.Y., Mekonnen, M.M., Chapagain, A.K., Mathews, R.E., and Richter, B.D. (2012) 'Global monthly water scarcity: Blue water footprints versus blue water availability', *PLoS One*, 7: e32688.

Mekonnen, M.M., Hoekstra, A.Y., and Becht, R. (2012) 'Mitigating the water footprint of export cut flowers from the Lake Naivasha Basin, Kenya', *Water Resources Management*, 26: 3725–42.

Mekonnen, M.M., Pahlow, M., Aldaya, M.M., Zarate, E., and Hoekstra, A.Y. (2015) 'Sustainability, efficiency and equitability of water consumption and pollution in Latin America and the Caribbean', *Sustainability*, 7: 2086–112.

Schyns, J.F., and Hoekstra, A.Y. (2014) 'The added value of Water Footprint Assessment for national water policy: A case study for Morocco', *PLoS One*, 9: e99705.

Zhang, G.P., Hoekstra, A.Y., and Mathews, R.E. (2013) 'Water Footprint Assessment (WFA) for better water governance and sustainable development', *Water Resources & Industry*, 1–2: 1–6.

Zhang, G.P., Mathews, R.E., Frapporti, G., and Mekonnen, M.M. (2014) Water Footprint Assessment for the Hertfordshire and North London Area, Report RESE000335, London: Environment Agency.

32

CARBON FOOTPRINT OF BIOFUELS

Harish Jeswani

Introduction

Many countries have developed policies and set mandatory targets to increase the use of biofuels in transport, in order to increase energy security, support rural development, and develop a low-carbon economy, in addition to mitigating climate change (Khanna et al. 2011). Owing to these policies, world ethanol production from biomass increased fivefold between 2000 and 2012, from 17 billion to more than 83 billion litres, and biodiesel production expanded from less than 1 billion to 22.5 billion litres during this period (Bringezu et al. 2009; IEA–ETSAP 2013). Biofuels currently contribute to about 4 per cent of global transport fuel consumption, and it is expected that their share will increase to 27 per cent by 2050 (IEA 2014).

As one of the key drivers for the growing interest in biofuels is their potential to reduce greenhouse gas (GHG) emissions from transport, the implementation of some of these policies depends on a life cycle assessment (LCA) of the GHG performance of the biofuel. Owing to the diversity of feedstock and inherently complex production systems, there are various methodological aspects that could lead to high uncertainty over GHG performances of biofuels. These include the definition of appropriate scope and system boundaries, selection of appropriate allocation methods, and accounting for direct and indirect land-use change (LUC) related to GHG emissions. In this chapter, these methodological aspects are discussed in detail.

Methodological approaches for assessing carbon footprint of biofuels

As mentioned above, concerns about climate change and the need to reduce GHG emissions are key drivers for policies and regulations concerning biofuels. The biofuel industry is therefore required to demonstrate that there are net GHG savings from biofuels over the whole life cycle, from 'crops to cars'. In order to determine whether a biofuel system results in a net reduction in GHG emissions, a life cycle-based approach is commonly used. The life cycle approach is also required by various legislative Acts related to biofuels, including the EU Renewable Energy Directive (EC 2009) and the US Energy Independence and Security Act (EISA; US 2007). LCA and carbon footprint methodologies are discussed in Chapters 10 and 11.

Approaches to estimate the GHG emissions from biofuels vary from one country to another (Azapagic and Stichnothe, 2010). In Europe, these approaches have been synchronised with the EU Renewable Energy Directive (RED). The US EPA has also developed the Renewable Fuel Standard to ensure that biofuels and

Table 32.1 Overview of the EU RED and US EPA GHG calculation methodologies for biofuels

	EU RED (EC 2009)	US EPA (EPA 2010)
Base year	2008	2005
Reference value for diesel (g CO_2 eq./MJ)	Most recent value or 83.8	91.4
Reference value for petrol (g CO_2 eq./MJ)	Most recent value or 83.8	91.4
System boundary	Cradle to grave	Cradle to grave
Capital goods/infrastructure	Excluded	Excluded
Pesticides	Included	Included
Fertilisers	Included	Included
Allocation method value*	Net calorific value	System expansion/economic
Direct LUC	Included (IPCC 2007)	Included (IPCC 2007)
Indirect LUC	Excluded	Included
GHG saving threshold (%)	35–60**	20–60***

Notes: * The allocation procedure used depends on the system; ** the current limit is 35%, which increases to 50% in 2017 and 60% thereafter; *** 20% on average for all renewable fuels, 50% for biodiesel, and 60% for cellulosic biofuels

fossil fuels are compared on an equivalent basis (EPA 2010). Table 32.1 provides an overview of the EU RED and US EPA GHG calculation methodologies. The key aspects covered in these methodologies, which include defining appropriate scope and system boundaries, unit of analysis (functional unit), selection of allocation methods, and consideration of LUC, are discussed below.

Scope and system boundaries

Selection of an appropriate system boundary depends upon the goals and scope of the study. The life cycle of biofuels, as outlined in Figure 32.1, encompasses planting, growing, and harvesting of biomass (if applicable), transportation of feedstock to biorefineries, conversion of biomass to biodiesel or bioethanol, distribution of biofuel, and its use in transport. Each stage in the life cycle of biofuel is associated with several environmental issues, depending on the type of feedstock and production route. For example, during cultivation of energy crops, GHG emissions are released from application of fertilisers and use of fuel in agricultural machinery. Emissions associated with the production of fertilisers, pesticides, and fuels also need to be taken into account in carbon footprinting. In cases where forest or grassland is converted to agricultural land for the production of feedstock for biofuel, LUC-related GHG emissions can significantly alter the GHG balance of biofuels. LUC aspects are discussed later. Processing of biomass into biofuels through various pathways also requires energy and material inputs, which may also influence the GHG performance of biofuels. GHG emissions associated with plant effluent can also be significant, especially in the case of biofuel from palm oil (Taylor et al. 2014). Combustion of biofuels releases biogenic CO_2 (captured during the plant growth), and, hence, it is not accounted. However, other combustion emissions, such as CH_4 and N_2O, are considered.

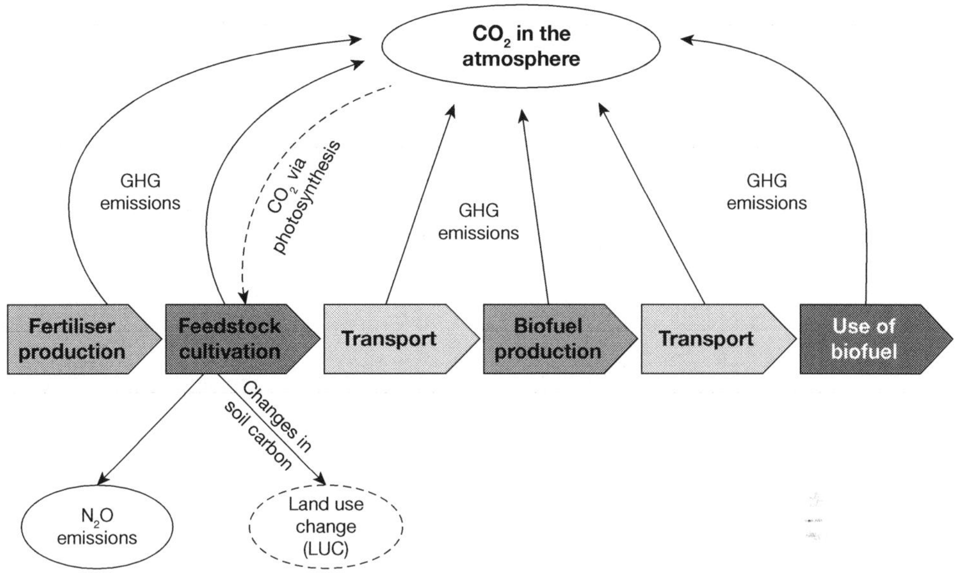

Figure 32.1
Life cycle of biofuels from 'cradle to grave'

Studies comparing different biofuels can employ a 'cradle-to-gate' or 'well-to-tank' approach and exclude the utilisation phase. However, if the purpose of the study is to compare biofuels with their fossil substitutes, the utilisation stage plays a crucial role, and a 'cradle-to-grave' or 'well-to-wheel' approach is used. Use of a 'cradle-to-grave' approach is recommended in regulations such as EU RED and US EPA.

Unit of analysis (functional unit)

In LCA, one of the main purposes of the unit of analysis is to provide a reference to which the input and output process data are normalised and a basis on which the final results are shown. Where the carbon footprint of biofuels is concerned, results should be expressed in terms of the same functional unit, to ensure that the comparison is based on delivery of the same service. The unit of analysis for comparison is often based on the equivalent energy content of the fuels and is usually defined (arbitrarily) as '1 MJ of fuel'. When the carbon footprint of transport fuels is assessed, the results can also be expressed on a per vehicle-kilometre basis. However, if the goal is to compare the land-use efficiency of different biofuel pathways, then results can also be expressed on a per hectare basis.

Allocation of emissions from co-products

Biofuel production systems generate various co-products (electricity, heat, animal feed, bio-chemicals), and, depending on the approach used to allocate the emissions and impacts between the biofuels and co-products, the results of LCA may vary significantly (Menichetti and Otto 2009). In LCA studies, two approaches are commonly applied: system expansion and allocation on the basis of either mass, energy content, or market value (price) of the products.

As shown in Table 32.1, the US EPA prefers system expansion. In this method, the system is credited for producing the additional output and displacing the need for another system producing the same or equivalent product. However, the methodological difficulty is in identifying the 'correct' way to credit the system. The choice of the allocation method and the 'credit' are of utmost importance, as often very different results are obtained using different approaches. In any case, this should be examined as part of sensitivity analysis.

The EU RED favours allocation based on the energy content of biofuels, although other allocation procedures, such as system expansion or economic value, might be more appropriate in particular cases. For example, energy allocation cannot be applied in systems where co-products do not have an energy value but have an economic value, for example animal feed, ash, and fertilisers. In these cases, allocation based on the economic value may be more appropriate. However, this produces volatile results in line with the economic values of commodities and should only be used where other allocation methods cannot be applied (ISO 2006).

LUC

LUC is probably the most controversial, uncertain issue associated with biofuels (Fargione et al. 2008). The main concern is related to possible additional GHG emissions when carbon stored in the soil is disturbed and released as CO_2 owing to LUC. Two types of LUC are considered: direct and indirect.

Direct LUC involves either the conversion of natural ecosystems (e.g. forest) into croplands, or changes in the management of already appropriated land, such as a shift from cultivating food crops or grazing animals to instead cultivating biomass feedstocks for biofuel production. As shown in Table 32.1, direct LUC is considered in both EU RED and US EPA approaches, and the IPCC factors are used for these purposes (IPCC 2007). To minimise LUC-related emissions, EU RED prohibits the use of certain lands, such as primary forests and wooded land, for biofuel production.

Indirect LUC refers to the changes in land use that take place elsewhere as a consequence of the biofuel production (Searchinger et al. 2008). This is often difficult to assess, owing to the uncertainties involved, particularly at the international level. Currently, only the US EPA approach considers indirect LUC.

It is important to note that agricultural and forestry residues and waste cooking oil do not necessitate LUC beyond what occurs for food production and existing forestry activities. Similarly, use of degraded or marginal lands for energy crop production could avoid impacts related to LUC and may have positive impacts.

In addition to LUC, crop management practices can also have a significant impact on the GHG balance of biofuels, owing to the GHG emissions from changes in soil organic carbon, litter, and dead wood (Cherubini 2010). Changes in soil organic carbon from LUCs are therefore considered especially important, and their effects can consistently reduce the GHG savings of bioenergy systems.

GHG saving potential of biofuels

According to EU RED, biofuels must achieve GHG savings of at least 35 per cent in comparison with fossil fuels. This saving requirement rises to 50 per cent in

2017 and 60 per cent thereafter. EISA requires a 20 per cent reduction in life cycle GHG emissions for any renewable fuel, a 50 per cent reduction in order to be classified as biodiesel, and a 60 per cent reduction in order to be classified as cellulosic biofuel. The GHG savings from biofuels compared with fossil fuels are calculated as follows (EC 2009):

$$GHG_{saving} = \frac{GHG_{fossil\,fuel} - GHG_{biofuel}}{GHG_{fossil\,fuel}} \times 100\ (\%) \quad (32.1)$$

Figure 32.2, which is based on data from a number of LCA studies, compares GHG savings for different conventional and advanced biofuels against fossil fuels (FAO, 2013; IEA, 2011; IEA-ESTAP 2013; Jeswani et al. 2015). The figure also compares GHG emissions with and without LUC. The results show a large variation in GHG mitigation potential for each biofuel, which reflects the diversity of potential feedstocks, production routes, the uncertainty regarding the technology, and the methodological differences (such as the allocation of emissions between co-products and LUC, as discussed above) in various studies. As shown in Figure 32.2, many conventional biofuels offer moderate savings in GHG emissions compared with their fossil equivalents. Among conventional biofuels, ethanol produced from sugar cane can offer a 70–90 per cent reduction in the life cycle GHG emissions compared with fossil gasoline, if no LUC occurs. This is because sugar cane crops require fewer fertilisers and offer high yields. Also, extraction of sugar is relatively easy, and the heat and power demands of the process can be met through combustion of bagasse. On the other hand, crops such as wheat, corn, and sugar beet require higher fertiliser inputs, and the cultivation process is energy intensive. As a result, biofuels from these feedstocks offer lower GHG savings. For biodiesel feedstock, such as palm, oilseed rape, and soya beans, the GHG emissions could be much higher than fossil diesel if grasslands or forests are converted for plantations. The use of waste oil as a feedstock shows significant potential for GHG mitigation, but its availability is limited.

Emerging and advanced technologies for producing biofuel from lignocellulosic feedstocks, such as forest and agricultural residue, and dedicated energy crops (miscanthus, poplar) in integrated biorefineries look more promising, as they can reduce emissions by 70–95 per cent (Figure 32.2) when co-products are used to produce heat and power, replacing fossil fuels, for example. However, uncertainty in the estimates for these processes is higher, as such plants are still under development and have yet to reach the commercial stage.

Other environmental impacts

Besides GHG emissions, other environmental impacts can occur from biofuel production. These include water use, biodiversity, acidification, eutrophication, and toxicity (Jeswani and Azapagic 2012; Jeswani et al. 2015). These impacts depend on various factors, such as feedstock, cultivation location and practices, and processing technologies. Also, large-scale production of biofuels from first-generation feedstocks competes with food production for arable land and water (Gasparatos et al. 2013) Therefore, it is important that, when regional policies on biofuels are developed, all relevant environmental impacts along the whole life cycle should be considered, to avoid shifting the burdens and making unsustainable choices.

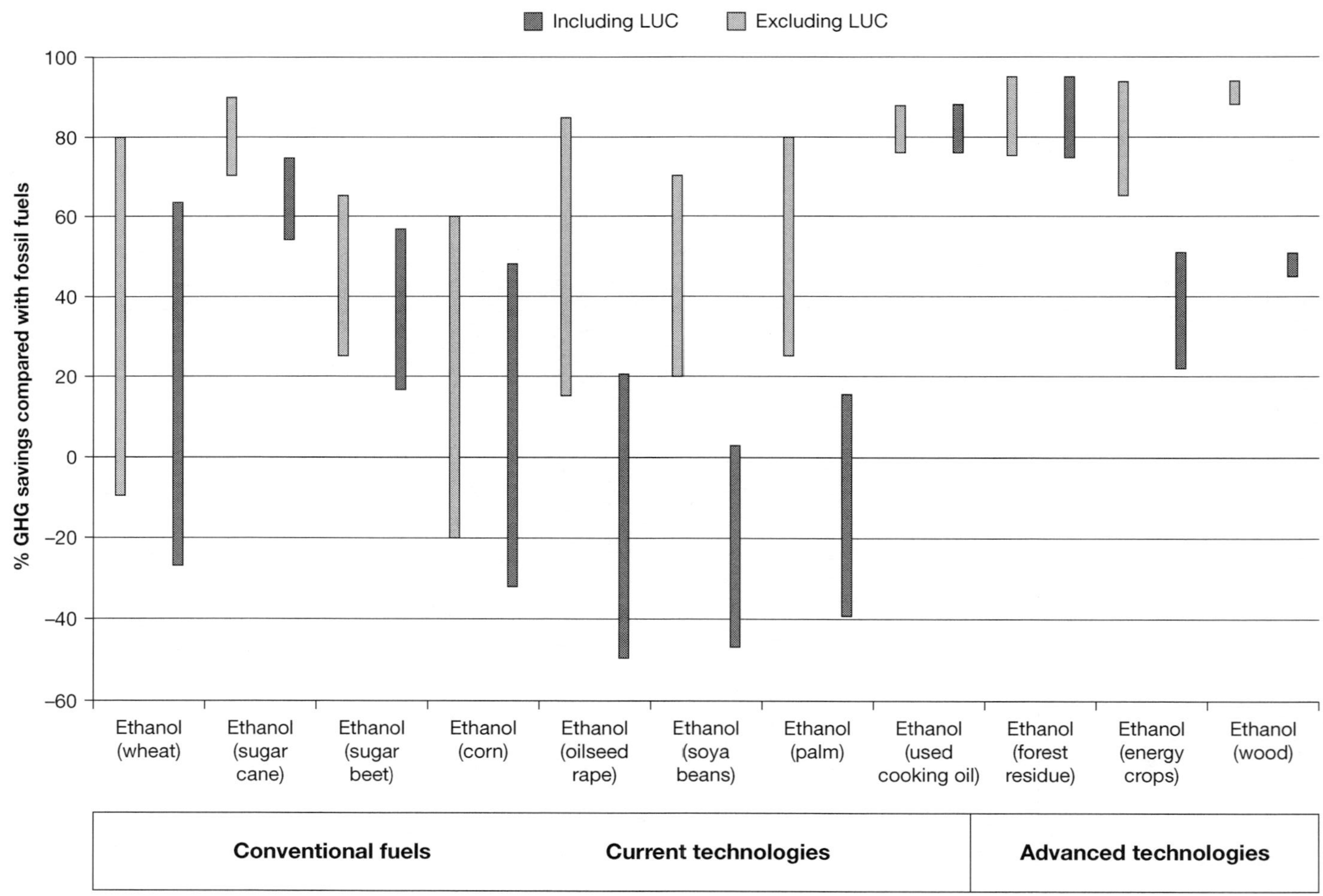

Figure 32.2
GHG savings for different conventional and advanced biofuels

Conclusions

Biofuels could play a strong role in the development of a low-carbon economy for regions. The sustainability of large-scale industrial development of biofuels will depend on a range of factors, including GHG emissions savings over fossil fuels, impact on biodiversity and water resources, availability of land for dedicated energy crops, and impacts on food security. Various LCA studies demonstrate that producing ethanol from sugar cane or biodiesel from waste oil shows significant potential for GHG mitigation. The levels of mitigation associated with other conventional biofuels are more modest and highly influenced by LUC. The GHG reduction potential of biofuels suffers significantly if grasslands or forests are used for the cultivation of feedstocks. The emerging technologies for producing biofuel from lignocellulosic feedstocks in integrated biorefineries look more promising. Biofuels can contribute to GHG mitigation strategies in regional transport policies if marginal/abandoned lands are used for energy crops, to avoid potential negative emissions from LUC. LUC-related emissions could be avoided by the production of biofuels from agricultural and forest residues.

References

Azapagic, A., and Stichnothe, H. (2010) 'Life cycle sustainability assessment of biofuels', in *Handbook of Biofuels Production: Processes and Technologies*, eds R. Luque, J. Campelo, and J. Clark, Cambridge, UK: Woodhead, pp. 337–60.

Bringezu, S., Schütz, H., O'Brien, M., Kauppi, L., Howart, R.W., and McNeely, J. (2009) Towards Sustainable Production and Use of Resources: Assessing Bio-Fuels, Paris: UNEP.

Cherubini, F. (2010) 'GHG balances of bioenergy systems – Overview of key steps in the production chain and methodological concerns', *Renewable Energy*, 35: 1565–73.

EC (2009) *Directive 2009/28/EC of the European Parliament and of the Council of 23 April 2009 on the Promotion of the Use of Energy from Renewable Sources*, Brussels: European Commission.

EPA (2010) *Regulation of Fuels and Fuel Additives: Changes to Renewable Fuel Standard Program; Final Rule, 40 CFR Part 80, Part II, March 2010*, Washington, DC: Environmental Protection Agency.

FAO (2013) Biofuels and the Sustainability Challenge – A Global Assessment of Sustainability Issues, Trends and Policies for Biofuel and Related Feedstocks, Rome: Food and Agriculture Organization of the United Nations.

Fargione, J., Hill, J., Tilman, D., Polasky, S., and Hawthorne, P. (2008) 'Land clearing and the biofuel carbon debt', *Science*, 319: 1235–8.

Gasparatos, A., Stromberg, P., and Takeuchi, K. (2013) 'Sustainability impacts of first generation biofuels', *Animal Frontiers*, 3: 12–26.

IEA (2011) Technology Roadmap – Biofuels for Transport, Paris: International Energy Agency.

IEA (2014) World Energy Outlook 2014, Paris: International Energy Agency, OECD/IEA.

IEA–ETSAP (2013) Production of Liquid Biofuels – Technology Brief, Paris: International Energy Agency – Energy Technology Systems Analysis Programme.

IPCC (2007) Climate Change 2007: Synthesis Report, Contribution of Working Groups I, II and III to the Fourth Assessment Report of the Intergovernmental Panel on Climate Change, Geneva: IPCC.

ISO (2006) *ISO 14044: Environmental Management – Life Cycle Assessment – Requirements and Guidelines*, Geneva: ISO.

Jeswani, H.K., and Azapagic, A. (2012) 'Sustainability assessment of second generation biodiesel', in *Advances in Biodiesel Production: Processes and Technologies*, eds R. Luque and J.A. Melero, London: Woodhead, pp. 13–27.

Jeswani, H.K., Falano, T., and Azapagic, A. (2015) 'Life cycle environmental sustainability of lignocellulosic ethanol produced in integrated thermo-chemical biorefineries', *Biofuels, Bioproducts & Biorefining*, doi: 10.1002/bbb.1558.

Khanna, M., Crago, C.L., and Black, M.J. (2011) 'Can biofuels be a solution to climate change? The implications of land use change related emissions for policy', *Interface Focus*, 1: 233–47.

Laborde, D. (2011) Assessing the Land Use Change Consequences of European Biofuel Policies, Final Report, Washington, DC: International Food Policy Research Institute.

Menichetti, E., and Otto, M. (2009) 'Energy balance and greenhouse gas emissions of biofuels from a life cycle perspective', in *Biofuels: Environmental Consequences and Interactions with Changing Land Use*, eds R. Howarth and S. Bringezu, New York: Cornell University, pp. 81–109.

Searchinger, T., Heimlich, R., Houghton, R.A., Dong, F., Elobeid, A., Fabiosa, J., Tokgoz, S., Hayes, D., and Yu, T.-H. (2008) 'Use of U.S. croplands for biofuels increases greenhouse gases through emissions from land-use change', *Science*, 319(5867): 1238–40.

Taylor, P.G., Bilinski, T.M., Fancher, H.R., Cleveland, C.C., Nemergut, D.R., Weintraub, S.R., Wieder, W.R., and Townsend, A.R. (2014) 'Palm oil wastewater methane emissions and bioenergy potential', *Nature Climate Change*, 4: 151–2.

US (2007) Energy Independence and Security Act of 2007, Public Law 110–140–DEC. 19, Washington, DC: US Federal Government.

33

SUSTAINABLE PROCUREMENT BASED ON LIFE CYCLE COSTING

Bernard Mazijn

Introduction

Although the focus of this chapter will be on sustainable *public* procurement, because there is an additional legal dimension, it should be clear that any organisation (including enterprises) can benefit from using life cycle costing (LCC) in its procurement practices. Furthermore, it is important to bear in mind that the procurement cycle is an important process within an organisation. It begins with the definition of needs, before goods or services to be purchased can be specified. Companies are selected as possible suppliers and are requested to submit bids. Evaluation of the bids leads to signing of a contract. Monitoring and follow-up are recommended, to be used (among other things) as input for the next procurement cycle.

Needs assessment

Needs assessment is the most important phase in the procurement cycle in terms of (future) impacts for all dimensions of sustainable development. Poor assessment will, in most cases, lead to incorrect interpretation of an organisation's actual needs and could have profound (cost) implications. The notion of needs assessment in the procurement cycle is related to the fact that procurement 'contracting authorities' are legal entities created with the specific purpose of meeting needs that are in the general interest. The purchase of goods or services should therefore start with an explicit (or at least implicit) needs assessment. However, the approach is generally not (well) developed and/or regulated. In fact, research has indicated the importance of first setting up a management system in the organisation, preferably within a context of sustainable development (Demeulenaere and Mazijn 2012). This can have a considerable impact on an organisation's demand for products or services and, therefore, on its budget requirements.

LCC

The calculation of costs all along the supply chain was used for the first time at the beginning of the twentieth century in the US in decision-making on huge investments (e.g. defence, agriculture, etc.). Since the 1960s, it has also become quite common practice in construction, aviation, and so on. The practice is now known as 'total cost of ownership' (TCO) or life cycle costing, the terms often being used synonymously. However, a distinction should be made between looking at the costs from a producer's perspective and from that of a consumer: see Table 33.1.

Table 33.1 All costs along the value chain from producer or consumer perspectives

Organisation (incl. enterprises) as a producer	Organisation (incl. enterprises) as a consumer
R&D costs	Purchase costs
Production costs	Installation costs
Distribution costs	Working costs
Service costs	Maintenance costs
(EoL costs)	Service costs
	EoL costs

In that sense, TCO/LCC helps the producer to use the results as a sales argument, to define the service offering around a product, to optimise the product design, and/or to look for alternative business models. The main argument for the consumer relates to the costs after purchasing. In both cases, it helps to observe more than the tip of the iceberg.

According to Hunkeler et al. (2008), the definition of environmental LCC is as follows:

> The assessment of all costs associated with the life cycle of a product that are directly covered by one or more actors in the product life cycle (supplier, manufacturer, user or consumer, and/or EoL actor), with the inclusion of externalities that are anticipated to be internalized in the decision-relevant future.

The first step in LCC is to map the life cycle of the product (or products), starting from a functional unit. The different life cycle stages and their costs are therefore considered. However, the question arises: in the definition, what do we mean by 'all costs'? One could imagine a matrix like that shown in Figure 33.1, where a distinction is made between private costs and externalities, reflecting what is at stake when the value chain producing goods or services is assessed within the context of sustainable development.

> An externality occurs when a decision within the value chain imposes costs or benefits on others which are not reflected in the prices charged for the goods and services being provided by the value chain. Externalities are sometimes referred to as spill overs. An externality may also result in private costs, even though it might not be accounted for in the decision-making.
>
> (Benoît and Mazijn 2009)

The solid black line represents the private costs and benefits reflected in market prices. Sometimes, 'the inclusion of externalities that are anticipated to be internalized in the decision-relevant future' (cf. the definition) is taken into account: see the broken line. An example could be the increasing prices of CO_2 emissions within the context of an emissions trading scheme. However, it is illusory to think one can reflect all externalities, within the limits of the dotted line, in the costs of products. In fact, there are many problems in terms of a scientific method, acceptable to all stakeholders, for calculating the monetary cost of these externalities at each stage of the value chain.

First, all usual and hidden costs, risks and liabilities, and opportunity costs need to be identified. The first categories can easily be understood, but the latter

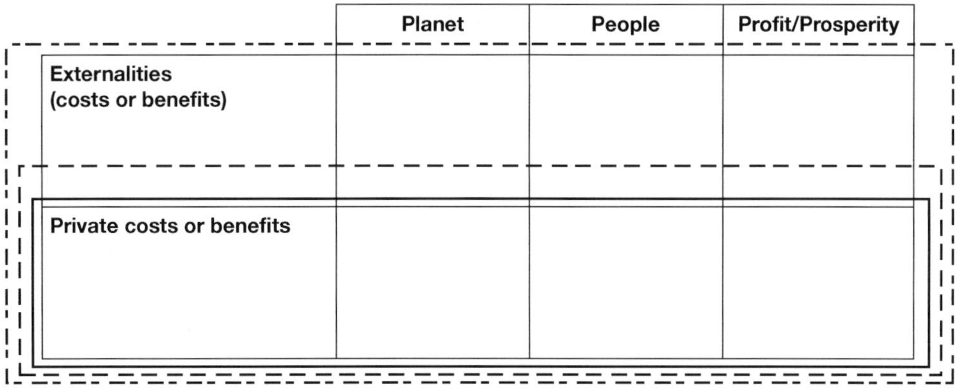

Figure 33.1
Detailing the assessment of the value chain producing goods and services within the context of sustainable development

needs some more detailed explanation. In fact, it is about the question: what is the cost for the customer when the product does not function or when it functions incorrectly? The consequence of this might be a loss of revenue or the cost of looking for the best alternative (e.g. outsourcing). It is clear that knowledge about different usage patterns, customer attitudes, and so on, will be essential. Within a context of LCA, one can refer to 'product utility' when defining the functional unit.

Next, as in every area of research, data collection for the estimation of costs is crucial. Who is entitled to deliver data (internal versus external, estimates versus observations, etc.)? What about historical data (cf. different product generations)? How are the differences between many components, cost factors, and so on, to be handled? Transparent and clear reporting on how data collection has been handled is important. But, the question arises as to whether all data will be communicated to the outside world, as there may be commercial or competition-related reasons, for example, for not publishing (parts of) a report.

Note that an important element in cost calculation over a product's lifespan is related to the net present value, a technique using discounting to convert costs or benefits occurring at different times to equivalent (net) costs today. Long-term interest rates and expected inflation are both important for this calculation.

Different software tools are available on the market, such as Oracle Crystal Ball, Palisade @Risk, VOSE Software – ModelRisk, and so on. Many of these include techniques to check reliability. It is also recommended that the variabilities and uncertainties in the calculation are looked for. The possibility of using a Monte Carlo simulation is embedded in these software packages.

In the end, the results provide a basis for drawing conclusions. However, they will very much depend on the aim of applying LCC. Overall, the impact in terms of eco-efficiency will be balanced against the overall costs of actions (research, development, design, etc.). An interesting example is the way the cost over the life cycle of refrigerators, freezers, and so on, has evolved. For example, the EU energy label, introduced in 1992, rated appliances in terms of a set of energy efficiency classes from A to G, A being the most energy efficient, and G the least efficient. Because consumers were well informed and could therefore look for the lowest life cycle cost – composed of purchase cost and lifetime operation costs – the market changed considerably over 10 years. The classes G–E were discontinued, and, in order to keep up with advances in energy efficiency, A+, A++, and A+++ grades were introduced.

Life cycle costing used in sustainable procurement

Within the context of sustainable (public) procurement (SPP), the interest in LCC is slightly different. The purchaser exerts only indirect influence over the costs related to the upstream stages, prior to purchasing the product. However, the downstream costs can represent a multiple of what was spent at acquisition.

It might start with costs related to the delivery or installation of the product. During operation, consumption of water, energy, and so on, or maintenance and the need for spare parts can figure prominently in terms of costs. Lastly, dismantling or treatment costs need to be taken into account or, conversely, the benefits from selling the product.

Using LCC in SPP equates to the quest for products with the greatest economic advantage over the life cycle, rather than the cheapest purchase price. This life cycle approach to costs makes a purchaser think about the function of a product. It takes into account the life cycle and the related costs for acquisition, operation, maintenance, and end of life. As a result, the purchaser knows the consumption and expenditure figures in detail, from the start.

The arguments in favour of LCC (as often formulated at the international level) are therefore obvious; that is, it helps the purchaser:

- plan future costs (and benefits);
- understand the influencing (often hidden) cost factors;
- compare offers in detail;
- assess future budget needs;
- evaluate the (economic) lifespan of a product;
- improve the support of the use of a product;
- balance costs against sustainability criteria.

It is clear that not all product groups have the same priority for use of LCC. First of all, a distinction is to be made between two major categories: goods and services used for an organisation's daily operation, and buildings. The latter form an important category with distinct specifications for the application of LCC. The focus here is on the first category.

The subcategory with the biggest potential for LCC is products that consume energy, water, and so on, during their operation, such as ICT in the office, vehicles, laboratory equipment, machines in a printing house.

Of lower priority are paper, fuel, postal services, software, and so on, but it should be borne in mind that there may be a relation with other products/ systems.

Furniture, office equipment, carpets, and so on, may also have no or lower priority. However, when there are offers on the market designed with a 'cradle-to-cradle' perspective (C2C), these products are leased and taken back by the supplier at the end of their life.

Note that (public) procurement that requires C2C solutions (i.e. in the format of product–service systems) is stimulating innovation and sustainable regional development.

In practice

In the public as well as in the private sectors, purchasers and sustainability experts are frequently seen as inhabiting different worlds. Those responsible for providing an organisation with goods and services have a practical approach to daily operation; their colleagues with expertise in sustainable development are capable of explaining environmental, social, and economic considerations.

A hands-on calculator for incorporating LCC in (public) procurement practices is therefore helpful. One of the few tools available for free is the SMART SPP tool (Smart SPP 2015), a programmed MS Excel sheet resulting from a project under the supervision of ICLEI and available in Dutch, English, French, and Spanish. There is a step-by-step guide. The tool is used worldwide in training in the use of LCC for SPP. Teams from different countries have been working together on the project and have developed and applied the tool in collaboration with local authorities. The city of Barcelona, for example, has identified the most innovative and most energy-efficient solution for the installation and management of on-street charging stations for electric vehicles, each with two charging docks. The public authorities in Bromley (London), Cascais (Lisbon), Cambridge (UK), and Kolding (Denmark) have all assessed the opportunity of the SMART SPP tool in advancing LED when procuring 'light at night' for public spaces, streets, and so on.

A final remark relates to the tool's possibility for monetising emissions such as CO_2. Research has shown that the real (negative) value of these emissions (e.g. the price of CO_2 per ton on the Emissions Trading System market) does not affect consumers' final selection of a product in terms of costs over the life cycle.

A couple of years ago, in Brussels (Belgium), the Centre d'Informatique pour la Région Bruxelloise started to challenge the market by including (inter alia) total cost of ownership in its tenders, in particular for desktop computers, laptops, printers, beamers, servers, and so on. With the exception of servers, this resulted in a reduction in the overall price of these devices of 10–25 per cent, and technical performance increased considerably compared with a previous tender.

Although it is hard to find case studies, research on the basis of real figures (Demeulenaere and Mazijn 2012) has revealed the possibilities for printers and cars. Printers with the same functional unit can be twice as expensive, and the total cost of ownership over a lifespan of 5 years can differ by three to eight times the purchasing price because of operation (electricity) and maintenance (toner) costs. When it comes to, for example, city cars, there can be a small difference in price between different brands and engine fuels (diesel, petrol). However, purchasers should be aware that the total cost of ownership over a lifespan of 6 years and 20,000 km/year can be double the purchase price.

Linking LCC in SPP to sustainable regional development

SPP is an instrument to stimulate regional development within a sustainable development context. Including environmental, social, and cost considerations when purchasing products (and/or services) is a first step, but it is also a valid approach for imported goods and services.

A further step is tendering with cradle-to-cradle approaches (see Chapter 14 in Part II on life cycle sustainability assessment). When this is the case, product–service systems will become more and more important. It provides opportunities for sustainable regional development. Within that context, LCC is an instrument that provides a picture of the wide range of costs.

Conclusions

LCC in SPP starts with a definition of needs. However, a public authority or a private organisation (including enterprises) setting up an (environmental) management system can reduce the material needs considerably before purchasing goods and services. Monitoring of resources and influencing people's behaviour are essential parts of this management system. As a result, impressive environmental and financial gains can already be expected. In addition, LCC can be advantageous if we reflect on the functional unit of the product – that is, we don't compare apples with oranges (e.g. when purchasing a printer). Furthermore, LCC reveals more than the tip of the iceberg, because it takes account of costs for purchase, installation, operation, maintenance, service and end-of-life handling. It delivers substantial information to inform purchase decisions. Moreover, tendering for product–service combinations stimulates sustainable regional development. However, LCC in sustainable procurement should be complemented by environmental and social criteria.

References

Benoît, C., and Mazijn, B. (eds) (2009) *Guidelines for Social Life Cycle Assessment of Products – A Social and Socio-economic LCA Code of Practice Complementing Environmental LCA and Life Cycle Costing, Contributing to the Full Assessment of Goods and Services within the Context of Sustainable Development*, Paris: UNEP-DTIE.

Demeulenaere, B., and Mazijn, B. (2012) *Manuel pour l'utilisation de l'analyse des coûts du cycle de vie (ACCV) dans les marchés publics (durables)*, Brussels: SPP Développement Durable.

Hunkeler, D., Lichtenvoort, K., and Rebitzer, G. (2008) *Environmental Life Cycle Assessment*, Boca Raton, FL: CRC Press.

Smart SPP (2015) Home. Available at www.smart-spp.eu/ (accessed 15 December 2015).

34
SUSTAINABILITY PERFORMANCE IN INDUSTRIAL PARKS

Jean-François Vallés

Introduction

As a common approach to industrialisation in the second half of the twentieth century, industrial parks marked a break in the way many businesses chose their locations. In response to the attraction of lower cost and to political and environmental pressures, companies moved out of cities to establish themselves in designated zones at the urban periphery. The apparent success of this strategy is demonstrated every day around the world, with regional administrations everywhere striving to create additional industrial zones and promoting them through structured marketing campaigns.

A quick glance at existing sites argues in favour of this strategy: we see many companies located side by side in well-organised industrial hives, with adequate space for vehicles and employees. There are undoubted benefits in establishing numerous businesses side by side, with increased overall efficiency and added value for the companies involved. But, seen from a wider perspective, the advantages for regional development are not always fully realised. In fact, few local authorities take advantage of business parks to create additional industrial synergies; too many still see such structures with the rather narrow focus of finding a convenient location to cluster tenant companies on land they have made available.

A definition of sustainable industrial performance

If we refer to the definition of sustainable development in Agenda 21, business parks should be a priority for public policies, as the grouping of businesses and jobs is at the heart of the triptych: economy, environment, and society.

Taking the historical definition further, we can expand the meaning of sustainable performance of an industrial park to the following: an industrial park must enhance the economic performance of businesses and regional authorities, while reducing their impact on the environment, and also contribute to the creation of jobs and the quality of life of employees. This is a broader vision than just management of a piece of real estate. It also goes further than the purely environmental purpose of the ISO14001 environmental management standard that is still all too often the sole sustainability objective of business management. This broadening of the agenda is one of the motives behind the attention now being paid in Europe and North America to ways in which to expand the concept of industrial parks, moving from economic productivity to also encompass the benefits that flow from collaborative action and circular economies.

Managing sustainable business parks

A subtle but important element is often overlooked when the business park concept is considered: how the park is managed. With the exception of imposing some administrative requirements, the manager of a business park usually has little capacity to intervene in the internal practices of a company. There are few initiatives aimed at proposing more efficient business solutions to companies located in parks.

In any case, there cannot be a single party responsible for the implementation of a holistic sustainable performance approach. Of course, the park manager has a potential role in initiating or proposing action, but managers must also work in conjunction with other institutional partners and, importantly, the businesses in the park. Other stakeholders can also be mobilised. Enhancement of the working environment through a 'sustainable business' approach is of interest to trade unions, for example, as already explained in Chapter 20 (Vallés and Van Meesche 2007).

The levels of intervention

How can a business park foster the emergence of a sustainable development policy when it has only limited influence over the individual tenant corporations? We can see a number of components in a collective sustainable performance strategy. Many of these were identified in a project carried out in the framework of the European project 'Invest in Med', which drafted the *MED Zones: For Sustainable Performance of Business Parks in the Mediterranean* guidelines (CCIMP 2010).

The first considerations in sustainable performance are the management and maintenance of a business park. Certain parks are managed very well. Those oriented towards foreign investors often have ISO 14001 or even ISO 9002 certification as proof of their quality credentials. Such certification is also found for private business parks (Croxley Green Business Park n.d.). Nevertheless, a large majority of industrial parks is not well managed, leading to dysfunction that discourages companies from undertaking collaborative ventures. Fewer than 500 of the 30,000 industrial parks in France are well managed; most do not take effective action on sustainability issues.

Although the implementation of a new management logic does not necessarily require significant work, it is important to have a proactive manager take charge of this area. Outside assistance is also often needed. For example, the framework of the SMIA project in Tunisia includes a component to strengthen the role of managers of business parks (Ksar Saïd n.d.). The training has focused on the evolution from a passive logic ('I am responding to requests') to a proactive logic ('I anticipate the needs and am seeking to improve my site'). The project is seen as a role model for other sites. In France, some business parks have developed newsletters for companies, discussing methods for making improvements, best practices, or new services, and so on.

Irrespective of leadership issues, it is essential to first develop an appropriate co-operation culture among the heads of companies. Regular meetings lead to a realisation that they have similar needs and can perhaps provide the motivation to develop mutually beneficial relationships. These meetings may also encourage the development of common venues, such as restaurants, research laboratories, or meeting rooms. In France, the Association Athelia regularly organises events to facilitate exchanges on waste, energy saving, transport for employees, and so on, as well as encouraging informal meetings between heads of businesses, inter-company sports competitions, and shared travel for employees (Athelia Association n.d.).

A recent improvement in industrial parks has been the increase in services geared towards businesses and their employees. Common-area upkeep, water and energy supplies, and waste treatment have been the focus of industrial parks around the world. This is now being complemented by services that help increase the competitiveness of companies, improve the quality of the work environment, and help with developing new solutions to reduce environmental impact. Services may include joint energy supply, laboratories, logistics, emergency and environment services, showrooms, and many others, including services meeting employees' needs. An industrial park in Turin, Italy (Environment Park n.d.), is an interesting illustration of this approach. This park proposes eco-activities as services: catering for employees, research laboratories, rooms for meetings, and green electricity, among others. The provision of such services naturally depends very much on the design of these areas.

Layout and land planning greatly influence the possibilities for collaboration. Traditional business parks usually have rectilinear avenues that serve large plots for established businesses. This simple approach is inefficient and a great consumer of land. In many parks in France, the land dedicated to tenants' buildings represents less than 20 per cent of the total surface area. More importantly, this type of layout is not conducive to the collaborative approaches presented earlier, as the proximity between firms is not taken into account in planning (IDEA35 2013).

More enlightened planning now includes public transport links and connections to urban areas or employees' places of residence. In the UK, Greenham Business Park offers a daily shuttle to Newbury city centre to reduce individual car use (Greenham Business Park n.d.). On the Eindhoven High Tech Campus, in Holland, the car parks are conveniently located in the heart of the business area, and the services are located so as to facilitate movement by foot and bicycle. Use of the car within the zone is forbidden. Beyond the reduction of impacts on the environment, this choice promotes exchanges between employees and provides opportunity for the development of synergies and innovation (Eindhoven High Tech Campus n.d.). When speaking of the setting up of the Blackberry© headquarters, near a university, the founder of the company was in the habit of saying that innovation is done in tennis shoes.

We must also take into account *external integration*, the ability of a site to facilitate the connection between businesses and their customers, partners, or suppliers. This includes ease of access to railway stations/airports and ports, but also to logistics platforms. Transport links are increasingly important to business parks dedicated to production for global markets. Some business parks offer logistics and warehousing services for companies (Industriepark Höchst n.d.).

It is also useful to consider *urban integration*, that is, the quality of the link between business parks and areas that surround them. This factor is particularly important, not only to reduce the potential nuisance to the bordering areas but also to create synergies for the region as a whole. An industrial plant in Morocco has strengthened its links with nearby residential areas by offering localised

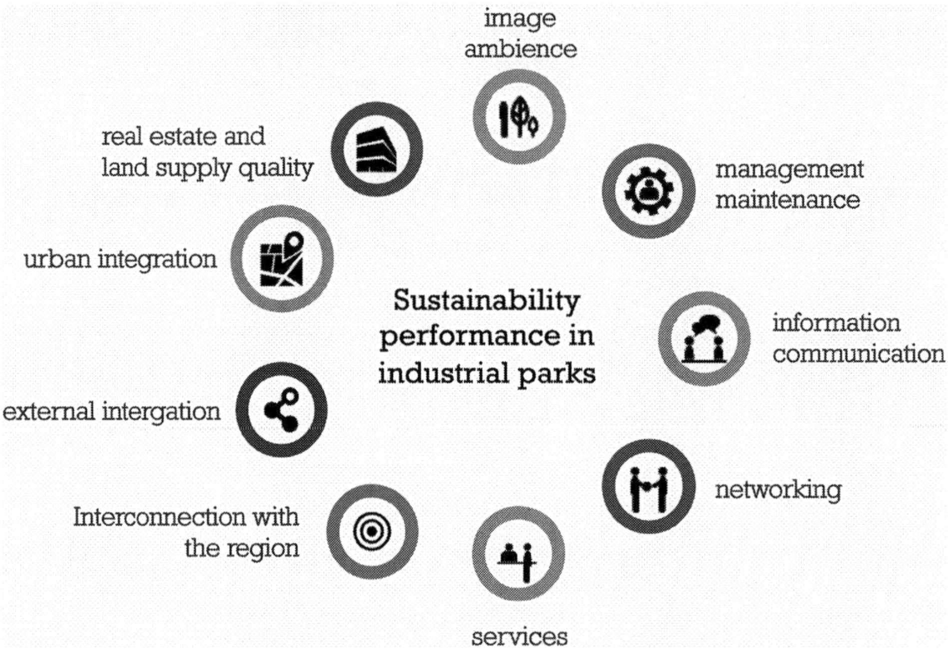

Figure 34.1 Sustainable performance of industrial parks

services in the vicinity (Bouskoura Industry Park n.d.). This approach helps reduce the negative effects of specialised urban zones on cities, something that has been the emphasis of numerous urban developers (Rogers 1997).

Addressing the above issues leads to a reconsideration of the nature of the land on offer to businesses. Is it only land or buildings? Does it propose future synergies? By their very nature, the content and quality of the offer may contribute to reducing the environmental impact of businesses and also help in facilitating the creation of cooperation between companies, thereby strengthening the regional industry fabric. A dynamic image will also attract a certain business clientele: an innovative park will be more attractive to companies (and employees) that present themselves as 'sustainable'.

Performance levels

Should all business parks look alike? Should they all seek excellence at each of the levels mentioned above? Each business park is unique in terms of its location, its size, the structures that are developed, and, especially, the nature of the businesses that are involved. Each company has different needs and expectations related to its operations, size, number of employees, area of business, and financial structure.

By its nature, each business park will focus on certain themes. The internal synergy will not be the same for a park dedicated to production and for one dedicated to innovation based on exchanges between employees. Similarly, the needs and expectations of services will differ. For parks with a focus on industry, services will aim to support the inner workings of the company, whereas, for parks dedicated to innovation, we can expect to find more intangible services (joint responses to calls for projects, development aid, collective presentations at trade fairs, etc.).

Intervention strategies

It is not a simple matter to further evolve the structure of an existing business park that already has many tenant companies. Evolution is most likely where we find a mechanism to motivate businesses to give a higher value to their park.

One of the strategies that seems to be bearing fruit in older business parks is to encourage a sustainable performance approach through staged action, where companies are initially invited to participate in resource efficiency and service sharing, before moving on to the more demanding step of collaboration on business opportunities. Each situation will be different, but the above examples demonstrate that, given sufficient leadership, good results can be expected. The role of the region in providing or supporting such leadership is clear where it is the owner of business parks, but regions have a role in facilitating the sustainable evolution also of private parks.

Conclusions

Working towards more sustainable performance of business parks is an effective response to the environmental, social, and economic issues that face a region. It can also be of benefit to individual companies. The results from various multi-stakeholder approaches being tried in regions around the world offer hope that these new approaches can be applied on a broader scale than has been the case so far.

References

Athelia Association (n.d.) Available at: www.atheliaentreprendre.fr (accessed 5 August 2016).

Bouskoura Industry Park (n.d.) Available at: www.gepib.ma (accessed 5 August 2016).

CCIMP (2010) MED Areas for a Sustainable Performance of Areas of Activities, Study no 19, CCIMP & ANIMA. Available at: www.animaweb.org/sites/default/files/investinmed-finalreport.pdf (accessed 5 August 2016).

Croxley Green Business Park (n.d.) Available at: www.croxleygreen.co.uk (accessed 5 August 2016).

Eindhoven High Tech Campus (n.d.) Available at: www.hightechcampus.com (accessed 5 August 2016).

Environment Park (n.d.) Available at: www.envipark.com (accessed 5 August 2016).

Greenham Business Park (n.d.) Available at: www.greenham-business-park.co.uk (accessed 5 August 2016).

IDEA35 (2013) Available at: www.idea35.fr/download.aspx?DocumentID=1083 (accessed 5 August 2016).

Industriepark Höchst (n.d.) Available at: www.industriepark-hoechst.com (accessed 5 August 2016).

Ksar Saïd (n.d.) Available at: www.gmg.com.tn (accessed 5 August 2016).

Rogers, R. (1997) *Cities for a Small Planet*, London: Faber & Faber.

Vallés, J.F., and Van Meesche, M. (2007) *The Sustainable Management of Areas of Activities*, Ghent, Belgium: Union CGSLB Liberal. Available at: www.google.fr/url?sa=t&rct=j&q=&esrc=s&source=web&cd=8&ved=0ahUKEwiUsqL34LvKAhVCjCwKHd6dCewQF.

ghHMAc&url=http%3A%2F%2Fwww.rise.be%2Ffiles%2Flibrary%2FPublications%2FPasser-a-l-action%2Fconcertation-sur-les-zone%2Fla-gestion-durables-des-zones-dactivite.pdf&usg=AFQjCNE1QBP2ycw6hStlrzLFAUcjns87qg&sig2=uT56jEoQ7Osf_9ZmrU-Gow&bvm=bv.112064104,d.bGg&cad=rja (accessed 5 August 2016).

35

SUSTAINABLE REDEVELOPMENT OF DEGRADED LAND AND LANDSCAPES

Peter Whitbread-Abrutat

Introduction

Combating land degradation is one of global society's major environmental challenges: the erosion of our terrestrial resources (and Earth's natural and social capital) directly and indirectly affects the lives of millions of people around the world (Nkonya et al. 2011; Bai et al. 2008; Millennium Ecosystem Assessment 2005). Restoring degraded landscapes (and ecosystems) is, therefore, a powerful way to rebuild ecological integrity and enhance the lives and livelihoods of people connected with them, for the long term – a sustainability concept that is gradually gaining traction with the advent of major new ecosystem-focused international initiatives such as IUCN's Commission on Ecosystem Management (Keith et al. 2013). In particular, it enables the 'recycling' of land for ongoing benefit to people, which, in the context of this chapter, may include the restoration of biodiversity, ecosystem functions and related services. In this sense, 'benefit' extends far beyond the realm of conventional economics and realises the critical importance of non-financial ecosystem goods and services and the human and social capital that can derive from them.

Defining landscape restoration

Whitbread-Abrutat (2012) defines landscape restoration as: 'The improvement of degraded land on a large scale that rebuilds ecological integrity and enhances people's lives'.

This definition is deliberately broad in order to encompass a diverse range of ambitions, activities, scales, environments, societies, and end-uses. It adapts the definition of forest landscape restoration, as defined by Maginnis and Jackson (2007), to non-forest landscapes and introduces the notion of scale. It also extends the Society for Ecological Restoration definition of ecological restoration (Society for Ecological Restoration International 2004) to include the concepts of scale and landscape (as opposed to ecosystem) and reinforces the human element – as recognised by Hobbs and Norton (1996) and Hobbs and Harris (2001) in their synthesis of the drivers for large-scale ecological restoration. The definition was also informed by comprehensive work on mining legacies by Whitbread-Abrutat (2008).

Key aspects of the definition are that landscape restoration activities should:

1 improve degraded environments through rebuilding ecological integrity;
2 operate on a large (temporal/areal) scale;
3 enhance people's lives.

Central to this definition is the consideration that, too often, social development programmes ignore the livelihood possibilities that restoring the natural environment may provide, and, similarly, environmentally focused rehabilitation programmes do not always consider the needs of people when determining final land-use options. Landscape restoration should aim to reconcile economic, social, and environmental concerns within a holistic framework, the best projects utilising environmental improvements as drivers for socio-economic regeneration. Less commonly, but more effectively, socio-economic development can be used to generate enduring environmental improvements. An excellent example of the latter is the Eden Project in Cornwall, UK. It was created in an exhausted 160-year-old china clay quarry at the turn of the century, in a region that had been suffering the consequences of a collapse of the mining industry for decades. Eden was perceived as a simultaneously environmental and socio-economic regeneration project, as the quarry was transformed into a botanical showcase and environmental education charity (see Figure 35.1). It employs 400 and has added value of over £1.2 billion to the local economy since opening in 2001 (Eden n.d.).

Such considerations are fundamental to many aspects of the sustainable development paradigm. It is this conceptual context, as well as the environmental and socio-economic potential that underlies it, that forms the backdrop to the landscape life cycle aspects of this chapter.

Exploring landscape restoration

Every landscape restoration project is a unique response to a specific set of challenges and opportunities, and yet there are common generic challenges addressed in all of them.

Figure 35.1a
China clay mining landscape, Cornwall, UK (top)

Figure 35.1b
Eden Project, created in a transformed china clay pit, Cornwall, UK (bottom)

The discussion below builds on a long-term research approach that has been pragmatic and practically focused. Over many years, it has considered a deliberately broad range of restoration projects across the world on landscapes degraded by mining/industry, agriculture, commercial forestry, urban development, and so on, and in a diversity of environments, geographical regions, and socio-economic/cultural contexts.

This work has attempted to draw generic conclusions in response to the following aims:

1. identifying the challenges to delivering landscape restoration projects;
2. understanding the elements of success required to deliver them;
3. developing a set of generic recommendations that should apply to almost any project, anywhere.

In particular, the achievement of sustainability objectives has been studied – particularly determining how the process of landscape restoration can rebuild ecological processes and ecosystem services, while simultaneously enhancing socio-economic priorities, particularly the restoration of livelihoods, improved health and well-being, community participation, and cultural enhancement.

Challenges to delivery and how they can be addressed

Every project will inevitably face external or internal challenges, and there is no guidebook to navigating them, but, if such challenges could be anticipated, a degree of preparation should be possible. The key challenges can be loosely assigned to four groupings, as outlined below, although the groupings are not mutually exclusive.

Participation

Meaningful stakeholder participation leads to more grounded sustainable development activities, justifying the additional time and resources required to ensure greater participation. A committed individual or small team working independently will have a limited positive effect locally, at best, but scaling will only be achieved by *building a constituency* with others, particularly of those with influence – local people, politicians, businesses, and conservation organisations – to support the landscape restoration initiative.

Engaging *local communities* is essential, but is not always easy; they may be isolated geographically and independent-minded, or may have been previously overlooked for being too thinly dispersed to warrant the effort, or may be blamed for environmental damage to satisfy their pressing needs when wider society offers no alternative. They may feel disenfranchised, apathetic, insular, independent, and wary of outsiders. Previous poor consultation experiences can persist in the collective memory, leading to distrust and a lack of constructive engagement. Cultural and language differences and poor education can be further barriers to engagement. These issues may be compounded by a lack of capacity. Such communities are not uncommon in degraded lands.

The greater the degree of participation, the greater the *sense of ownership*; however, there may also be a requirement for upfront investment in capacity-building in the short term. This is apparent, for example, in the management of Costa Rica's Area de Conservación Guanacaste (AGC) – an expansive area of

Figure 35.2
Dry tropical forest restoration – before and after 20 years of restoration effort

tropical forest largely regrown from previously cleared grazing land – which instituted a policy to employ only Costa Ricans and promotes park staff from within, thus building local ownership and pride (see Figure 35.2).

It is important to be aware that an individual's or organisation's objectives may well be different from those of the overall project. The key to building a constituency is, therefore, to develop a common agenda that satisfies a range of expectations and often involves addressing people's *perceptions* of the land and how they have interacted with it in the past. An example is provided by the Reserva Ecológica de Guapiaçu (REGUA) project, 80 kilometres north-east of Rio de Janeiro, in the upper watershed of the Guapiaçu river. REGUA has an enlightened and brave policy of employing and training up local people to work on its project – the main thing it looks for in potential employees is the 'right attitude'. A fascinating example is offered by the guides employed to lead wildlife-watching trips into the forest. Each of them originally worked illegally in the forest as hunters/poachers, and REGUA offered them the first formal employment of their lives as guides. They learned on the job and by informal teaching by local national park staff. Adilei is one of those inspiring guides, having learned the calls and behaviours of local birds, and was able to attract birds closer for better viewing by bird-watching tourists. He had little formal education, but explained that he is now learning English in his spare time so that he could interact better with his tourist charges (Whitbread-Abrutat 2012).

Governance

Attention paid to the details of how the restoration project and its aftermath are managed and governed greatly improves outcomes and the chances of attracting further funding and in-kind contributions, ultimately leading to greater project sustainability. Good governance is essential for ensuring the long-term viability of projects that go beyond simply restoring and walking away and is particularly important where projects are set up to deliver long-term socio-economic benefits. The development of sound project goals is critical, based on a multi-stakeholder process that also addresses the inevitable institutional barriers and, where feasible, seeks to influence policy and the legislative process.

In this sense, good governance includes clear decision-making, financial/fiduciary responsibilities, and efficient administration. *Transparency* is also a key element of good practice here. Decision-making should be guided by principles established at the outset. A critical activity of the governing body is to facilitate the regular monitoring and evaluation of project activities, to ensure quality is maintained and funds are spent wisely and effectively. The governing body should also plan for the long term – the team that establishes and delivers the project is probably not the same governing team that should be managing land and its socio-economic aspects indefinitely.

Controlling the land is critical, and actual title is important; most straightforward is to simply buy the land outright, as in the restoration of farmland and development of Chilean Patagonia's Pumalin and Future Patagonia Parks by Doug and Kris Tompkins of Tompkins Conservation (Tompkins n.d.). But, according to Janzen, after his long experience in the restoration of dry tropical forest in Costa Rica, equally important is that ownership needs to be 'psychologically and socially visible if conservation is to work over the longer term', meaning that outsiders need to physically see and understand that an area of land earmarked for conservation is actually being beneficially used by people (Janzen 2010, personal communication).

Sustainability

The point in time when the landscape restoration project ends is too late to consider implementing ongoing sustainability initiatives to ensure that the restored landscape remains viable and that the people associated with it flourish. Transitional thinking between the project and post-project scenarios needs to be engrained within the project goals from the start. The transitional phase should be as smooth as possible for stakeholders and local communities and should be embedded within the governance system, including the incorporation of changes to governing bodies to increase stakeholder participation over time and development of funding and project management and delivery mechanisms that are sustainable financially as well as ecologically and socially.

The restoration process can also provide training opportunities for those involved, building their own capacities for future benefit, either locally or further afield. In the AGC, for example, Janzen trains university leavers as 'parataxanomists' to identify and catalogue AGC species. In the restoration of defunct Patagonian farms in Chile, Tompkins has researched farming practices more sensitive to the local environmental conditions and has developed a viable, growing honey and preserves industry as part of a long game in transforming the regional economy from one of extraction and exploitation to sensitive production and environmental tourism.

A prime example of landscape restoration sustainability in this sense is provided by one the earliest major landscape restoration projects: the re-greening of Sudbury, Ontario. The forests and lakes of Sudbury – a major nickel mining centre – were ecologically devastated by smelter emissions in the early to mid twentieth century. Air quality began to improve in the 1970s, when the technologically advanced 'Superstack' was commissioned. In 1973, the provincial government, primarily concerned about joblessness among local miners, employed them to plant trees around communities and major transport routes (Lautenbach 2005, personal communication). Other stakeholders involved, including ecologists at the local Laurentian University, deduced how best to restore the forest ecologically, and, ultimately, a major re-greening effort began, including all levels of government, the university, mining companies, schools and local communities. To date, well over 10 million trees have been planted, which has had substantial knock-on benefits for the city's economic diversification, improved health and well-being of its citizens, and vast ecological enhancements (see Figure 35.3). The re-greening programme has been the

Figure 35.3
Regrowing forest trees in the shadow of two closed smelter stacks, Sudbury, Ontario, Canada

educator of countless students and school children and is used for training foresters and land managers. The project continues.

'The oil in the machine'

Delivering a successful landscape restoration project is about far more than ticking a checklist of stages and processes. In many ways, it is an organic thing, built on human relationships and evolving over years and decades, with changing personal circumstances, societal expectations, and the personalities involved. Inevitably, unpredictable events and opportunities occur that will need to be carefully considered and reacted to. The more flexibility and adaptability that a project has engrained into its modus operandi and collective philosophy, the more durable and ultimately successful it will be. There are several cross-cutting themes that are critical to every successful project everywhere, which I call, for want of a better term, 'the oil in the machine'. The key elements of it are leadership, communication, collaboration, knowledge, and creativity and beauty.

Often overlooked, yet of critical importance, is the notion of *creativity and beauty*. The success of the Eden Project is, in large part, due to the integration of engineering, science, and art to deliver something that people want to see – whether they are interested in plants or not!

The challenge of acid mine drainage – one of the most widespread and difficult-to-treat environmental impacts of the global mining industry – was addressed innovatively and creatively by the AMD&ART project at an abandoned colliery in Vintondale, Pennsylvania (Vintondale 2011). In Vintondale, an environmental reclamation and community enhancement initiative brought broad public participation to the design and construction of acid drainage treatment systems. The AMD&ART team of scientists, engineers, and artists transformed the site into new wetlands and recreational space for the community, while also creating a large-scale passive water treatment system. The underlying premise of this approach is that it is the combination of good science with good design and historical perspective that brings community understanding and support.

Conclusions

Post-project sustainability needs to begin at the outset of project planning to ensure that future environmental and community opportunities are built into the collective mindset. The suggested changes in approach will require engagement with appropriate experts – maybe artists, historians, landscape restoration experts, academics, horticulturalists, politicians, business people, wildlife experts, engineers, entrepreneurs, and so on – those with knowledge of a range of successful approaches in different circumstances.

The plan should outline the strategy that creates the foundation upon which to build a new environmental and socio-economic future for the surrounding communities. As much detail should be provided on the participation, governance, sustainability, and 'oil in the machine' issues as on the technical reclamation aspects.

In a twenty-first-century, climate-changing world of more than 7 billion people, we need to be cleverer than thinking that landscape restoration is just about planting trees and walking away; the best in this field are able to see the potential of a landscape and its people and alter their own perceptions and lives accordingly to achieve this potential.

For most of the people involved in such work, there is a frustration that they wish to do more, but that a human lifetime is inevitably limiting. We may not know all the answers, but we know enough now to make substantial improvements to degraded lands and their people.

Ultimately, a landscape restoration project will only ever be a chapter in the narrative of a landscape. It is incumbent on those connected with it to ensure that the end of the project signals the start of a new chapter, and not the end of the story.

References

Bai, Z.G., Dent, D.L., Olsson, L., and Schaepman, M.E. (2008) 'Proxy global assessment of land degradation', *Soil Use & Management*, 24: 223–34.

Eden (n.d.) Available at: www.edenproject.com (accessed 5 August 2016).

Hobbs, R.J., and Harris, J.A. (2001) 'Restoration ecology: Repairing the Earth's ecosystems in the new millennium', *Restoration Ecology*, 9(2): 239–46.

Hobbs, R.J., and Norton, D.A. (1996) 'Towards a conceptual framework for restoration ecology', *Restoration Ecology*, 4(2): 93–110.

Keith, D.A., Rodríguez, J.P., Rodríguez-Clark, K.M., Nicholson, E., Aapala, K. Alonso, A., Asmussen, M., Bachman, S., Basset, A., Barrow, E.G., Benson, J.S., Bishop, M.J., et al. (2013) 'Scientific foundations for an IUCN Red List of Ecosystems', *PLoS ONE*, 8(5). Available at: http://dx.doi.org/10.1371/journal.pone.0062111 (accessed 5 August 2016).

Maginnis, S., and Jackson, W. (2007) 'What is FLR and how does it differ from current approaches?', in *The Forest Landscape Restoration Handbook*, eds J. Rietbergen-McCracken, S. Maginnis, and A. Sarre, London: Earthscan.

Millennium Ecosystem Assessment (2005) *Living Beyond Our Means: Natural Assets and Human Well-Being*. Washington, DC: Island Press.

Nkonya, E., Gerber, N., von Braun, J., and de Pinto, A (2011) Economics of Land Degradation. The Costs of Action versus Inaction. Bonn, Germany: International Food Policy Research Institute, Center for Development Research, University of Bonn. IFPRI Issue Brief 68, September.

Society for Ecological Restoration International (2004) *Primer on Ecological Restoration*, Version 2. Available at: www.ser.org/resources/resources-detail-view/ser-international-primer-on-ecological-restoration (accessed 5 August 2016).

Tompkins (n.d.) Available at: www.tompkinsconservation.org (accessed 5 August 2016).

Whitbread-Abrutat, P.H. (2008) *Mining Legacy Survey (for the ICMM-IUCN Dialogue on Mining and Biodiversity)*. Post-Mining Alliance, Eden Project. Available at: www.postmining.org/documents/surveysummaryreportFINAL25Feb08.pdf (accessed 5 August 2016).

Whitbread-Abrutat, P.H. (2012) Exploring World Class Landscape Restoration. Travelling Fellowship report for the Winston Churchill Memorial Trust.

Vintondale (2011) Available at: www.youtube.com/watch?v=F8yjArnrxXg (accessed 5 August 2016).

Part V

INTEGRATED USE OF LIFE CYCLE APPROACHES IN KEY ECONOMIC SECTORS

36

SUSTAINABLE AGRI-FOOD CHAINS AND LCA

Bruno Notarnicola, Pietro Alexander Renzulli, and Giuseppe Tassielli

Introduction

As explained in Chapter 10, the environmental impact of private consumption is a cause of environmental damage worldwide (Tukker et al. 2006), and food production and consumption are no exception. This has led to the development of methods used to study sustainability issues surrounding agri-food chains. Among these, life cycle assessment (LCA; see Chapter 10) is a powerful tool that can clearly identify the environmental burdens of a product system throughout its life cycle (Notarnicola et al. 2015).

LCA-based initiatives and labels for the agri-food sector

The quest for sustainable production and consumption in the agri-food sector has stimulated the creation of many international initiatives and strategies designed to reduce the environmental impacts of food production and consumption. These include numerous LCA-based initiatives and labels related to the agri-food chain.

Labelling and declaration programmes

This section summarises the labelling and declaration programmes that are used worldwide and that represent a means of encouraging sustainable food and beverage production and consumption at the regional level (for an introduction to eco-labels, see Chapter 18).

- The China Environmental Labelling programme: this includes two types of voluntary label. The first of these is for products within the scope of existing national technical standards. Currently, the only standard applicable to the agri-food sector is one applying to soft drinks. The second is based, essentially, on self-declaration verified by a national certification centre.
- Ukrainian Living Planet label: this label, created by the non-governmental organisation Living Planet, and supported at governmental level, is a Type 1 ISO 14024 scheme (see Chapter 18), with specific standards already specified for twelve food products and six types of beverage.
- Russian Vitality Leaf: this is the only Russian label; it is based on the internationally recognised ISO 14024 standard. There are currently specific standards for vegetables, drinking water, and alcoholic beverages.
- Indian Ecomark: this eco-labelling scheme, created in the early 1990s as a means for easy identification of environmentally friendly products, provides

specific standards for many types of food product, but has not achieved the expected appeal to consumers or industry over the years.

- At the European level: EU Ecolabel certification is applicable to twenty-six categories of products/services, none of which belongs to the food and drink sector. This is symptomatic of the difficulties in creating a unified labelling system that satisfies both the consumers' need for information on the sustainability of products and the needs of the food industry for a labelling system that is easy to apply.

There are currently five environmental product declaration (EPD) labelling schemes based on ISO 14025 (see Chapter 18), namely:

- the international EPD system: the most widespread, with specific product category rules (PCRs) for more than twenty types of food and drink product;
- EPD Norge: there are currently no food- and drink-related PCRs for this kind of Norwegian EPD;
- Earthsure®: the first North American EPD programme, compliant with ISO 14025; the two PCRs available for the agri-food sector apply to meat and beer;
- Ecoleaf environmental label: a Japanese scheme currently only implementing one PCR regarding the agri-food sector (meat products);
- sustainability measurement and reporting system: this North American initiative, based on Type 3 declarations, is currently under development.

The numerous international labelling and declaration programmes mentioned above that communicate footprint results (particularly the EPD schemes) lack a standardised methodology for the publication of LCA-based information on the sustainability of products. This has prompted the European Commission to create a new common methodology, known as the Product Environmental Footprint (PEF), for the evaluation and communication of product environmental performance. The PEF scheme is undergoing pilot testing, and there are currently no definite PCRs related to the agri-food sector (see Chapter 23 for more information on PEF).

LCA-based initiatives and protocols

Various actors in the food and drink supply chain have produced numerous initiatives and protocols based simply on LCA, as per ISO 14040 and ISO 14044 (see Chapter 10). A brief, non-exhaustive description of some of these is given below.

For the livestock and dairy sector, the Food and Agriculture Organization (FAO) has launched an initiative named Livestock Environmental Assessment and Performance, to provide stakeholders with a robust, harmonised footprint methodology for food and feed products. Similarly, the International Dairy Federation and the FAO have conducted a joint study of greenhouse gases (GHGs) in the dairy sector, with the aim of providing useful, specific information on the dairy food chain in the world's main regions and farming systems. Other similar initiatives, on national dairy sectors, have also been carried out in Australia, Canada, Chile, South Africa, the UK and US.

The Beverage Industry Environmental Round Table, a coalition of beverage industry companies created to quantify and encourage the reduction of GHG emissions within the beverage sector, has published a protocol entitled 'Beverage

Industry Sector Guidance for Greenhouse Gas Emissions Reporting'. This is a sectoral application of the carbon footprint method introduced in Chapter 11. In addition, a report entitled 'A Practical Perspective on Water Footprinting in the Beverage Sector' was published as a guide to the application and development of existing water footprinting (WF) tolls for the drink sector. More information on WF can be found in Chapter 12.

Lastly, some initiatives regarding the creation of specific carbon footprint standards for products have been developed by retailers such as the Tesco Group in the UK and Casino in France, mainly as a basis for product labelling.

Typical LCA methodological issues for agri-food systems

Agri-food systems, in comparison with other industrial systems (e.g. energy, transport sector, building sector), have the particular feature of involving, not only an industrial system, but also an agricultural one, making the product system to be analysed via LCA more complex, as it will include, not only technical processes, but also biological ones. As a consequence, system boundaries have to be chosen carefully. Moreover, system expansion and/or allocation procedures will have to be judiciously defined, as, in many cases, agri-food systems can generate large quantities and types of co-products. This means that establishing the allocation procedure solely on mass criteria or on the economic value of the co-product can lead to erroneous results. In such cases, allocation should take account of both mass-related aspects and economic value, through a weighted average of these, in order to balance the large quantities of co-products with their low economic value.

The choice of the functional unit (FU) is of vital importance for an LCA. Typically, FUs are based on mass or volume. However, in agri-food systems, such an FU may not be suitable for the scope and goal of the LCA, or it may not correctly represent the function of the product. Furthermore, an agricultural process may yield the same types of product but with different qualitative characteristics (for example, extra-virgin olive oil or just virgin olive oil), in which case the two FUs (e.g. 1 litre of oil) would not be comparable. The production of high-quality wines involves product systems with the main function of producing wines with a certain hedonistic value. In this case, a mass- or volume-based FU will make low-quality wines environmentally outperform high-quality wines that require much more complicated, long-lasting production processes. An FU based on the hedonistic value of the wine will invert the results (Notarnicola et al. 2003).

For most agri-food products, the agricultural phase is the one with the greatest impact, owing primarily to the use of pesticides and fertilisers, as well as the need for land, water resources, and energy. There are many dispersion models concerning pesticides and fertilisers (none of which is universally accepted) that have to take into account site- and time-specific aspects of the product system under examination (see Chapter 13 for more information on dynamic and spatial modelling in LCA). For example, meteorological conditions may influence how a pesticide is dispersed into soil, air, and groundwater. The choice of dispersion model can thus greatly influence the results in terms of impact categories. Furthermore, there are no specific characterisation factors for many types of pesticide. In such cases, assumptions need to made and clearly stated, as these could significantly alter the final results of an LCA study (Notarnicola et al. 2012).

As pointed out in Chapter 10, there is no single methodology for evaluation of the change in biodiversity resulting from occupation of land relating to product systems. Similarly, there is no single, universally accepted approach for evaluation of the degradation of the quality of soil in terms of organic carbon content, erosion, microbial biomass, or salinisation. Aspects such as these, of paramount importance for the sustainability evaluation of food product systems, are often ignored in LCA studies.

BOX 36.1 LCA APPLIED TO QUALITY WINE PRODUCTION IN CHILE

Environmental impacts occur all the way along the extended life chain of wine production. Reducing these impacts requires a total life cycle management approach. Figure 36.1 shows the results of an LCA of a 750 cc bottle of Chilean premium wine, undertaken by a pilot study directed by the Chilean Clean Production Council. The objective was to include life cycle approaches in Clean Production Agreements, a voluntary instrument led by the Council. The project identified environmental 'hot spots' for seven wine products and then established goals and actions aimed at adequate sustainability management. A cradle-to-gate scope was used that included the impacts related to orchard establishment, grape production, vinification, and packaging. The most significant stages were grape production and packaging. For grape production, the relevant impacts concern eutrophication from fertiliser run-off, agrochemical toxicity, land-use change, and water use for irrigation. The packaging stage is associated with mineral and fossil fuel depletion arising from the use of glass bottles. The study allowed the setting of reduction goals for fertiliser and agrochemical application, as well as water use during grape production. Results also identified actions for packaging procurement.

Andrés Rolon (Fundación Chile)

Figure 36.1
LCA of a 750 cc bottle of Chilean premium wine

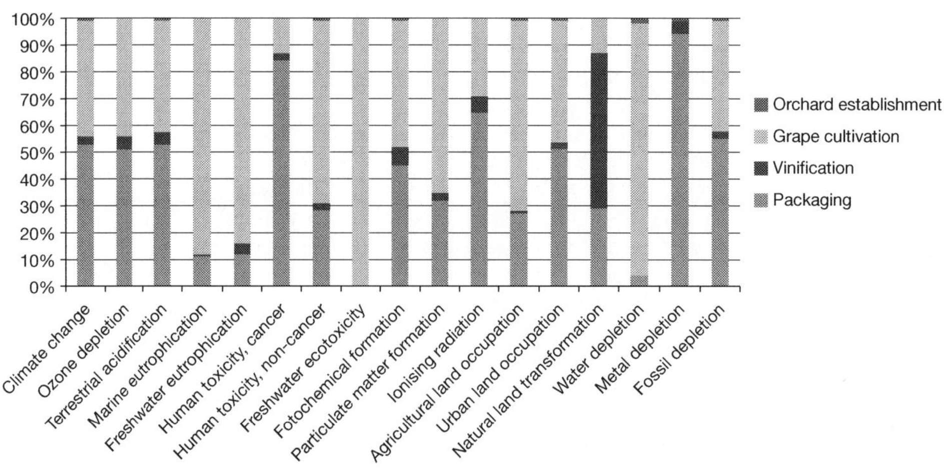

Many agri-food LCAs do not include the use and end-of-life phases. In some cases, these phases are responsible for a large portion of certain types of potential impact (FAO 2011). It is, therefore, important, whenever possible, to include as much of the total life cycle as possible, with a correctly chosen FU that best represents the scope of the assessment, in order to also consider the possibilities of making a product system more recyclable or disposable, within the context of a circular economy.

As already highlighted in Chapter 13, site specificity is an issue that can cause great variability in LCA results. Commercial LCA database producers are starting to provide regionalised data sets in order to account for site dependency. However, the data are still not sufficiently detailed. It is therefore important, when using commercial software, to have a clear picture of what data are being used and of the geographical area they represent; otherwise, the LCA results could be misleading. Lastly, when evaluating carbon sequestration during a food product's life cycle, it is important, not only to carefully evaluate all the carbon sequestered during the photosynthesis of a plant, but also to account for the carbon released during plant degradation, biomass combustion, and human digestion and excretion. This remains one of the least-studied life cycle stages of agri-food products.

Typical LCA methodological issues of agri-food systems

A drive from regional governments for dissemination of information or for subsidising of the use of the above-mentioned types of label for the typical products of a region (e.g. pasta, olive oil, wine, cereal-based products, fruit and vegetables in the case of southern Italy) will not only increase producers' and customers' sustainability awareness, but will also, in the long run, reduce the environmental burden of such product systems. For example, wine LCAs (e.g. Notarnicola et al. 2013) have shown that electrical energy consumption and packaging are some of the main causes of the environmental burden of this sector. Similarly, LCAs on olive oil (e.g. Salomone and Ioppolo 2012) have shown that the agriculture phase and the use of pesticides are the main hot spots of the olive oil supply chain. Such information can be used by producers to reduce the environmental impact of their products and can also be used concurrently to promote these agri-food products via labelling systems, thereby making them more competitive on the market and, hence, stimulating regional sustainable development.

Similarly, encouraging the use of LCA jointly with life cycle costing and social LCA (see Chapters 10, 14, and 36) to support the study of organic/biological food production can help clarify which agricultural practices are effectively more sustainable (some literature seems to indicate that organic practices are not always more sustainable).

Furthermore, the literature indicates that there is a lack of education of farmers on the correct use of pesticides, fertilisers, and water (which determines a large share of the overall environmental burden of food products; Notarnicola et al. 2013). Systematic use of LCA-based methodologies can help farmers to define the minimum amounts of pesticides, fertilisers, and water needed for their crops. This is a crucial aspect for sustainable regional development, as it enables prevention of future degradation of soil quality and of transfer of the environmental burden to other sectors (such as eutrophication of waters affecting tourism) and, in the case of a region limiting water use, makes it possible to free water resources that can be used elsewhere in the regional economy to generate economic growth.

LCA can also be implemented at a macro level as a means of assessing the sustainability of diets of regions or areas of the world. Specifically, statistical data on food consumption, representing the typical food basket of a region, can be evaluated to give an idea of the impact associated with such consumption. This is often done by coupling LCA to environmentally extended input–output analysis (see Chapter 8). If the results indicate that a typical diet of a given region is more sustainable than others, then this information could, for instance, be used to promote sustainable tourism in such areas and the export of those food products.

LCA at macro-system level (see Chapter 10) can also be used to evaluate a region's food logistics and distribution scenarios. For example, a project for food distribution in a new eco-district, implemented with the principles of industrial symbiosis (with, for example, the sustainable reuse of waste food within the region, together with other synergistic exchanges between industrial sectors operating in the area), could be evaluated using an LCA system that would highlight the advantages of such a scenario and make the development of the symbiotic project more appealing. Moreover, LCA can be used to implement regional programmes to reduce food waste and to encourage social programmes aimed at reutilising food waste, as opposed to sending it to landfill or for composting.

Conclusions

In conclusion, an increasing number of LCA studies supporting the regional sustainable development of agri-food chains exist. Some methodological challenges need to be overcome to strengthen those studies further. As a perspective, linking the results of those studies with the implementation of the Eco-Management and Audit Scheme for the environmental auditing and management of organisations involved in the agri-food sector is an option to increase sustainability awareness and encourage better practices and development at a regional level.

References

FAO (2011) *Global Food Losses and Food Waste – Extent, Causes and Prevention*, Rome: FAO.

Notarnicola, B., Salomone, R., Petti, L., Renzulli, P.A., Roma, R., and Cerutti, A.K. (2015) *Life Cycle Assessment in the Agri-food Sector. Case Studies, Methodological Issues and Best Practices*, Heidelberg, Germany: Springer.

Notarnicola, B., Tassielli, G., and Nicoletti, G.M. (2003) 'LCA of wine production', in *Environmentally-Friendly Food Production*, eds B. Mattsonn and U. Sonesson, Cambridge, UK: Woodhead, pp. 306–26.

Notarnicola, B., Tassielli, G., and Renzulli, P. (2012) 'Modeling the agri-food industry with life cycle assessment', in *Life Cycle Assessment Handbook*, New York: Wiley, pp. 159–84.

Notarnicola, B., Tassielli, G., and Renzulli, P.A. (2013) 'La variabilità dei dati nella LCA della produzione olivicola, VII Convegno della Rete Italiana LCA', in *Life Cycle Assessment e ottimizzazione ambientale: esempi applicativi e sviluppi metodologici*, Rome: ENEA, pp. 29–35.

Salomone, R., and Ioppolo, G. (2012) 'Environmental impacts of olive oil production: A life cycle assessment case study in the province of messina', *Journal of Cleaner Production*, 28: 88–100.

Tukker, A., Huppes, G., Guinée, J., Heijungs, R., de Koning, A., van Oers, L., Suh, S., Geerken, T., Holderbeke, B., Jansen, B., and Nielsen, P. (2006) *Environmental Impacts of Products (EIPRO). Analysis of the Life Cycle Environmental Impacts Related to the Total Final Consumption of the EU-25*, Seville, Spain: European Commission, Joint Research Centre (JRC), Institute for Prospective Technological Studies (IPTS).

37

INTEGRATION OF LCA, LCC, AND SLCA METHODS FOR ASSESSING A BIOECONOMY REGION

Alberto Bezama, Anke Siebert, Jakob Hildebrandt, and Daniela Thrän

Introduction

In some industrial sectors (energy, chemical, materials), biomass resources in general, and wood in particular, are seen as a major element for the transition towards a greener economy. Concepts such as 'bioenergy systems' and 'biorefineries' have been the starting point for applied research in these areas. However, as biomass resources are limited, there is a need for a coherent approach to optimise the different biomass uses in a more value-oriented way, as stated in several national bioeconomy strategies. An advanced approach has been developed by the German government, which has defined bioeconomy as, 'the knowledge-based production and use of renewable resources, in order to provide products, processes and services in all areas of the economy, within the framework of an economic system that is viable for the future' (BMEL 2014). The current challenge is, therefore, how to implement such a strategy, as the transformation towards a bio-based economy can be managed from many different perspectives. In this regard, from a resources conversion perspective, the following tasks should be accomplished in order to ensure satisfactory implementation from the regional perspective: (1) sustainable resource provision, (2) strategic implementation of new facilities in bio-based industrial clusters, and (3) fostering of transsectoral cooperation.

Indeed, the estimated intensification of wood harvesting processes and the implementation of novel processing facilities for wood resources will affect the environment as well as communities living near to biomass catchment areas and production sites. In this context, a socially sustainable transition – together with consideration of environmental and economic impacts – constitutes a central element for the success or failure of a future development. Actually, these three impact dimensions (social, economic, and environmental) are highly interrelated, as together they influence the shaping of an overall societal perception of the impacts and benefits of a product or process. Integrated evaluation of these dimensions should therefore be considered during an attempt to provide information to local, regional, and national decision-makers.

Achieving an integrated assessment of these dimensions of sustainability, especially from a regional perspective, is a challenging task. Life cycle assessment (LCA) and life cycle costing (LCC) are standardised tools for product assessment that can be adapted for the assessment of regional effects. Social life cycle assessment (SLCA), on the other hand, is a method still under development and is a tool that could provide knowledge about potential social effects resulting from products' life cycles. Moreover, for the particular case of evaluating bio-based value chains in a regional context, a common denominator for integration of the tools is also still lacking.

To address this need, this chapter introduces the development of both an SLCA method and an integrative sustainability index for a wood-based bioeconomy region. Although the models are still in the last steps of development, we discuss their applicability and the major lessons learned during the development process.

SLCA and its applicability to bioeconomy

To date, SLCA has been used for two major purposes: hot spot assessments and comparative assessments. The first of these is intended to identify severe social issues within countries and sectors attributed to a production system, with the aim of improving social conditions for affected stakeholders throughout the product's life cycle. Generic data at national level are, therefore, the most commonly used proxies to describe general performance of the production system examined, meaning that detailed information on the value chains involved is not provided. Comparative assessments, on the other hand, aim to identify to what extent one product is more sustainable than another. These studies, therefore, have a more limited scope (e.g. focus on one unit process or on a single stage of its/their life cycle) and use site-specific data generated, for example, from company interviews.

SLCA is defined here as a technique that aims to assess the relative social performance of an organisation's behaviour that has implications for a product's life cycle in the light of performance reference points. We argue that, in order for wood-based products from bioeconomy regions to actually be assessed, the relative social performance of all organisations involved should be evaluated on a regional level. We therefore aim to identify social hot spots (e.g. incidences of relative social performance that can be improved) within regional wood-based production systems. A framework to carry out this task has been developed and it is shown in Figure 37.1.

Figure 37.1
Proposed framework for a product-oriented SLCA for the evaluation of wood-based products from bioeconomy regions

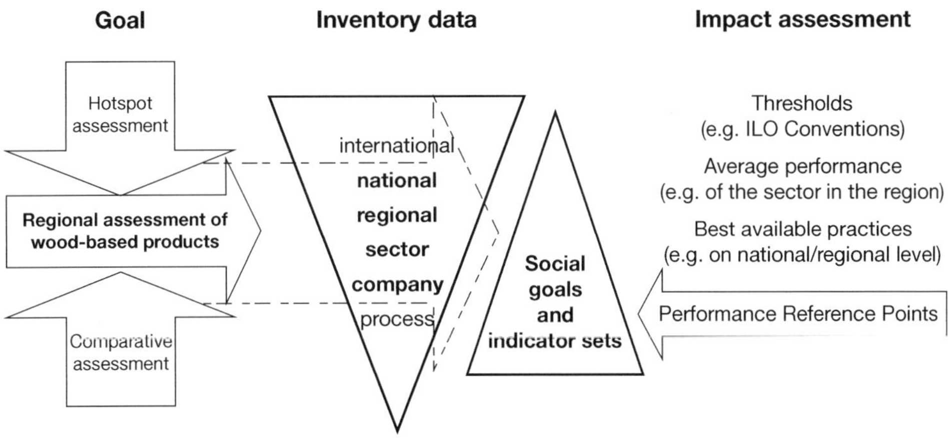

The *Guidelines for Social Life Cycle Assessment of Products* (referred to below as the Guidelines), published by UNEP in collaboration with SETAC in 2009, provide a general framework for carrying out such assessments (Benoît et al. 2009); the method's structure was oriented by the ISO 14040 framework. In the initial stage, it is important to define the goal and scope of the evaluation tool. The overall objective of the SLCA approach is to provide a socio-economic profile of wood-based production systems in a bioeconomy region in order to identify potential social hot spots. We argue that social effects are caused by the behaviour of organisations rather than by the processes themselves, the latter traditionally being evaluated with technical, environmental, and economic LCA-based tools (see Part IV for further examples). The system studied is thus defined by actors along the value chain (e.g. forestry enterprises, solid wood-based industries, wood logistics or chemical industries).

A first step in evaluating the relative social performance of the most relevant organisations involved in the value chain is to identify the social indices that are relevant to assessment of wood-based products from a regional perspective. Significant social themes for our SLCA approach are proposed in the Guidelines. Human rights, working conditions, health and safety, governance, cultural heritage, and socio-economic repercussions are identified, among other issues, and are characterised globally as main social impact categories. However, for the assessment of wood-based products in a regional bioeconomy context, these themes had to be adapted to suit the characteristics of the wood-production and wood-related industries in Germany, so that all potentially relevant social issues were included. In a previous article (Siebert and Bezama 2014), we argued for the use of a combined top–down and bottom–up approach to identify relevant social aspects, integrating broad societal goals and the perspectives of affected stakeholders. On the one hand, we reviewed international sustainability standards (e.g. Global Reporting Initiative, ISO26000, SA8000, FSC etc.) and SLCA approaches with regard to their applicability to our study context. On the other hand, we made use of stakeholder interviews to identify key socio-economic issues and opportunities on a regional scale, as well as potentially affected stakeholders' preferences. This enabled development of social indices (health and safety, fair salary, discrimination, stakeholder participation, etc.). The status of each social index is measured by one or more indicators. So that indicators such as *company-based fixed-term job rate*, *gender pay gap*, or *the availability of information days for local residents* can be quantified, standardised questionnaires are used to gain information from the organisations involved in a product's life cycle. In this way, data are collected from relevant foreground organisations such as companies involved in wood harvesting, transport, and production of wood products. Although this analysis focuses on foreground regional activities, social effects outside the region should also be assessed using, for example, generic data for hot spot screening.

To lend meaning to the indicator values, additional information is needed to characterise relative social performance along the value chain. Performance reference points are used in the literature to identify whether or not an organisation complies (and to what extent) with a particular reference (Parent et al. 2010; Ciroth and Franze 2011). In the SLCA framework proposed here, indicators are calibrated with utility functions (such as those used for balanced scorecards) so that the indicators can be assessed (for example) against the average performance or the best available practice in the wood sector on a national or regional level, depending on data availability. These reference points will be used as thresholds to distinguish between the relatively 'good' and 'bad' social performances of activities attributed to the production system. Generic sector-specific data are, therefore, collected from official statistics to calibrate the different utility functions. This provides an actual benchmarking framework within the bioeconomy system, with potential for reproducibility to other industrial regions involved in the production of wood-based goods, including industries and service companies.

Sustainability index for the bioeconomy on a regional scale

The integration of sustainability assessment methods and multi-criteria analysis methods is an advancing field, with several robust, established methods for technical, economic, and environmental/ecological product-oriented assessments. Moreover, a series of methods is under development for an appropriate product-based socio-economic assessment and for the integration of all of these dimensions (Benedetto and Finkbeiner 2011). However, when bioeconomy systems are assessed on a regional scale, such as the case of industrial bio-based clusters, the sustainability performance of single products, even though very relevant for individual producers, is not the most relevant aspect to consider. In this case, one of the most relevant aspects to evaluate is the actual sustainability potential of (partial or whole) regional product portfolios with highly integrated by-product valorisation and process energy (heat and electricity) supply services, and/or the potential value added of these systems compared with individual production chains.

The evaluation not only has to encompass the complex production networks, it must also include the coupled use and end-of-life (EoL) management options, such as cascade utilisation of biomass resources. However, an increasing number of assessed production systems, co-production options, and secondary raw material utilisation options will result in an increasing number of potential reference points for benchmarking (Ishizaka 2013; Walther 2010). The assessment of EoL management options is operationalised by means of qualitative indicators for ranking different product recycling alternatives. A consequence of this is an extension of the system boundaries for internal and external benchmarking. For monitoring of the sustainable development of regional production systems, an aggregated sustainability index could deliver a single scoring method to trace performance improvements achieved along the value-added chains. This could help in assessing whether or not novel or mature technology options could further help to address identified sustainability deficits in the overall system. A methodological framework for such a sustainability index is presented in Figure 37.2.

This methodology is conceptualised to support the sustainability management of material flows in cross-sector industrial cooperation by monitoring the resulting progress and/or deficits from induced changes in technological pathways. Besides the regionalised impacts of the industrial metabolism, the characterisation factors for assessing the regional resource regime must be included to account for the regional impacts of the actual biomass production activities, as stated by O'Keeffe et al. (2016).

In the same way as for SLCA, a first step was to identify and characterise the most appropriate sustainability goals. For that reason, the preferential and alternative options for various resource use alternatives were defined, and the stakeholders were attributed vertically and horizontally throughout the added-value chains identified. Through the involvement of these stakeholders

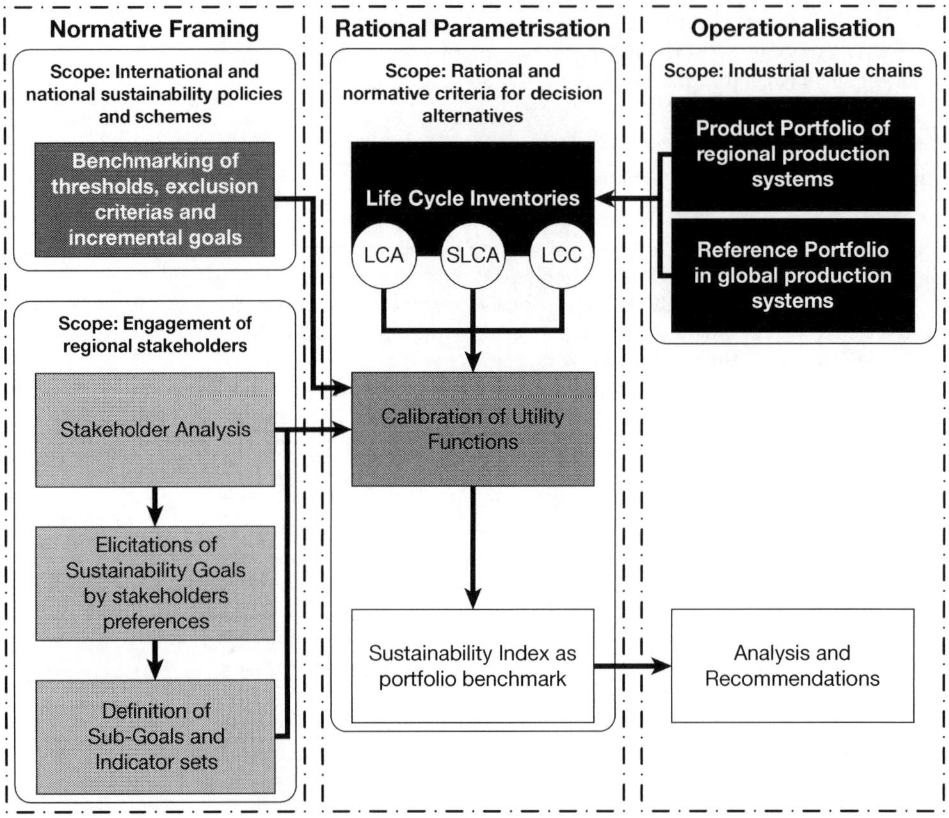

Figure 37.2
Analytical framework for a sustainability index methodology with generalised data aggregation procedures

(consulted through questionnaires, workshops, and direct interviews), a series of goals and selective and distinctive indicators were identified, selected, and prioritised. Elicited from these stakeholder consultations, normative framing was carried out for validation of thresholds, for the identification of exclusion criteria, and for the definition of incremental optimisation goals. With this information, it is possible to derive the appropriate utility functions that characterise the case study system (Hildebrandt et al. 2014a; Hildebrandt et al. 2014b).

Operationalisation of the method requires, initially, derivation of the different life cycle inventories, which are then normalised by means of reference-point approaches. Life cycle inventories for the value chains of the bio-based industries involved are derived for the regional product portfolio, the *business-as-usual*, end-of-life recycling paths, and the most adequate reference system(s), and are aggregated with a utility basket method for comparing products with same utilities but different production processes. A second step of the operationalisation is interpretation of the results and sensitivity analysis.

To carry out a sensitivity analysis, both the production phase and the end-of-life management options are assessed with various scenario assumptions for the *business-as-usual* case, for the *best-practice* case, and for the *most conservative estimations*. This enables testing of the robustness of the results and making of recommendations to anticipate future developments.

To estimate uncertainties and the degree of freedom in future developments of production processes, the technological maturity of plant concepts in the R&D stage is used for the scenario-based sensitivity analysis to anticipate future changes in the calibration of reference points, which are benchmarking the distance from the best performers. Moreover, for extrapolation of the impacts related to the products' use phase and their end-of-life management options, the *business-as-usual* case is compared with *best-practice* reference scenarios derived from emerging design and technology trends on international and national scales. The major advantage of this approach is the good regional resolution of assessed material flows resulting from the combined characterisation and quantification of the regional resource regime and the socio-industrial metabolism of industrial networks.

The aggregated sustainability index is therefore capable of monitoring regional energy and material flows and the socio-economic and techno-ecological impacts and benefits caused by upscaling of the regional production capacities of bio-based industry sectors.

Conclusions

The development of an integrated approach such as SLCA for assessing the sustainability of products or production systems is a necessary yet challenging step. Particularly in the case of novel bio-based industrial clusters, such an approach could serve to support appropriate material and energy flow management within a bioeconomy region.

In this regard, the proposed SLCA framework provides knowledge for producers about their social responsibility compared with their competitors and delivers information on where to improve their behaviour with regard to affected stakeholders. The approach relating organisation-based primary data to regional sector-specific characterisation factors provides the opportunity for direct assessment of wood-based products and enables accurate evaluation of regional value chains and a comparison of production systems in this context.

The sustainability index methodology is now being applied to the case study of a wood-based bioeconomy region in central Germany. The system compartments assessed for the regional case study include the engineered timber, biorefinery, composite manufacturing, and insulation and saw mill industries. The benchmarking is performed in relation to established processes, for example, in the chemical, steel, and cement industries.

Regional assessments should take account of the local conditions and development expectations. Key regional issues should, therefore, be reflected in the structure of the defined goal systems. For this reason, adopting a bottom-up approach to include regional stakeholder preferences and their objectives is very important for both of the methods presented here. In addition, a regional assessment must take into account the regional status quo, that is, an adequate location-specific reference level to enable a relativised social responsibility assessment of organisations.

Considering the above, the overall frameworks of both the SLCA approach and sustainability index can be adapted for application in other wood-based bioeconomy regions in Germany, if the characterisation factors are developed for the region. With the necessary adaptations, including refinement of regional and sectoral conditions, these methods could be applied to assess other production systems and industrial sectors with a regional scope.

References

Benedetto, E., and Finkbeiner, M. (2011) *Towards Life Cycle Sustainability Management*, 1st edn, Berlin: Springer.

Benoît, C., Mazijn, B., and Andrews, E.S. (2009) *Guidelines for Social Life Cycle Assessment of Products*, Paris: United Nations Environment Programme.

BMEL (Federal Ministry of Food and Agriculture) (2014) *National Policy Strategy on Bioeconomy – Renewable Resources and Biotechnological Processes as a Basis for Food, Industry and Energy*, Berlin, Federal Ministry of Food and Agriculture, Division 531 – Strategy and Coordination of the Directorate-General 'Biobased business, sustainable agriculture and forestry'.

Ciroth, A. and Franze, J. (2011) *LCA of an Ecolabeled Notebook – Consideration of Social and Environmental Impacts along the Entire Life Cycle*, Berlin: GreenDeltaTC GmbH.

Hildebrandt, J., Bezama, A., and Thrän, D. (2014a) 'Establishing a robust sustainability index for the assessment of bioeconomy regions', in *Proceedings of the International Conference and Utility Exhibition 2014 on Green Energy for Sustainable Development*. Bangkok: Asian Institute of Technology, pp. 1–4.

Hildebrandt, J., Bezama, A., and Thrän, D. (2014b) 'A hybrid-MCDA tool for assessment of added value networks in a woodbased bioeconomy', *Bioenergy from Forests Conference*, Helsinki.

Ishizaka, A. (2013) *Multi-criteria Decision Analysis: Methods and Software*, Chichester, UK: Wiley.

O'Keeffe, S., Majer, S., Bezama, A., and Thrän, D. (2016) 'When considering no man is an island – Assessing bioenergy systems in a regional and LCA context: a review', *International Journal of Life Cycle Assessment*, 21(6): 885–902.

Parent, J., Cucuzzella, C., and Revéret, J.P. (2010) 'Impact assessment in SLCA: Sorting the sLCIA methods according to their outcomes', *International Journal of Life Cycle Assessment*, 15(2): 164–71.

Siebert, A., and Bezama, A. (2014) 'An analytical framework for a regional sLCA goal system: Application in a wood-based bioeconomy region in Germany', in *DepoTech 2014 – Abfallwirtschaft, Abfallverwertung Und Recycling, Deponietechnik Und Altlasten*, Leoben, Austria: Lehrstuhl für Abfallverwertungstechnik Und Abfallwirtschaft, pp. 657–60.

Walther, G. (2010) *Nachhaltige Wertschöpfungsnetzwerke: Überbetriebliche Planung und Steuerung von Stoffströmen entlang des Produktlebenszyklus*. Wiesbaden, Germany: Gabler Verlag/GWV Fachverlage.

38

RESPONSIBLE MINING – FROM A LIFE CYCLE TO A STEWARDSHIP APPROACH

Ross Hamilton and Andrew S. Parsons

Introduction

Conventional life cycle approaches have enabled mining operations to understand the inputs (e.g. water, energy) and outputs (e.g. waste, greenhouse gas emissions) of their processes in order to minimise impacts, manage costs, and meet regulatory requirements. With the growth in importance of broader sustainability issues, an approach based on eco-efficiency has nevertheless proven insufficient to insulate businesses from many risks. A more expansive stewardship approach is increasingly being applied. This chapter examines some of the trends, challenges, and opportunities associated with life cycle approaches in the mining sector. The examples provided largely have an environmental focus, but the authors recognise the imperative of including social and other dimensions in a stewardship approach.

Life cycle management

Minerals are the start of the life cycles of many products in use today. All metals, industrial minerals, and building materials must either be recovered from redundant products or mined (we include 'quarrying' here also). Everything, from steel reinforcing to the components found in electronic devices, jewellery, gravel, and energy minerals such as coal and uranium, starts life in the ground. Looking forward, the mining industry contributes to the building of a lower-carbon economy through the provision of the metals and minerals used in products, from fuel cells and photovoltaic panels to wind turbines and insulating materials.

The life cycle starts where a mineral is located. A miner has some flexibility in mine design or may choose not to mine, but, unlike some other sectors, cannot move the operation to a more convenient location. Processing of concentrates by smelting and refining may take place elsewhere, depending on local conditions, infrastructure, and transport.

The way in which mining companies manage environmental, social, and economic issues continues to evolve in response to changing conditions. The traditional management approach to many sustainability issues concentrated on operational efficiency, most recently using a life cycle approach, typically within the operational boundary. This approach has been an important first step in improving the efficient use of inputs and in reducing wastes and discharges and, coincidentally, in gaining a broad understanding of a mine's impacts on the environment and on neighbouring communities. The use of efficiency measures as a management tool has also been advanced by corporate sustainability reporting standards, such as the Global Reporting Initiative's Guidelines (www.globalreporting.org/standards) and its Mining and Metals Sector Supplement (www.globalreporting.org/standards/sector-guidance/sectorguidanceG4), which have been widely adopted. An example of company action is energy-efficiency initiatives, such as those described in a case study reported by the South African company Gold Fields (www.goldfields.com/case_study.php?caseID=23).

Such approaches presuppose a linear relationship between quantity and cost or impact. But efficiency is not necessarily the end game for all issues, and, hence, such an approach does not necessarily insulate a company against risks that manifest outside the local operational site. In recent years, leading companies have started moving from a management (efficiency) approach to a broader approach that takes into account the regional context and product life cycles and can be better described as a stewardship approach.

From management to stewardship

A management approach based on stewardship concepts considers the priorities and concerns of communities and of the environment beyond the operational site boundary, in the wider catchment, the region, and even at the national or global level. Aspects such as water connect a mining operation with the surrounding landscape and local communities.

Central to the increased adoption of a stewardship philosophy is that a 'one-size-fits-all' management approach to complex issues is seldom effective, as issues relating to water and community relationships are location-specific. Every situation is different. For instance, even the most efficient operation can have detrimental impacts if it is situated within a sensitive environmental and/or social setting.

Regional considerations are vital when environmental and social impact assessments and other studies are undertaken, rather than conditions within the given project area being the focus. The concept of strategic impact assessment (SIA) is important here – SIAs are usually carried out by or for government agencies, rather than companies, so that regional-scale impacts can be assessed across multiple actors and timeframes. Closely related concepts are strategic environmental assessment (Minerals Council of Australia 2015) and cumulative effects assessment (Canter and Ross 2010; Canadian Environmental Assessment Agency 2014). All these assessment tools take longer timescales and more extended life cycle chains into account than are of immediate concern to a local production site.

Regional approach – Water

Businesses and governments are increasingly recognising the inherent complexities and interdependencies associated with management of aspects such as water and biodiversity. For instance, water can be a human rights, environmental, cultural, regulatory, recreational, and economic issue.

Inadequate consideration of water stewardship can result in costly delays, cancellation of licences, community conflicts, and reputational damage. Examples include the Conga Project in Peru – its environmental impact assessment was approved by the relevant government agency in 2010, but the project was suspended in 2011 following community protests, chiefly over water (www.newmont.com/resources/Case-Study-Library/Case-Study-Library/2015/El-Agua-Primero-La-Mina-Despus-Water-First-Then-Mine).

The result has been a proliferation of water stewardship initiatives (Alliance for Water Stewardship 2014; CEO Water Mandate 2015, ceowatermandate.org; WWF 2013) to promote the responsible sharing and use of fresh water that are socially and economically beneficial, as well as environmentally sustainable.

The interpretation and application of water stewardship concepts have been a challenge for the mining sector owing to the inherent complexities associated with collaborating or partnering with others to mitigate shared risks and address shared impacts. Stewardship can transcend individual companies. In the Olifants Catchment in South Africa, collective action groups comprise a wide variety of organisations. These groups are established in law to: (a) manage water resources in a defined water management area, (b) co-ordinate the functions of other institutions in water matters, and (c) involve local communities in water management (Department of Water and Sanitation n.d.).

The International Council on Mining and Metals (ICMM) has launched a framework (International Council on Mining and Metals 2014) to make water stewardship meaningful for the mining industry. It comprises four core elements and associated supporting activities (Figure 38.1), with solutions that work for businesses as well as other water users within the surrounding region.

The framework adopts a catchment/watershed approach that considers water requirements beyond those of the mining operation, so that companies are better positioned to mitigate shared risks by partnering with others within a given catchment.

The associated guide (International Council on Mining and Metals 2015) outlines how companies can identify, evaluate, and respond to catchment-based water issues. The guide assists companies to develop water plans in accordance with the local social, cultural, regulatory, economic, environmental, and hydrological context.

In Peru, for example, a public–private partnership was initiated between Freeport-McMoRan's Cerro Verde operation, the local water utility company, local authorities, and the central government, to ensure access to water for mining, the sustainable delivery of potable water, and the treatment of the city's wastewater. Engagement of local, regional, and national stakeholders found that treating wastewater for use at the mine was the best of several alternatives. The mine constructed a potable water treatment plant and a water storage and distribution network.

Regional approach – Closure planning

Closure planning is another area in which the industry has recognised the necessity of a life cycle approach (International Council on Mining and Metals 2008). All mines eventually exhaust their economically viable resources and subsequently close. Effective closure planning starts at the beginning of the mining life cycle, taking into account all aspects of pre-operational planning, operational management, and post-closure activity, including any residual social

Figure 38.1
ICMM water stewardship framework outlines a regional water approach

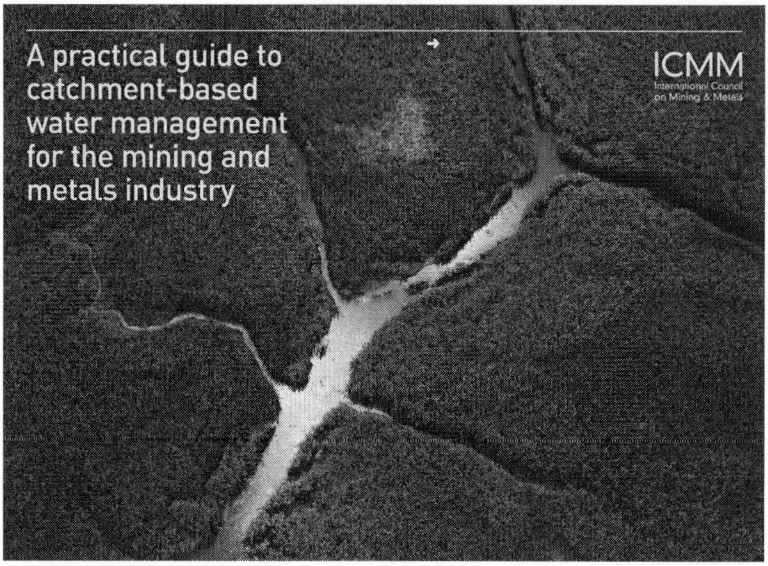

Be transparent and accountable

Publicly report material water risks, management activities and performance.

Engage proactively and inclusively

Engage stakeholders in an open and transparent manner to understand their priorities, share plans and collaborate on solutions.

Effective water resource management

Manage operational water (quantity and quality) inputs, use and outputs to maximize resource sustainability, operational flexibility and economic benefit.

Adopt a catchment-based approach

Understand the social, cultural, economic and environmental value of water at the catchment scale to identify material water stewardship risks and provide context for corporate and operational water management.

and environmental issues that arise from closure implementation. This process has been a challenge for many companies because sustainability considerations often change over time and confront near-term mine production priorities and traditional valuation techniques.

Because communities and governments remain in place after a mine closes, engaging them early in the process of closure planning is important, despite the inherent sensitivities in regions where mines are the major source of employment and income. Extending mine life cycle considerations into the post-closure phase to look at subsequent economic and social development is important to many regional stakeholders.

Biodiversity

All stages of the mining and metals life cycle have the potential to impact biodiversity. Biodiversity conservation is important, not only to protect species, but also because of the ecosystem services fauna and flora provide, including cultural values, livelihoods, and medicinal properties. Conditions relating to biodiversity protection stipulated by governments and project financiers for extractives projects continue to evolve. The International Finance Corporation Performance Standard 6 (2012; www.ifc.org/wps/wcm/connect/topics_ext_content/ifc_external_corporate_site/ifc+sustainability/our+approach/risk+management/performance+standards/environmental+and+social+performance+standards+and+guidance+notes), for example, requires prospective projects to implement a mitigation hierarchy to aid the conservation of biodiversity.

The mitigation hierarchy is a sequence of four actions ('avoid', 'minimise', 'restore', and 'offset') to reduce the impacts on biodiversity and ecosystem services that could arise from a project's routine activities related to the mining cycle of exploration, construction, operation, and closure. These actions aim, in the first instance, to anticipate and *avoid* causing impacts wherever possible; where avoidance is not possible, the next consideration is to *minimise* any potential impacts of the project; where impacts have occurred, *restoration* should be considered to assist with the recovery of the impacted ecosystem; finally, a biodiversity *offset* or *offsets* may be considered to compensate for impacts that cannot be avoided, minimised, and/or restored. To be effectively applied, the mitigation hierarchy must consider the uniqueness, prevalence, and vulnerability of biodiversity aspects, not just within the operational footprint, but also within a regional and even national/international context.

To support the consistent implementation of the mitigation hierarchy, guidance developed by the Cross-Sector Biodiversity Initiative (CSBI) was released in 2015 (Cross-Sector Biodiversity Initiative 2015). CSBI is a unique collaboration between the mining and metals (ICMM), oil and gas (IPIECA), and finance (Equator Principles Association) industries.

Materials stewardship: Sustainable procurement and product stewardship

Mining companies are important consumers of materials and services. As well as the technical specifications underpinning such purchasing, the choices should ideally reflect the corporation's sustainability objectives. Most large mining companies extend their sustainability reach through setting requirements that suppliers must meet (see, for example, those of Rio Tinto: procurement.riotinto.

com/ENG/doingbusiness/407_policies.asp). At the same time, they have a significant potential to use their procurement budgets to enhance local, regional, and national development. In the past, companies operating in developing countries tended to source inputs from their home markets, but they have recognised the benefits achieved through sourcing materials locally. The Zimele programme of Anglo American in South Africa associates procurement with development of small and medium enterprises (SMEs) to provide a ready market for the products (southafrica.angloamerican.com/our-difference/zimele-enterprise-development.aspx). In some cases, changes have also come about as a result of local sourcing requirements in legislation.

At its formation in 2002, the ICMM recognised that the sustainability agenda for the mining industry included stewardship of the industry's products: its 8th Sustainable Development Principle requires that members 'Facilitate and encourage responsible product design, use, re-use, recycling and disposal of [their] products' (www.icmm.com/en-gb/about-us/icmm-10-principles). ICMM's materials stewardship work programme has developed an online toolkit and several guidance documents (www.icmm.com/page/84173/materials-steward ship). Since then, several sector-specific programmes have been formed: for example, the Global e-Sustainability Initiative (gesi.org), Responsible Jewellery Council (www.responsiblejewellery.com), Aluminium Stewardship Initiative (aluminium-stewardship.org), and the Copper Alliance (copperalliance.org/core-initiatives/sd/lifecycle).

Challenges and opportunities for taking a regional approach

A life cycle approach to regional development requires multiple actors to have a common appreciation of appropriate planning and decision-making, and the need to collaborate on shared solutions. Although landscape-level planning and collective action approaches are still evolving, initial experience within the mining industry has identified a number of challenges.

In many regions, local decision-makers do not hold critical data sets (they are often held by other actors), nor do they have the supporting systems such as geographic information systems to interpret the data. Often, data needed to inform stakeholders and related decision-making processes are fragmented, not accessible, or not comparable.

The MAP-X initiative recently initiated by the World Bank, UNEP, and the G7+ aims to improve transparency in the extractives sector by consolidating relevant information into one open data platform for all stakeholders. Data critical for decision-making relating to the Extractive Industries Transparency Initiative (eiti.org), as well as contextual environmental, social, and economic data, will be hosted for fragile states on the platform (extractivesfragilestates.github.io/ExtractivesFragileStates/about).

A number of initiatives and programmes encourage a consistent approach to measuring, monitoring, and reporting within the industry. For example, the Minerals Council of Australia has developed a standardised approach to water accounting that is increasingly being adopted by the industry beyond Australia (Minerals Council of Australia 2014). CSBI recently released a guideline (Gullison et al. 2015) for conducting environmental baseline assessments.

Biodiversity conservation requires a regional or landscape approach (Pope and Moore 2013). The Western Australian Biodiversity Institute is a platform established to facilitate the acquisition and interpretation of biodiversity-related knowledge. Foundation members include representatives from research agencies, government advisory agencies, and industry (www.creatingcommunities.com.au/breaking-down-the-barriers-to-collaboration-in-biodiversity). Collaborative initiatives are useful as they often reduce concerns about sharing data with competing companies or other sectors.

However, collective initiatives seldom proceed smoothly. Any approach that relies on collaboration across sectors has its challenges, and collective action groups must build trust in order to work together. The CEO Water Mandate has published a *Guide for Managing Integrity in Water Stewardship Initiatives* (CEO Water Mandate 2015) to enable participants to face up to this challenge.

Small- and medium-scale mining

Most of the methods described above have been applied by large companies that need to take a longer-term, broader-scale view. Some instruments may also be appropriate to smaller companies. The Cement Sustainability Initiative (www.wbcsdcement.org) and the Responsible Jewellery Council include companies across a broad range of sizes. The Fairtrade Gold (www.fairgold.org) and Fairmined (www.fairmined.org) initiatives have set standards for artisanally mined gold. Flexibility and a bottom–up approach have been key to the success of the latter initiatives.

Conclusion: Towards a life cycle view of sustainable mining

Mining and metals are essential for modern life. Responsible mining companies recognise the need to integrate sustainability into business planning and decision-making. Data collected using conventional efficiency-based life cycle approaches provide an important yet limited input. Stewardship approaches have a broader geographic, timescale, stakeholder, and issue perspective and, hence, better support landscape and regional-level planning.

Mining companies and governments must be cognisant of community views and the need to balance environmental and social considerations. A dynamic approach to life cycle analysis and risk assessment is required in recognition of the changes in the social, environmental, economic, political, and regulatory context over the multi-decadal timeframes in which they are situated. This means that many of the approaches described here cannot be conducted as one-off exercises; they require ongoing, iterative engagement and interaction with actors at both the operational and regional level. Further, such approaches require, not only an understanding of historical trends, but also the ability to forecast dynamics such as water supply, climate change, user demand, and social changes. Identifying, communicating, and integrating uncertainty and the limitations of technical knowledge throughout life cycle approaches, while better understanding social and community needs and dynamics, will help ensure government and business decision-making, management responses, and assumptions are practicable and adaptable to the range of scenarios that may eventuate over the life cycle of a mine.

Regional government agencies should promote regional, landscape-level approaches. Companies seek consistent application of the law, prosecution of offenders, and predictable legislation. Agencies can assist companies through provision of environmental and demographic databases and prompt consideration of licence and permit approvals. Governments have a unique convening power that can be used to bring opposing parties together to find long-term solutions to the age-old challenge of development that is truly sustainable.

References

Alliance for Water Stewardship (2014) *The AWS International Water Stewardship Standard*. Available at: www.allianceforwaterstewardship.org/aws-standard-system.html#aws-standard (accessed 7 August 2016).

Canadian Environmental Assessment Agency (2014) *Cumulative Effects Assessment Practitioners' Guide*. Available at: www.ceaa-acee.gc.ca/default.asp?lang=En&n=43952694-1&offset=6 (accessed 7 August 2016).

Canter, L., and Ross, W. (2010) 'State of practice of cumulative effects assessment and management: The Good, the bad and the ugly', *Impact Assessment & Project Appraisal*, 28(4): 261–8.

CEO Water Mandate (2015) *Guide for Managing Integrity in Water Stewardship Initiatives: A Framework for Improving Effectiveness and Transparency*. Available at: ceowatermandate.org/integrity (accessed 7 August 2016).

Cross-Sector Biodiversity Initiative (2015) *A Cross-Sector Guide for Implementing the Mitigation Hierarchy*, London. Available at: www.csbi.org.uk/tools-and-guidance/mitigation-hierarchy (accessed 7 August 2016).

Department of Water and Sanitation (n.d.) *Catchment Management Agencies*. Available at: www.dwa.gov.za/IO/Docs/CMA/WMI%20Poster%20Booklets/CMAposter.pdf (accessed 7 August 2016).

Gullison, R.E., Hardner, J., Anstee, S., and Meyer, M. (2015) *Good Practices for the Collection of Biodiversity Baseline Data*, prepared for the Multilateral Financing Institutions Biodiversity Working Group and Cross-Sector Biodiversity Initiative. Available at: www.csbi.org.uk/tools-and-guidance/biodiversity-data-collection (accessed 7 August 2016).

International Council on Mining and Metals (2008) *Planning for Integrated Mine Closure: Toolkit*, London. Available at: http://hub.icmm.com/document/310 (accessed 7 August 2016).

International Council on Mining and Metals (2012) *Water Management in Mining: A Selection of Case Studies*, London, 16–17. Available at: www.icmm.com/www.icmm.com/water-case-studies (accessed 7 August 2016).

International Council on Mining and Metals (2014) *Water Stewardship Framework*, London. Available at: www.icmm.com/water (accessed 7 August 2016).

International Council on Mining and Metals (2015) *A Practical Guide to Catchment-Based Water Management for the Mining and Metals Industry*, London. Available at: www.icmm.com/publications/water-management-guide (accessed 7 August 2016).

Minerals Council of Australia (2014) *Water Accounting Framework for the Australian Minerals Industry*. Available at: www.minerals.org.au/leading_practice/water_accounting_framework_for_the_australian_minerals_industry (accessed 7 August 2016).

Minerals Council of Australia (2015) *Cumulative Environmental Impact Assessment Industry Guide*. Available at: www.minerals.org.au/news/cumulative_environmental_impact_assessment_industry_guide (accessed 7 August 2016).

Pope, J., and Moore, S.A. (2013) *Planning and Assessment for Biodiversity Conservation at a Landscape-Scale: An Evaluation of Current Approaches and Opportunities in Australia*, Hobart, Tasmania: University of Tasmania. Available at: www.nerplandscapes.edu.au/publication/planning-and-assessment-biodiversity-conservation-landscape-scale-evaluation-current (accessed 7 August 2016).

WWF (2013) *Water Stewardship: Perspectives on Business Risks and Responses to Water Challenges*. Available at: wwf.panda.org/what_we_do/how_we_work/conservation/freshwater/water_management (accessed 7 August 2016).

39

TRENDS AND CONSIDERATIONS OF RENEWABLE ENERGY DEVELOPMENT

A regional perspective

Mark Radka, Giulia Ferrini, Agathe Laure, and Ravina Brizmohun[1]

Introduction

Thousands of regions, cities, and towns around the world have adopted policies and targets to promote renewable energy (RE). Motivations include reducing GHG emissions, developing local skilled industry and creating local jobs, reducing energy demand, achieving security of supply and possibly independence from the national grid, and becoming more resilient to climate change (Renewable Energy Policy Network for the 21st Century (REN21) 2014). The benefits can be large, but environmental and social problems from RE development are also a possibility. Life cycle analysis and social impact analysis can help regional and local planners realize the many benefits of RE, while reducing risks.

Renewable energy is generally understood to mean energy that comes from resources that are naturally replenished on a human timescale, such as sunlight, wind, rain, tides, waves and geothermal heat. Renewable energy technologies are technologies that use – or enable the use of – one or more RE sources.

Global perceptions of RE have shifted considerably in the past decade. Continuing technology advances, cost reductions, and experience in deploying RE systems have contributed to their growing attractiveness, not least because of the large role RE can play in reducing emission of GHGs. Renewable energy installations are currently cost competitive with fossil fuel alternatives in good locations and are increasingly displacing conventional fossil fuels in four areas: electricity generation, heating and cooling of air and water, transport, and rural (off-grid) energy services (REN21 2014). All projections point to continued reductions in equipment and installation costs, and so the trend in RE expansion is likely to accelerate.

Excluding large hydroelectric projects (those greater than 50 MW capacity), global investment in renewable power and fuels was US$270 billion in 2014, nearly 17 per cent higher than the previous year. By early 2014, at least 144 countries had RE targets, and 138 countries had RE support policies. Developing and emerging economies have led the expansion in recent years and account for ninety-five of the countries with support policies, up from fifteen in 2005 (REN21 2014). Although national governments are principally driving the transformation of the energy market, policies enacted at the subnational level are important. The experience to date is that complementary policies developed at different levels of government have helped bring about a consistent market for clean energy (Busche 2010). Policies relevant at the subnational level include targets, subsidies, public investment, innovative financing, bulk procurement, green power purchasing, building codes, transport fuel mandates, and municipal utility regulation (REN21 2013).

Local and regional governments are closer to citizens and often have greater flexibility than national governments in making energy planning decisions, allowing them to develop policies that better meet specific local needs. In the US, state governments are increasingly implementing policies to drive market transformation in the clean energy sector (Frankfurt School–UNEP Centre/Bloomberg New Energy Finance 2015). The same is true in Europe, where the transition is underway from a centralized, largely fossil fuel- and nuclear-based power system towards a more decentralized system that relies on electricity from RE sources and combined heat and power (European Parliament 2010). Many jurisdictions have also benefitted from policies promoting the use of solar hot-water systems in residential and commercial buildings (REN21 2015). By the end of 2014, European cities had submitted 734 sustainable energy action plans under the EU Covenant of Mayors effort, bringing to 3,333 the number of European local governments with action plans aiming to reduce emissions through the use of energy efficiency and renewables (REN21 2014).

Because RE technology is decentralized, it depends on different infrastructure, market, and ownership models. The participation of a broader range of stakeholders, including local citizens, is typically necessary to realize projects (World Future Council 2013). Whether the policy push for RE comes from the national level or is more local, the role of regional and local governments in realizing RE projects is critical, because they often have jurisdictional authority over transportation, land-use planning, building codes, and other areas that affect community-wide energy use. Greater decentralized policy development is providing the opportunity for wide policy experimentation, as more and more countries begin a transition to clean energy, with regional and local governments uniquely positioned to frame the discussion in a way that connects it to the values of local citizens (Galarraga et al. 2011; Busche 2010). Policy development at the regional level is proving in many instances to be more effective, because it is easier to achieve social acceptance on a smaller scale. Framing energy choices as a local issue increases citizen involvement in the clean energy discussion, because people can be more directly involved in decision-making processes (Busche 2010).

LCA of RE technologies

As described earlier in this book, the basic goal of LCA is to determine the environmental impact resulting from the production, distribution, and disposal of the products or systems of interest: in this case, either the individual RE technologies or the RE *system* as a whole (Hertwich et al. in UNEP 2015). Like all energy technologies, RE technologies have impacts on the environment. These may not be immediately evident, however, if they are distributed. LCA can help local authorities identify in advance any adverse impacts that the choice of a particular technology might cause and suggest ameliorative actions that – if taken – would mitigate the impacts (European Parliament 2010; UNESCAP 2012; IRENA 2015).

Unlike conventional power plants – where the operating phase usually dominates – environmental impacts of RE mostly arise from the manufacturing, installation, and disposal of the power conversion devices – for example, the wind turbines or solar panels (Figure 39.1). The production of liquid biofuels and burning of biomass for heat or electricity are the exception, particularly if production of the agricultural feedstocks is considered part of the operations. Impacts from RE systems may arise from exploitation of natural resources, local pollution, or changes to the integrity and stability of the local ecosystem (Del Río and Burguillo 2008). For a particular RE system, the impacts depend on the type of energy source being tapped, the conversion technology used, whether the system is connected to the grid or is 'off grid', whether energy storage is included, the size of the system, and other factors.

Areas to examine in an LCA include metal and material use or depletion, land-use change, habitat change, and human health. Although most of these impacts (on human health, for example) are typically low for RE technologies, their widespread deployment can put additional pressure on ecosystems and biodiversity (US National Renewable Energy Laboratory 2013). A detailed assessment would consider all impacts, starting with the extraction of the materials used to construct or operate the conversion technology, through to its final decommissioning and disposal, although this might be beyond the scope of a regional or local LCA exercise (Hertwich et al. in UNEP 2015). Social impact assessment (SIA) can be a useful complementary tool; opposition to RE facilities such as wind farms can often be reduced or eliminated through consultations with affected groups. SIA can also help identify distributional aspects of RE development.

UNEP's International Resource Panel has recently examined the life cycle impacts of different RE technologies, compared with fossil fuel alternatives at the global scale, and concluded that they show clear environmental advantages, including

Figure 39.1
LCA of energy systems

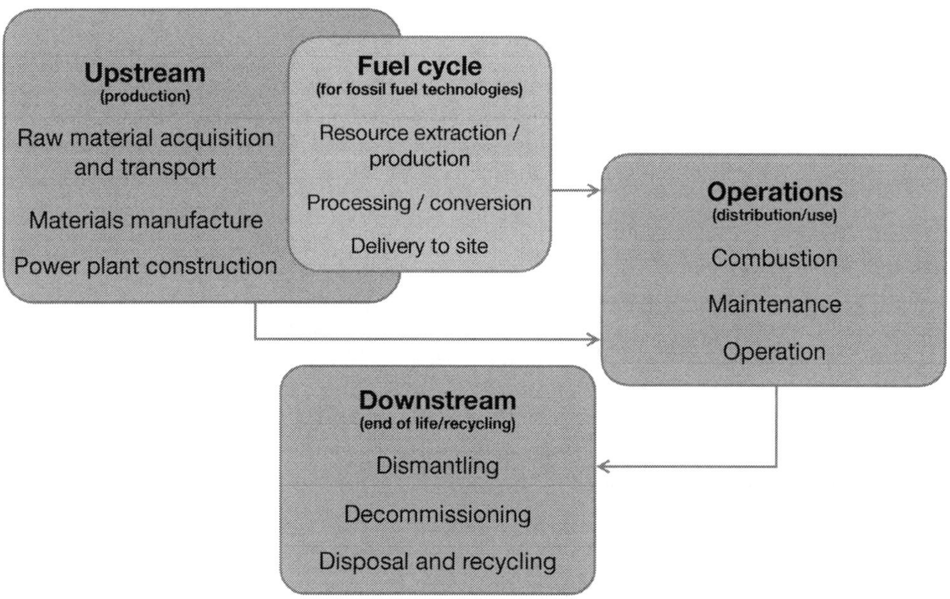

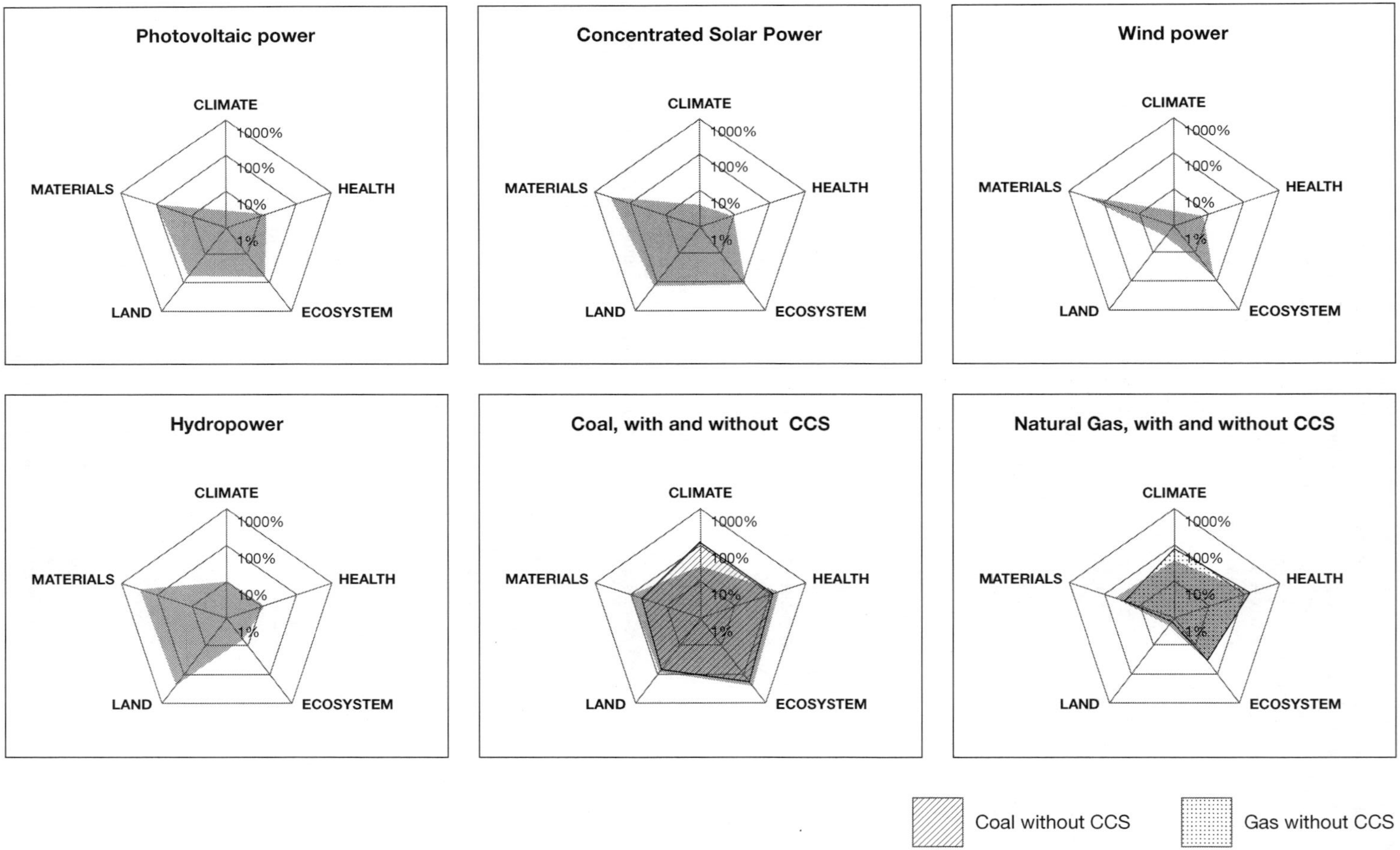

Figure 39.2
An overview of the life cycle results of different technology groups compared with the global average mix of electricity production technologies

a 'substantial decrease of the life cycle greenhouse emissions of grid electricity . . . as well as a decrease in the extraction of fossil resources' (Hertwich et al. in UNEP 2015). Figure 39.2 compares, for six different fossil-fuel and RE technologies, the magnitude of GHG emissions and impacts on human health, ecosystem health, land use, and materials consumption. The indicators are the use of the bulk materials (cement, iron and steel, copper, and aluminium), land occupation, GHG emissions, and the end-point indicators for human health and ecosystems. The overall conclusion is that RE substantially reduces impacts on the environment and human health, with the exception of material use. This points to the need to systematically examine end-of-life issues and ensure recycling of products, a responsibility that may fall disproportionately on local and regional authorities.

Environmental aspects of different RE technologies – Regional and local concerns

Wind power

Early wind farms had open lattice towers that encouraged bird nesting and were often sited in areas that attracted raptors and other large birds. Bird collisions with spinning turbine blades were consequently a concern. Better tower designs, lower turbine speeds, and more attention to siting have reduced bird mortality associated with wind farms. More recent studies have implicated wind farms in bat deaths. If poorly sited, wind farms can cause habitat disruption, but the 'footprint' of a wind farm, occupied by the turbines, access roads, and associated structures, is small compared with the area over which the turbines are distributed, meaning the land is usually available for other uses, such as farming. Sound and visual impairment has led to opposition in some localities, although local ownership of the generating assets by cooperatives or other distributed ownership structures has reduced this friction in countries such as Germany and Denmark.

Solar power

Two main solar technologies are solar photovoltaic (PV) and concentrating solar power (CSP). Potential impacts from solar power projects include land use and habitat loss, water requirements, and toxic emissions in the manufacturing stage.

Unlike wind facilities, large-scale solar projects cannot easily share land with agricultural uses. Impacts from large systems can be reduced by their siting at lower-quality locations, such as old industrial sites, abandoned mining land, parking lots, or existing transportation and transmission corridors. Some developers have placed panels over irrigation canals, and Japanese companies are experimenting with floating solar farms in order to avoid competition with other uses of land. Smaller-scale PV systems are often affixed to roofs of homes or commercial buildings; making use of existing structures in this manner requires no additional land. The same is true of solar hot-water systems.

Water is used in cleaning mirrors of CSP plants and PV panels, and this might be an issue in areas where it is limited. Dry areas often have fragile ecosystems that can be easily disrupted during construction or operation of larger utility-scale solar facilities. Concentrating solar thermal plants also require water for cooling; requirements depend on the plant design, location, and type of cooling system. Both CSP and PV plants are optimally located in areas that have high insolation, but such areas are often water stressed. A good LCA conducted at the regional level should indicate whether water availability is likely to be a problem.

The PV cell manufacturing process uses hazardous materials, most of which are used to clean and purify the surfaces of the semiconductor materials used in the cell itself. Thin-film PV cells contain more toxic materials than those used in conventional silicon PV cells; elements of concern include gallium, arsenic, indium, selenium and cadmium, although these are not used in the pure metallic form (US National Renewable Energy Laboratory 2013). Toxic emissions from PV solar cells and CSP mirrors are a possibility if they are disposed of improperly. In smaller PV systems, such as those installed on individual rooftops, toxic emissions might also arise from the 'balance of system' – inverters, controllers, batteries, and other electronic components that, with the PV cells, make up the PV system. For local or regional authorities, it is important to consider end-of-life issues, particularly if approaches for dealing with e-waste, batteries, or other electronic components of used systems are weak.

Biomass for electricity and fuel

Biomass power plants are similar to fossil fuel power plants, and biomass is sometimes co-fired with coal or municipal waste in order to reduce net GHG emissions. Hence, biomass plants have air emissions and water requirements that are similar, but not identical, to fossil fuel power plants. The difference is that, if feedstock of a biomass power plant is produced in a sustainable manner, the net GHG emissions can be close to zero.

Biomass power plants require approximately the same amount of water for cooling as coal power plants, but the actual water use depends on the individual facility's cooling technology and local conditions. Water is also needed to produce the biomass feedstock in many locations. Some feedstock sources are 'waste'; examples include agricultural waste biomass, such as straw or corn stover, forestry waste, and municipal solid waste. These raw materials require no additional water to be produced, but energy crops intentionally grown to be burned or converted into liquid biofuels can have water demands as well as compete with other crops for agricultural land.

As with any combustion process, burning biomass to produce electricity can impact air quality. Impacts depend on the feedstock, combustion technology, pollution control technology, and local meteorology. Pollutants of concern include nitrogen oxides, carbon monoxide, and particulates.

Land-use impacts from biomass production largely depend on the feedstock. If the biomass is a by-product of logging or farming, there is no overall increase in land use. Additional effort is required to collect and transport the material, however, and, if this is not done well, the use of such materials can lead to adverse social and environmental impacts. Dedicated production of biomass feedstocks destined for liquid biofuels, electricity, or heat may have wider social implications.

Geothermal energy

Geothermal power plants can have impacts on both water quality and consumption. Hot water pumped from underground reservoirs often contains high levels of sulphur, dissolved salts, and other minerals. Most geothermal facilities have closed-loop water systems in which extracted water is pumped directly back into the geothermal reservoir after it has been used to produce heat or electricity.

The distinction between open- and closed-loop systems is important with respect to air emissions. In closed-loop systems, gases removed from the well are not exposed to the atmosphere and are injected back into the ground after giving up their heat, and so air emissions are small. Open-loop systems, however, can emit hydrogen sulphide, carbon dioxide, ammonia, methane, and other gases if these are present in the water (Galarraga et al. 2011).

The amount of land required by a geothermal plant varies depending on the properties of the resource reservoir, the amount of power capacity, the type of energy conversion system, the type of cooling system, the arrangement of wells and piping systems, and the substation and auxiliary building needs (Galarraga et al. 2011).

Other RE technologies

Other sources of RE include running water (hydropower), tides, and waves. Projects using technologies to tap these sources can have environmental and social impacts if they are improperly sited and operated. This is particularly true of hydropower projects. Space precludes an in-depth discussion here, and the interested reader is urged to consult the literature.

For an illustration of how countries may regard their current power mix with respect to renewables, the reader is referred to Box 39.1 concerning the situation in Mauritius.

BOX 39.1 RENEWABLE ENERGY COMPARED WITH CONVENTIONAL SOURCES IN MAURITIUS

Background

Many countries still rely on conventional electricity production. This example shows how LCA can reveal the sustainability characteristics of a national supply scheme. Such studies can reveal the most suitable energy sources to support regional sustainability policies.

Country case study – LCA and energy supply

The energy mix of Mauritius is predominantly based on fossil fuels. Coal and fuel oil account for 80 per cent of the mix, and the remainder is from renewable sources such as bagasse (derived from sugar cane) and hydropower. An LCA determined the respective environmental impacts of these sources of electricity. The methodology was based on the ISO 14040–14044 standard. Mid-point impact categories were global warming, acidification, abiotic depletion, eutrophication, toxicity, ozone depletion, and photochemical ozone. A cradle-to-consumer approach was used as the boundary.

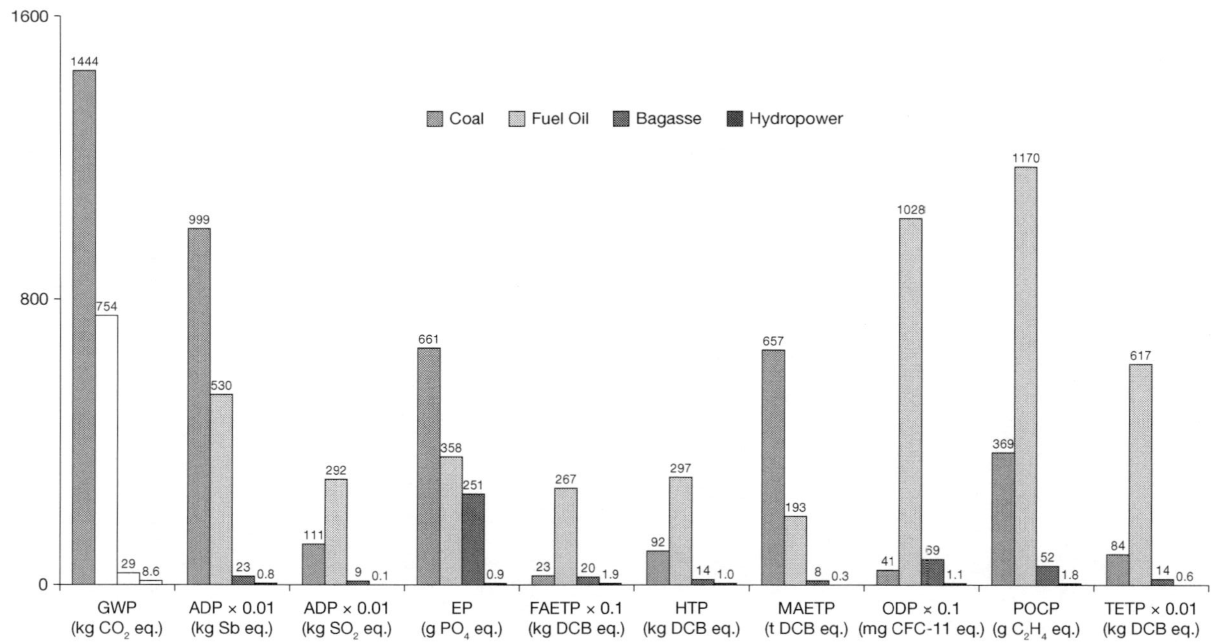

Figure 39.3
Midpoint impacts per MWh for major fuel sources in Mauritius

Figure 39.3 compares the impacts attributed to each fuel source. As expected, coal and oil have a significantly higher impact on nearly all effect categories than the two renewable sources. For global warming, there is a factor of more than 45 when electricity from coal is compared with bagasse, for example. The study also shows that the operating phase of renewable power generation results in significantly lower impacts over the life of the installation compared with fossil fuel.

The LCA study was based on the present energy mix. It could be extended to include other renewable sources that are now being considered worldwide in energy policies. Policies to increase the share of renewables have already been formulated in Mauritius, with a declared national target of 35 per cent by 2025.

Ravina Brizmohun

Conclusion

The world is undergoing an accelerating transition to an energy system that is more intelligent, distributed geographically, and based on RE sources. There are many benefits for regions, but, as is the case with any energy transition, problems may arise in the change to new technologies and operating practices. LCA can help identify potential problems early enough that corrective actions are possible.

Note

1 The authors are, respectively, head of the Energy, Climate, and Technology Branch in the UN Environment Programme and consultants working in the Branch. They alone are responsible for the views expressed, which do not necessarily represent the decisions or policies of UNEP.

Further reading

OECD (2012) *Linking Renewable Energy to Rural Development*, OECD Green Growth Studies, OECD: Paris.

Taylor, M. (2008) 'Beyond technology-push and demand-pull: Lessons from California's solar policy', *Energy Economics*, 30(6): 2829–54.

United Nations Environment Programme (2014) *District Energy in Cities: Unlocking the Potential of Energy Efficiency and Renewable Energy*, Paris: UNEP.

Union of Concerned Scientists (n.d.) *Benefits of Renewable Energy Use*. Available at: www.ucsusa.org/clean_energy/our-energy-choices/renewable-energy/public-benefits-of-renewable.html (accessed 7 August 2016).

References

Busche, S. (2010) *Clean Energy Policy Analyses: Analysis of the Status and Impact of Clean Energy Policies at the Local Level*, Golden, CO: US National Renewable Energy Laboratory.

Del Río, P., and Burguillo, M. (2008) 'Assessing the impact of renewable energy deployment on local sustainability: Towards a theoretical framework', *Renewable & Sustainable Energy Reviews*, 12: 1325–44.

Edenhofer, O., Pichs-Madruga, R., Sokona, Y., Seyboth, K., Matschoss, P., Kadner, S., Zwickel, T., Eickemeier, P., Hansen, G., Schlömer, S., and von Stechow, C. (eds) (2012) *IPCC Special Report on Renewable Energy Sources and Climate Change Mitigation*, Cambridge, UK: Cambridge University Press.

European Parliament (2010) *Decentralized Energy Systems*, Brussels: European Parliament.

Frankfurt School–UNEP Centre/Bloomberg New Energy Finance (2015) *Global Trends in Renewable Energy Investment 2015*, Frankfurt, Germany: Frankfurt School-UNEP Centre.

Galarraga, I., Gonzalez-Eguino, M., and Markandya, A. (2011) 'The role of regional governments in climate change policy', *Environmental Policy & Governance*, 21: 164–82.

IRENA (2015) *Off-Grid Renewable Energy Systems: Status and Methodological Issues*, Abu Dhabi: International Renewable Energy Agency.

Leidreiter, A., et al. (2013) *From Vision to Action: A Workshop Report on 100% Renewable Energies in European Regions*, Hamburg, Germany: World Future Council.

Pehnt, M. (2006) 'Dynamic life cycle assessment (LCA) of renewable energy technologies', *Renewable Energy*, 31(1): 55–71.

REN21 (Renewable Energy Policy Network for the 21st Century) (2015) *Renewables 2015 – Global Status Report*, Paris: REN21.

REN21 (Renewable Energy Policy Network for the 21st Century) (2014) *Renewables 2014 – Global Status Report*, Paris: REN21.

REN21 (Renewable Energy Policy Network for the 21st Century) (2013) *Renewables Global Future Report*, Paris: REN21.

Sathaye, J., Lucon, O., Rahman, A., Christensen, J., Denton, F., Fujino, J., Heath, G., Kadner, S., Mirza, M., Rudnick, H., Schlaepfer, A., and Shmakin, A. (2012) 'Renewable energy in the context of sustainable energy', in *IPCC Special Report on Renewable Energy Sources and Climate Change Mitigation*, eds O. Edenhofer, R. Pichs-Madruga, Y. Sokona, K. Seyboth, P. Matschoss, S. Kadner, T. Zwickel, P. Eickemeier, G. Hansen, S. Schlömer, and C. von Stechow, Cambridge, UK: Cambridge University Press, Figure 9.7.

UNESCAP (2012) 'Decentralised energy systems', in *Low Carbon Green Growth Roadmap for Asia and the Pacific*, United Nations Economic and Social Commission for Asia and the Pacific.

UNEP (2015) *Green Energy Choices: The Benefits, Risks and Trade-Offs of Low-Carbon technologies for Electricity Production*, Report of the International Resource Panel, E. Hertwich, T. Gibon, S. Suh, J. Aloisi de Larderel, A. Arvesen, P. Bayer, J. Bergesen, E. Bouman, G. Heath, C. Peña, P. Purohit, and A. Ramirez, Paris: UNEP.

US National Renewable Energy Laboratory (2013) *Life Cycle Greenhouse Gas Emissions from Electricity Generation*, Golden, CO: US National Renewable Energy Laboratory.

US National Renewable Energy Laboratory (2015) *Best Research-Cell Efficiencies*, Golden, CO: US National Renewable Energy Laboratory.

World Future Council (2013) *From Vision to Action: A Workshop Report on 100% Renewable Energies in European Regions*, World Future Council.

40

CLOSING THE LOOP

The example of packaging

Massimo Lombardi

Introduction

This is the story of a project born in Sabox (n.d.), a small manufacturer of cardboard boxes that developed sustainable products and founded a network of companies to give effectiveness to the proximity recycling principle and to offer, even to big players, packaging solutions sourced by their own waste-paper, thus creating sustainable value for any stakeholder involved.

Sabox sustainable packaging: The GreenBoxX®

Sabox is a company that specializes in the design and manufacture of corrugated-cardboard packaging. The main offices and the production plant are 30 km south of Naples, Italy. In 2008, Sabox mainstreamed sustainability and social responsibility into the core of its operating business, adopting the GreenProject, a sustainability programme aiming to reduce product and manufacturing environmental impact and CO_2 emissions; Sabox was the first in Italy to achieve the Carbon Trust Standard in 2011. Refocusing R&D efforts into finding greener solutions, an entire range of ecopackaging, GreenBoxX®, was developed: it is made of recycled paper from the local differentiated waste collection and is processed by very close manufacturing plants. GreenBoxX® features high physical and mechanical performance, is lighter and easier for shipping, has a lower impact on the environment, being manufactured using water-based inks, and uses plasticizer-free glues. Having carried out a product's life cycle assessment (LCA), Sabox has specific data about the carbon footprint related to its packaging life cycle. Bearing an EPD, Sabox (n.d.) communicates certified data for its product's carbon footprint, along with its commitment to reducing it: in this way, clients can be provided with detailed figures about the packaging's environmental impact, which thus integrates customers' sustainability plans and positively differentiates from competitors.

Further, any paper reduction or weight lightening can be easily translated into CO_2 equivalent figures.

The network 100% Campania for sustainable packaging

In 2012, Sabox guided the process that led to the foundation of the network *100% Campania* for sustainable packaging (Campania is the name of the region in which the city of Naples lies). The founding companies of the network are all local companies operating mainly in the paper industry and managing the entire value chain (from collection and selection platforms to paper mills, from corrugators to box plants) – representing about €200 million turnover and with more than 400 employees.

The network was established to achieve the following objectives:

- implement the proximity recycling principle through managing a territorial, integrated, sustainable collection of secondary raw materials for the production of packaging and other paper-based products;
- reduce environmental, economic, and social impacts of the whole collecting and recycling cycle;
- utilize the same customers' waste-paper to produce their packaging, a closed loop that reduces natural resources exploitation, emissions, and waste generation;
- research and development of new types of material to produce sustainable packaging and to identify new applications for the locally recycled waste-paper;
- develop sustainability projects with the goal of the gradual reduction, in terms of CO_2 equivalent, of the environmental impact of the network's companies and their products;
- communicate to citizens and institutions the environmental and economic benefits of a local processing cycle to boost the quantity and quality of differentiated waste collection.

In Campania, every year, 150,000 tons of waste-paper are collected, and just a small portion is recycled in the same area: most of it is sold all over Italy and also abroad.

To understand the value of a proximity manufacturing cycle, just consider that 100,000 tons of waste-paper recycled locally corresponds to 300 jobs and more than €80 million of added value.

In general, paper-waste generated from a manufacturing plant is taken by an authorized operator, transported to a platform where it is selected and palletized, and then placed on the market to be sold, even abroad, without the company that generated it knowing or deciding on its destination.

100% Campania guarantees that all the waste-paper entrusted to the network will be recycled in proximity to produce packaging and other paper-based products and to be distributed to companies in the same territory, while the traceability and sustainability of the entire manufacturing *proximity principle* cycle are ensured.

Directive 2008/98/EC of the European Parliament establishes a legal framework for the treatment of waste in the EU. It sets out the basic concepts and definitions related to waste management, such as the 'waste hierarchy' (reduce, reuse, recycle) and the 'proximity principle': the principle states that waste should be disposed of in one of the nearest appropriate installations.

Wherever possible, the proximity principle should be applied. This recognizes that the transportation of waste has environmental, social, and economic costs, so, as a general rule, waste should be dealt with as near to the place of production as possible. This has the added benefit of raising awareness about waste and encouraging ownership of the problem at the local level. Another benefit is to create and keep a local value-creating manufacturing cycle.

The Ferrarelle project

This project illustrates how the 100% Campania network worked to create and manage a closed-loop supply chain to produce corrugated packaging sourced from waste-paper from the Ferrarelle manufacturing plant. This is an example of a circular economy creating sustainable value along the supply chain (other examples of a circular economy can be found in Chapter 26). Ferrarelle is a well-known Italian mineral water producer dating from the end of the nineteenth century. Today, Ferrarelle employs around 370 people and exports to more than forty countries worldwide. The marketing and sales offices are in Milan, the operational headquarters are in Riardo (30 km north of Naples), and Parco delle Sorgenti hosts the natural area with Ferrarelle's water sources.

This project, elaborated by the consulting firm Greener Italia with Sabox, was designed to be consistent with the Ferrarelle sustainability targets for packaging and emissions, as illustrated in their corporate sustainability report.

The closed-loop project aims to produce a new kind of corrugated-cardboard packaging produced from waste-paper from the Riardo Ferrarelle plant; this packaging will be then adopted by the same plant for its products.

Switching to the EPD-certified GreenBoxX®, Ferrarelle will be able to communicate reliable and comparable data about savings and CO_2 reductions, improve stakeholders' relationships, and give effectiveness to the proximity recycling principle.

The project started with analysis of the current situation and focused on the cardboard interlayer, which accounts for 90 per cent of the paper-based packaging used by Ferrarelle. According to the project, waste-paper will be recycled in a local paper mill and, following a special manufacturing process, will become a new kind of paper fit for industrial purposes, *greenpaper®*. The new packaging, GreenBoxX®, produced using the greenpaper®, will be manufactured by the current packaging provider (Sabox), who will also take care of the switch.

To date, more than 6 million interlayers have been produced by the closed-loop project.

Phases of the closed-loop chain project

1. Waste-paper generated by the Riardo Ferrarelle plant is stocked in containers located inside the manufacturing plant.
2. Periodically, containers are emptied, and paper waste is moved by the usual operator to its local platform.
3. The network, by virtue of a contractual agreement, purchases all the waste-paper generated by the Riardo plant from the operator. The platform's operator certifies the waste-paper's origin.
4. Waste-paper is then transferred to the local paper mill where, according to detailed manufacturing specifications, it is used to produce the greenpaper®.
5. Rolls of greenpaper® are then moved to the network corrugator and box plant to be transformed into GreenBoxX® interlayers, ready to close the loop by being sent back to the Riardo plant.

Complementary phases of the closed-loop supply chain

The complementary phases of the closed-loop supply chain are as follows:

1 determination of production potential based on the amount of waste generated by the plant's facilities;
2 evaluation of alternative, more efficient paper's grade;
3 calculation of the environmental impact in terms of CO_2 equivalent per box;
4 printing layout and elaboration of the environmental claims 'on product';
5 supply test;
6 production start-up.

Benefits

Waste-paper generated by the Riardo plant is no longer collected to be sold on the market (even abroad), but is destined for a local recycling cycle that creates sustainable value for the territory and optimizes and reduces the use of resources, costs, and environmental and social impacts.

- The proximity recycling principle is fully and effectively realized.
- The entire GreenBoxX® production cycle is FSC-certified.
- The GreenBoxX® life cycle environmental impact is EPD-certified.
- All companies in the supply chain are committed to seeking solutions to reduce the overall environmental impact of GreenBoxX®.
- There are integration and consistency with Ferrarelle's targets for sustainability.
- There is communication on and off product.
- The switch to GreenBoxX® interlayers saved, in one year, more than 170 tons of CO_2e and about 172 tons of paper, according to Figure 40.1.

Calculations are based on an estimated amount of waste-paper (1,000 tons/year) generated by the Riardo manufacturing plant. An example of the use of recent software to calculate the impact of such changes in the packaging industry generally is shown in Box 40.1. Other methods may also be applicable.

BOX 40.1 CASE STUDY FROM ANOTHER REGION (AUSTRALIA) – THE EXAMPLE OF PIQET

The Australian Packaging Covenant requires all industry sectors to reduce packaging's environmental impact. In response to the need for an independent and credible evaluation methodology, the former Sustainable Packaging Alliance has developed the cloud-based Packaging Impact Quick Evaluation Tool – PIQET©.

PIQET is a streamlined, intuitive, rapid LCA, providing environmental performance indicators to support decision-making in packaging development and innovation, selection of materials, and specification of packaging for procurement of goods. PIQET emphasizes environmental factors as well as including functional performance and social indicators such as litter and consumer convenience.

PIQET uses data sets such as Ecoinvent and others, supplemented where necessary by company-specific data. The life chain of packaging is

Ferrarelle Spa: switching to greenboxx® interlayers

Interlayers are used to seperate and stabalize layers of product placed on a pallet

Starting from previous interlayer in corrugated used by Ferrarelle, we identified a greenboxx® alternative able to generate consistent savings in weight and CO_2 emissions

comparison table	board grade	g/sqm	mm	sqm	g/pcs	pcs/year	toatal weight (tons)	△tons	△%	CO2e tons	△CO2e tons
previous packaging	TBFF/2322/B	409	800x1200	0,96	399	3mio	1.197	-172	-14,35	1.186	-170
greenboxx®	CGLCGLCGL/111/B	356	800x1200	0,96	341	3mio	1.025			1.016*	

Calculations by Massimo Lombardi based upon an estimated amount of 3mio pcs/year - weight and CO2e calculations based on GIFCO-FEFCO technical specifications for packaging in corrugated and *greenboxx®LCA 2011 and annual updates - Bicocca University - Milan - EPD n. 000221SP - www.environdec.com

Figure 40.1
Comparison between GreenBoxX® and the previous Ferrarelle packaging

increasingly complex; as new converting processes come online, these are added to PIQET.

The tool is used by major consumer goods companies, retailers, and packaging manufacturers around the world to evaluate the impact of packaging and to design lower-impact products.

PIQET reports on the following indicators:

- climate change (kg CO_2 eq);
- cumulative energy demand (MJ LHV);
- minerals and fuels (MJ surplus);
- photochemical oxidation (kg $C_2 H_4$ eq);
- eutrophication (kg PO_4^{3-} eq);
- land use (Ha*a);
- water use (kl H_2O);
- solid waste (kg).

Performance on the above indicators is of direct relevance to regions, as, for example, in resource use, energy consumption, land use, environmental control, and waste management.

For further information, see www.piqet.com. Technical details can be found in Verghese, K., Lewis, H., and Fitzpatrick, L. (eds) (2011) *Packaging for Sustainability*, London: Springer. This contains a chapter on tools and provides a detailed overview of PIQET (www.springer.com/us/book/978 0857299871)

Prepared by F. Balkau, with permission from PIQET.

Conclusions

The project shows some 'active' ways to utilize an LCA study. First, Sabox carried out the study to determine where and how great were the environmental impacts of its product. Subsequently, starting from the LCA results, Sabox set up reduction projects in collaboration with its suppliers, establishing a network (Rete 100% Campania n.d.) of companies sharing sustainability targets and including most of Sabox's product life cycle actors (waste-paper platforms, paper mills, corrugators, logistics, box plants).

Besides integrating its commercial offer with the product's environmental performance, Sabox elaborated a project proposal that involved the customer (Ferrarelle) and embodied the principles of circular economy and proximity recycling, also being consistent with the customer's supply chain reduction targets.

In conclusion, the project examined here shows how an LCA can:

1 provide the basis on which a company decides to develop projects to reduce and optimize the use of resources needed to manufacture its products;

2 be integrated with the participation of other companies along the product life cycle to develop projects to reduce their contribution to the overall environmental impact.

Regions should promote collaboration among companies along the supply chain and promote the use of life cycle tools, and above all they should give the right value to the efforts made by enterprises towards sustainable development.

Further reading

Castellani, V., and Sala, S. (2011) *Atlante dell'ecoinnovaziane*, Milan, Italy: Franco Angeli.

Iraldo, F., and Melis, M. (2013) *Green Marketing*, Milan, Italy: Edizioni Sole 24 Ore.

Zamboni, S. (2012) *L'Italia della green economy*, Milan, Italy: Edizioni Ambiente.

References

Sabox EPD (n.d.) Available at: www.environdec.com (accessed 7 August 2016).

Rete 100% Campania (n.d.) Available at: www.packagingsostenibile.com (accessed 7 August 2016).

Sabox (n.d.) Available at: www.sabox.it (accessed 7 August 2016).

The Carbon Trust (n.d.) Available at: www.carbontrust.com (accessed 7 August 2016).

41

LIFE CYCLE ASPECTS OF THE WALLOON WOOD SECTOR

Nelle Thyssen and Vanessa Zeller

Introduction

Most environmental footprint studies are carried out at national level. Also many process LCAs and environmental databases present national averages. However, the regional perspective is important when regional differences in the economic structure of a country exist and when diverging technology and energy mixes are used. Belgium is such a case, where regional variation in the economic structure (production) occurs, and where regional differences in income level (consumption) can be found. Furthermore, the Belgian regions are responsible for most environmental policies, and the decision-making often takes place at the regional level.

Many studies have determined the environmental impacts of consumption and production in a specific region and identified hot spots among all economic activities. In a review of studies carried out in the European context, it has been found that, together with the food and transport sectors, the building sector is responsible for the highest environmental impacts (Tukker and Jansen 2006); also, in other areas of the world, the building sector can make a relevant contribution to the environmental impact, as shown in Chapter 42. In order to decrease these impacts, a reduction in operational energy use is the focus of environmental policies for the building sector (e.g. the EC's *Energy Performance of Buildings Directive*; European Commission 2003). Additionally, variations in building fabric (material choices) are proposed as measures to increase environmental performance. In this context, wood buildings have been discussed as an alternative to conventional buildings (e.g. concrete, steel, and brick; Ortiz et al. 2009; Thormark 2002; Gong et al. 2012; Guardigli et al. 2011; Gerilla et al. 2007; Perez-Garcia et al. 2005). The majority of those articles show lower energy consumption and climate impact for wood-based buildings.

Regional context: The Walloon wood construction sector

For Belgium, the building structure and building occupancy, as well as personal transport, are important contributors to environmental impacts such as resource depletion, acidification, global warming potential (GWP), and waste creation (more than 60 per cent; Jansen and Thollier 2006). More specifically, 40 per cent of the CO_2 emissions are released by the residential sector (Buyle et al. 2012).

On average, around 1,800 wood-based residential buildings are constructed per year in Belgium, which represents about 7.5 per cent of the number of new dwellings built (2010–13; Houtinfobois 2014). The wood products sector is the sector that fabricates wooden construction products, from sawn wood to prefabricated buildings, and also wood products with non-construction applications (European Commission 2008). This sector represents a 2.5 per cent share of the total economic production from manufacturing sectors in the Walloon region (Bureau fédéral du Plan 2008) and has the forestry sector as its main supply sector. In Belgium, forestry activities can be found mainly in the Walloon region, where the forest has a surface of around 555,000 ha, with a covering of 57 per cent hardwood and 43 per cent softwood (Office économique wallon du bois 2015). Every year, about 3,700 m^3 of industrial and round wood are harvested. In total, about 7,900 enterprises are active in the forestry–wood chain (including production of paper, furniture, etc.), of which 40 per cent are in joinery (Office économique wallon du bois 2015).

Objectives and methodologies of the case studies

The ENECOBOIS project is a research project executed by WOOD.BE with the University of Brussels, financed by the Walloon region. The overall goal of the project is to analyse the environmental impacts of the Walloon wood construction sector by considering the complete forestry–wood chain.

Within the project, the environmental impacts of the wood-based construction sector are studied on three levels: the product, the building, and the sector level. For the first two, process LCA is applied, from cradle to gate and from cradle to grave, respectively (cf. Chapter 10). Specific case studies, including primary data collection in Walloon enterprises, are executed. The European standards EN 15804, EN 16485, and EN15978 (in accordance with the international standards ISO 14040/44) are applied. The LCA at product level aims to analyse hot spots for the current production of wood building products in Wallonia and to study options for improvements to the environmental profile. The analysed products are sawn wood, glued laminated timber, MDF, parquet, cladding, and windows. At building level, the four most common types of residential wood-based building in the Walloon region are analysed: wood framework, beam column, log building kit, and solid wood panel structures. In particular, the LCA study aims to investigate the relation between impacts resulting from the production of building products (embodied energy) and the energy that is consumed during the use phase (operational energy use).

At sector level, LCA is applied based on environmentally extended input–output tables (cf. Chapter 8). The analysis aims to identify and reduce the environmental impacts of the wood construction sector. The sector is compared with other manufacturing sectors that produce construction products, such as the concrete and steel sectors.

Case study results

Two important contributors to the environmental impact of wood products are the energy consumption (heat and electricity) of the production process (e.g. sawn wood, MDF, and parquet) and transport of Nordic wood by truck to the Walloon region (e.g. glued laminated timber and cladding). Heat is often produced on site based on the co-products, and so is under the direct sphere of influence of the enterprise. For the electricity consumption, for example, the installation to evacuate the co-products along the processing chain represents an important share.

At *building level*, the production of construction wood was identified as an important contributor to several impact categories. The relative contribution depends on the wood type and quantity used (10–30 per cent of the GWP in the product stage). However, the high amount of biogenic carbon stored in the wood positively influences the greenhouse gas (GHG) balance. The use stage was identified as the life cycle phase making the biggest contribution, with an average contribution of 77 per cent of the overall impacts. The relative importance of the use phase decreases in buildings with high insulation performance. The construction and end-of-life phases show a relatively small contribution for most impact categories, owing to specific advantages of wood constructions such as lower impacts from demolition and landfill and high recyclability and energy recovery potential.

At *sector level*, the study identifies the general environmental hot spots of the wood building sector by determining the average contribution of process inputs for several impact categories. Regarding GWP, for example, transport (18 per cent), energy (17 per cent), and chemicals (10 per cent) are identified as the most important contributors. The results also show the impacts from sectors that are usually not considered in process LCA, such as the impacts from service sectors, which contribute to 5 per cent of the total GWP. The analysis identifies where the wood sector is situated within the overall economy. By comparing the sector with the average GHG emissions from manufacturing sectors, it is demonstrated that the wood construction sector is not one of the most contributing ones in the regional economy (79 per cent lower GHG emissions than the average).

LCA results in decision-making processes

The results of the product LCA support the development of recommendations for the specific enterprises under study and give insight into the impacts along the forestry–wood chain (in the sector). When LCA results are discussed with enterprises and actors in the forestry–wood chain, economic concerns, legal and practical limitations, and even social aspects are mentioned as the most important barriers to implementing improvement options.

The improvement of one environmental hot spot of the production could consist in lowering the energy consumption, for example, by optimising the evacuation system for co-products. Anyhow, owing to the high investment cost, an economic analysis (e.g. life cycle costing, as described in Chapter 33) of the potential environmental improvements would put the LCA results in a more realistic industrial scope and could convince the industrial partner to undertake action.

Truck transport of Nordic wood, responsible for an important share of the impacts of products based on this wood, is, economically and practically speaking, the most appropriate means of transport. Based on a preliminary analysis, boat transport would improve the environmental profile, but extra custom duties, risk of damage, and transport duration make it economically unsuitable. It becomes clear that, with environmental LCA, only a part of the industrial reality is represented. Transport optimisation is an issue that individual enterprises can't cope with, and it should be managed on a bigger (sector or regional) scale – for example, by promoting and enabling (practically and economically) more environmentally friendly transport methods.

The results at *building level* address the industrial decision-maker, more precisely, the wood construction enterprises that have the ability to influence choices made by the clients/final consumers and the architects. If LCAs that cover the whole life cycle of a building are conducted, the awareness within wood construction enterprises of the environmental consequences of material and design choices can be raised. For example, the results from a replacement scenario demonstrate the environmental effects of durability of construction products. The results from operational energy use show how the technical performances of insulation products and of technical installations determine the energy balance, but also other impact categories. In general, the LCA results help to broaden the view, from impacts that can be controlled by the enterprises (supply chain management, optimisation of transport distances, etc.) to impacts that occur further downstream in the process chain, especially in the end-of-life phase. If the most important contributors are known, the most effective improvement options can be implemented. The results at building level can also be used to determine reduction potentials for the whole region. Therefore, the results from comparative LCAs have to be combined with information about the building stock.

The results at *sector level* are relevant for political decision-makers, because representative results about environmental hot spots at sector level can be provided. With this information, priority areas for actions can be selected more easily, so that the environmental impacts of the wood construction sector can be reduced. Furthermore, as the trade flows towards different kinds of consumer are included in the background data, the links between production and consumption can be studied. Sustainable consumption scenarios can help the regional decision-maker to understand the implications of potential improvements and to evaluate the effectiveness of sustainable consumption measures. As the regional sector database/LCA model comprises all economic activities in the region, comparisons between sectors can be carried out in a systematic, consistent, and cost-efficient manner. In this way, priority sectors for regional environmental policies can be identified.

Conclusions

The fact that the environmental impacts of the wood-based construction sector in the Walloon region are studied simultaneously with impacts at the enterprise level gives a global view of the performance of the sector. It helps to put the performance of one enterprise or one sector of the region into perspective. Economic and social aspects are not directly studied during this project, but appear to be an important complementary aspect.

Individual enterprises become aware of the hot spots of their production thanks to the LCA at product level, although they can't easily influence all processes of the whole production chain, because most of the upstream and downstream processes are outside their direct sphere of influence. Additionally, the investment costs for on-site optimisations can be an economic barrier. This is where sector federations and regional decision-makers could develop joint solutions and implement actions to help individual enterprises.

Even though the wood-based building products and construction show a relatively good environmental performance, the LCA results for the building level show high impacts in the use phase. Therefore, priority setting in the use phase by the regional decision-makers can still be justified. As results also demonstrate the ongoing development towards increased importance of the product phase, it is also necessary, at the same time, to continue monitoring and optimising the impacts of the products used and of the other life cycle stages.

On the one hand, the LCA at sector level reveals the hot spots of the wood construction sector that should be prioritised when it comes to regional action plans for the sector. On the other hand, it shows that the wood-based construction sector shows better environmental performance than other (construction) sectors. From this perspective, wood-based construction could be promoted as a good example.

The project revealed the importance of the regional perspective for different levels of analysis for the wood-based construction sector. Analyses at product, building, and sector level are complementary, and the bigger picture should be considered when conclusions are reached and decisions are made based on the results.

References

Bureau fédéral du Plan (2008) *Données régionales input–output élaborées par le Bureau fédéral du Plan dans le cadre du projet,* 'Modèle environnemental-IO pour la Flandre' commandité par la Région flamande.

Buyle, M., Braet, J., and Audenaert, A. (2012) 'Review on LCA in the construction industry: Case studies', in *Mathematical Modelling and Simulation in Applied Sciences,* Finland, pp. 98–104.

European Commission (2003) 'Directive 2002/91/EC of the European Parliament and of the Council of 16 December 2002 on the Energy Performance of Buildings', *Official Journal of the European Communities,* L(1): 65–71.

European Commission (2008) *NACE Rev. 2 Statistical classification of economic activities in the European Community, Methodologies and Working papers.* Luxembourg: Eurostat, European Commission.

Gerilla, G.P., Teknomo, K., and Hokao, K. (2007) 'An environmental assessment of wood and steel reinforced concrete housing construction', *Building & Environment,* 42: 2778–84.

Gong, X., Nie, Z., Wang, Z., Cui, S., Gao, F., and Zuo, T. (2012) 'Life cycle energy consumption and carbon dioxide emission of residential building designs', *Beijing Journal of Industrial Ecology,* 16(4): 576–87.

Guardigli, L., Monari, F., and Bragadin, M.A. (2011) 'Assessing environmental impact of green buildings through LCA methods: A comparison between reinforced concrete and wood structures in the European context', *Procedia Engineering,* 21: 1199–206.

Houtinfobois (2014) *Etat de la construction en bois en Belgique en 2011–2014 (Enquête),* Brussels: Hout Info Bois.

Jansen, B., and Thollier, K. (2006) 'Bottom–up life-cycle: Assessment of product consumption in Belgium', *Journal of Industrial Ecology,* 10: 41–55.

Office économique wallon du bois (2015) *Panorabois Wallonie édition 2015,* Marche-en-Famenne, Belgium: Office économique wallon du bois.

Ortiz, O., Castells, F., and Sonnemann, G. (2009) 'Sustainability in the construction industry: A review of recent developments based on LCA', *Construction & Building Materials,* 23: 28–39.

Perez-Garcia, J., Lippke, B., Briggs, D., Wilson, J.B., Bowyer, J., and Meil, J. (2005) 'The environmental performance of renewable building materials in the context of residential construction', *Wood & Fiber Science,* 37: 3–17.

Thormark, C. (2002) 'A low energy building in a life cycle: Its embodied energy, energy need for operation and recycling potential', *Building & Environment*, 37: 429–435.

Tukker, A., and Jansen, B. (2006) 'Environmental Impacts of products: A detailed review of studies', *Journal of Industrial Ecology*, 10: 159–82.

42

GLOBAL WARMING ASSESSMENT OF HIMALAYAN BUILDINGS

Silu Bhochhibhoya and Raffaele Cavalli

Introduction

In every country, the building sector is a major contributor to emissions as well as a huge user of natural resources and energy (Asif et al. 2007; Fan and Junfeng 2001). Environmental impacts arising from the production and transport of building materials vary according to regions and countries (Buchanan and Honey 1994; Pandit 2013). Developing countries may experience greater environmental impacts per unit of material as a result of lower production efficiencies and also, in part, because of the smaller size of manufacturing plants (Huberman and Pearlmutter 2008; Pittet and Kotak 2010). Furthermore, the inputs for production of materials and for their subsequent transport are very different, with potential consequences for the overall environmental impacts (Huberman and Pearlmutter 2008; Pittet and Kotak 2010).

Information about the environmental impact of building materials is very limited in developing countries, even though they represent some of the most vulnerable areas in the world in terms of the hazards associated with climate change (Gentle and Maraseni 2012; Pandit 2013).

With the aim of better understanding of the environmental performance of buildings in developing countries, a study has been performed in the Sagarmatha National Park (SNP), Nepal, which attracts about 30,000 tourists each year (Census 2014) and is vulnerable to the consequences of climate change. The government policy to attract more tourists in the near future will contribute to a growth in construction, with negative effects on an already critical situation in terms of environmental pollution and other impacts (Salerno et al. 2010). Given the projected high tourist flow, lodge owners are modifying their traditional building style based on local materials to favour modern buildings. These use materials transported by air from the capital city, as transport by road is difficult. To date, most traditional building materials have been found near to where they are used. In this situation, life cycle assessment (LCA) could assist decision-making about technologies and materials so as to minimise the environmental impact of future buildings (Zabalza Bribián et al. 2011; Passer et al. 2012).

This chapter provides an overview of an LCA study of building materials, identifying major emission sources and potential areas of improvement based on a comparative carbon footprint study of the three main types of building pattern found in the Himalayan region: traditional, semi-modern, and modern.

Traditional buildings (Figure 42.1a), known as Sherpa houses, follow ancestral tradition in using mainly materials such as timber, stone, and mud. Semi-modern types of building attempt a transformation to a more modern type of functioning (Figure 42.1b) but still have limited or no insulation materials besides wooden planks, dry stones, and a minimal amount of cement and mud as plaster. Conversely, modern buildings for tourists rely heavily on imported construction materials such as cement, glass wool, and polystyrene sheets (Figure 42.1c).

Figure 42.1a
Traditional building

Figure 42.1b
Semi-modern building

Figure 42.1c
Modern building

Materials and methods

The globally recognised LCA methodology was used to assess the environmental impact of buildings over their lifespan. LCA for building materials provides quantitative and comparative values of the environmental impacts of various building technologies (Singh et al. 2011). This study follows the ISO 14040/44 methodology that addresses the environmental aspects and potential environmental impacts over a product's life cycle. The LCA modelling was carried out in GaBi V 6.0. The CML 2001–April 2013 method was used for the environmental impact assessment.

Goal and scope definition

The study examines the use, maintenance, and replacement phases of the three building types. The end-of-life phase includes impacts arising from demolition of the building, reuse of some materials, transportation of the waste to a treatment site, and different treatment processes. The building materials end-of-life phase was, however, not considered, because there is limited information on options for building demolition and recycling in the Himalayas. The functional unit was thus 'construction and occupation' (Cuéllar-Franca and Azapagic 2012) over a 50-year lifespan.

For this study, the construction phase comprises the environmental impact (carbon footprint) associated with extraction of raw materials, processing of these into building products, their transportation to the construction site, and then assembly of the products. The operation phase considers the impacts of energy consumption in different household activities, such as space heating, water heating, cooking, lighting and use of electrical appliances. The maintenance and replacement phase considers the impacts associated with replacing building materials and maintenance over the 50-year lifespan. The life of the building materials was calculated using standard techniques (ATD 2015).

The inventory analysis phase

The life cycle inventory covers the collection of data in accordance with existing standards on building materials and energy use. The Eco-invent database was used for the emission factors as a secondary source of data. The detailed information is presented in Table 42.1. Data were collected from inventory reports, direct observations and measurement, and from interviews with contractors and local people.

Impact assessment and interpretation of results

The impact assessment phase uses both qualitative and quantitative approaches to understand how use of raw materials, energy generation, and emissions to the atmosphere affect the environment. This study focuses on the global warming potential (GWP) attributed to different processes and materials used in buildings in SNP. GWP is a common indicator in LCA studies (Knauf 2015), usually expressed in terms of carbon dioxide equivalents (CO_2eq). Although the time frame for the assessment recommended by the Kyoto Protocol (1997) and by the Intergovernmental Panel on Climate Change (IPCC 2013) is 100 years, this study has chosen a time frame of 50 years to reflect the local circumstances of building life.

Results and discussion

The GWP of the three different building types is presented in Figure 42.2. The results show that the highest carbon footprint component during a building's

Table 42.1 Profile of the three types of building under study

Type	Traditional	Semi-modern	Modern
Elevation (m)	3,800	3,800	3,800
Operational season	7 months	7 months	7 months
Net area (m^2)	210	244	301
Gross volume (m^3)	1,953	2,868	3,897
Construction method	Load bearing-framed structure	Load bearing-framed structure	RCC
No. of floors	3	4	3
No. of guest rooms	30	35	16
Occupancy (% of rooms occupied)	42	64	66
Number of guests per annum	3,675	4,183	3,910
Walls	Mud plaster inner and outer side of dry stone, with wooden plank & plywood in internal wall	Cement pointing in dry stone, with wooden plank & plywood on internal wall	Cement pointing in dry stone, with insulating materials in the space, wooden plank & plywood on internal wall
Insulation	Mud plaster	Polystyrene	Glass wool & polystyrene
Windows	Wooden frame with single-glazed glass	Wooden frame with single-glazed glass	Wooden frame with double-glazed glass, 4-mm thick each, and air space of 6 mm
Roofing	Galvanised sheets (CGI)	Galvanised sheets (CGI)	Galvanised sheets (CGI)
Floor	Wooden plank	Wooden plank	Wooden plank
Door	Wooden	Wooden	Wooden

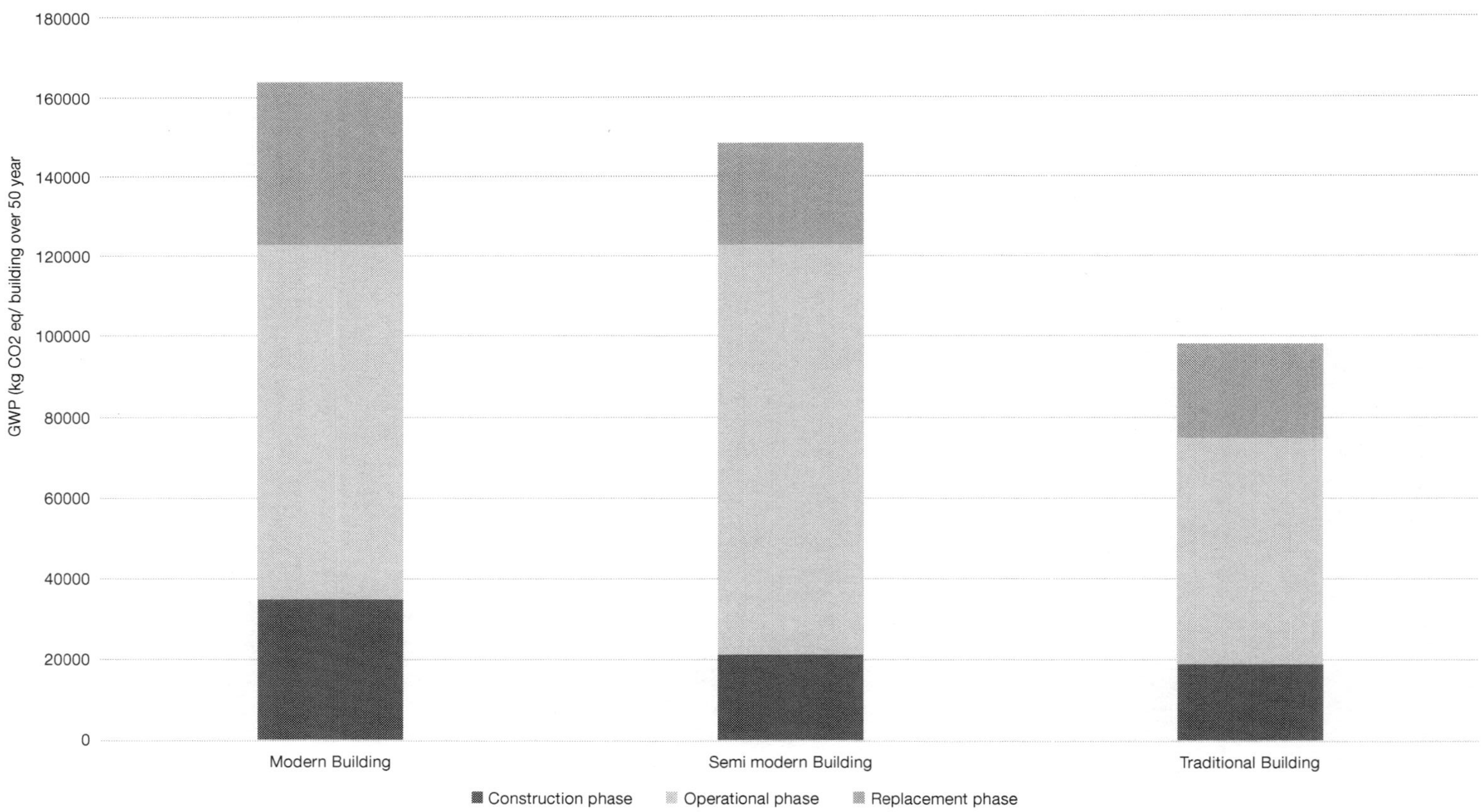

Figure 42.2
Comparison of GWP in different life cycle phases of buildings

life is the use phase, accounting for approximately 53 per cent (88 t CO_2eq) of total life cycle impact. In the case of modern buildings, the construction phase accounts for 21 per cent (35 t CO_2eq), and the replacement phase accounts for 25 per cent (41 t CO_2eq). For semi-modern buildings, the use phase accounts for 65 per cent (97 t CO_2eq), the construction phase accounts for 14 per cent (21 t CO_2eq), and replacement accounts for 20 per cent (30 t CO_2eq). Traditional buildings have the lowest GWP, as they are built mainly from local materials (stone, mud, and wood), with few commercial materials (mostly corrugated galvanised iron (CGI) sheets and plywood). Emissions during the operation phase are, however, high, at 57 per cent (56 t CO_2eq) of total emissions; the construction phase accounts for 17 per cent (17 t CO_2eq), and the replacement phase accounts for 26 per cent (25 t CO_2eq).

Emissions from different sectors in each building phase were also analysed. GWP from the manufacturing of materials accounts for approximately 85 per cent of total GWP; transportation of materials by lorry and aircraft accounts for 14 per cent.

A breakdown of the carbon footprints from different materials during the lifespan is presented in Figure 42.3. The results show that CGI roof sheets and plywood for walls are the main contributors in all building types. CGI sheets have the highest impact: 22 t CO_2eq for a modern building, against only 7 t CO_2eq for semi-modern and traditional buildings. Plywood occupies the second position in terms of environmental impact, with 8 t CO_2eq for modern buildings, 12 t CO_2eq for semi-modern, and 8 t CO_2eq for traditional. A reduction in the carbon footprint of Himalayan buildings, therefore, requires attention to be paid to finding alternatives for roofs and wall envelopes.

The operation phase is, nonetheless, the highest contributor to total GWP: a modern building generates 86 t CO_2eq or more, a semi-modern one produces 96 t CO_2eq, and the traditional building generates 56 t CO_2eq over 50 years. Kerosene and firewood are the main contributors in the operational phase, contributing, respectively: 44 per cent from kerosene and 22 per cent from firewood in modern buildings, 48 per cent from kerosene and 36 per cent from firewood in semi-modern buildings, and 30 per cent from kerosene and 34 per cent from firewood in traditional buildings.

As in the construction phase, CGI sheets and plywood are the major contributors to GWP in the replacement phase. Over a 50-year period, a CGI sheet is typically replaced once, and plywood is replaced twice. CGI sheets account for 55 per cent of total GWP, and plywood accounts for 41 per cent in modern buildings. In semi-modern buildings, CGI sheets contribute 25 per cent, and plywood contributes 74 per cent, and, in traditional buildings, CGI is responsible for 30 per cent and plywood is reponsible for 68 per cent of total GWP in the replacement phase. The study shows that local materials are more environmentally sustainable than modern commercial ones.

Carbon footprint is just one element of the social, economic, and environmental indicators of sustainable development. All these aspects are relevant for the construction industry in all countries (Cabeza et al. 2014; Ortiz et al. 2008). This chapter, therefore, also gives a brief overview of the social and economic aspects of Himalayan buildings. The detailed methodology can be seen in Chapters 14 and 15, where social LCA and life cycle costing are concerned.

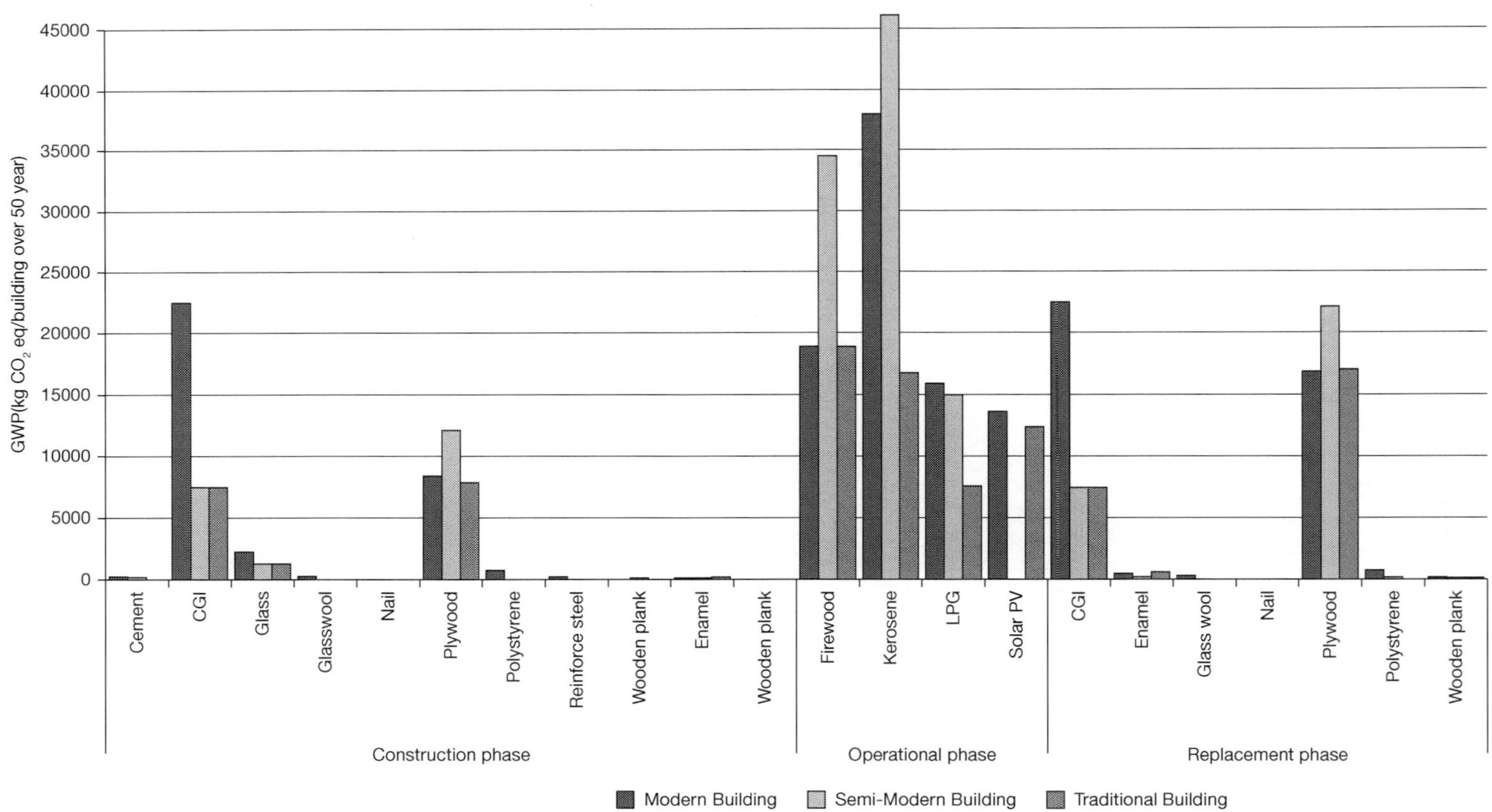

Figure 42.3
GWP of different materials in different phases of building life cycle

Traditional Sherpa houses have embodied social values linked to local materials such as stone, mud, and wood. Modern construction – based on commercial materials such as cement, glass wool, polystyrene, and so on – is more costly. The operational phase is also significant, being responsible for 88 per cent of total costs over 50 years. In semi-modern buildings, the operational cost is 85 per cent of the total, 1 per cent is construction cost, and 14 per cent is replacement cost. In a traditional building, 69 per cent of the total cost is attributed to the operational phase, 3 per cent to construction, and 29 per cent to replacement.

There is little difference regarding the social aspects of these three building types (for the same use), but workers' health and safety are an important social issue where Himalayan buildings are concerned.

Calculation of the relative contributions to global warming of the main greenhouse gases (GHG) from the three different types of building shows that carbon dioxide is the main contributor (> 90 per cent) for all types. The contribution of nitrous oxide is 1 per cent from all building types. Methane contributes 5 per cent for a modern building, 4 per cent for a semi-modern one, and 3 per cent for a traditional building. Non-methane volatile organic compounds represent only a minor contribution to the total GWP emissions for the three building types.

It is interesting to also compare the above calculation results with LCA initiatives in other countries. The case study shown in Box 42.1 concerns a building initiative in Spain where a software tool was used to reveal the best retrofitting options to improve energy performance, among other sustainability criteria.

BOX 42.1 STUDYING URBAN GHG REDUCTION STRATEGIES IN SPAIN

The ESSAI URBAIN project in San Sebastian (Spain) investigated different retrofitting strategies at the district/city level in the context of the city's sustainability objectives. NEST software was used to investigate both the current situation (baseline scenario) and a range of possible future strategies for three districts. Although principally focused on energy retrofitting, the strategies also included some other parameters, such as public lighting, mobility, and waste management.

The different strategies were assessed via the NEST tool to reveal primary energy consumption and associated GHG emissions. The project evaluated these against the San Sebastian GHG emissions objectives for 2020 and 2030. As illustrated in Figure 42.4, the currently envisaged options lead to a GHG emissions reduction of, respectively, 13 per cent and 20 per cent, in comparison with the 20.5 per cent and 30 per cent objectives for 2020 and 2030 (in comparison with 2007).

The results obtained within the ESSAI URBAIN project have helped the San Sebastian municipality to position the currently envisaged retrofitting strategies against its objectives, and to identify the most efficient ways to achieve them.

Prepared by Maxime Pousse, NOBATEK

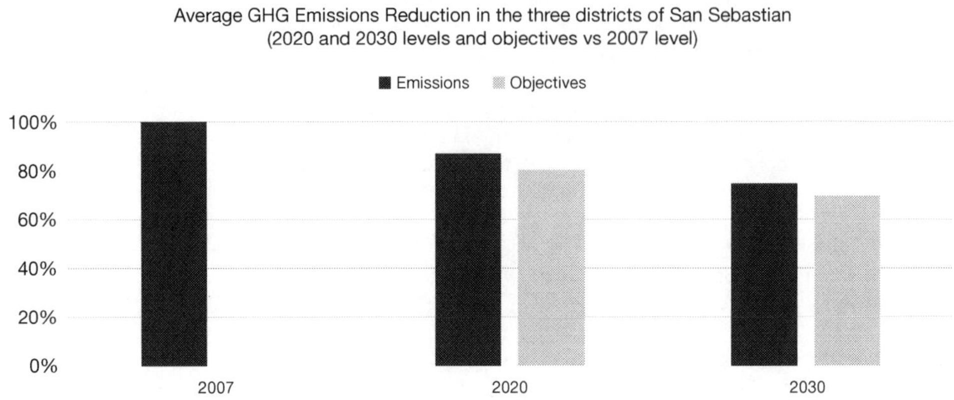

Figure 42.4 Average GHG emissions reduction in the three districts of San Sebastian

Conclusions

The total GWP over a 50-year lifetime is 164 t CO_2eq for a modern building, 148 t CO_2eq for a semi-modern building, and 98 t CO_2eq for traditional buildings. For all the three types of building, the use phase has the highest carbon footprint during the building's lifespan, as a result of energy use. The main opportunities for improvement in the building sector in the Himalayan region lie in the reduction of impacts in this use phase. Studies show that energy saving and emission reduction in households can be achieved by behavioural changes and via product innovations such as energy-efficient lighting and more efficient stoves (O'Neill and Spash 1999; Streimikiene and Ciegis 2010).

The study has outlined that the modern and semi-modern building types that use commercial materials have a GWP five times higher than traditional buildings, which are built from locally available materials.

The study also reveals that CGI and plywood generate higher CO_2eq emissions compared with other materials. To reduce the carbon footprint of buildings in the Himalayas, attention should, therefore, be focused on finding alternatives for roof and wall envelopes. Locally available materials are more environmentally friendly, especially those that use wood from sustainably managed forests.

Building with local materials is more environmentally friendly than building with equivalent commercial materials, because of the higher impact associated with the production and transport of such materials (Morel et al. 2001). From this standpoint, use of wood and wood products could be a good alternative (depending on the source of the timber). An energy-efficient building constructed from local materials is a highly recommended, sustainable building design for the Himalayan region.

References

Asif, M., Muneer, T., and Kelley, R. (2007) 'Life cycle assessment: A case study of a dwelling home in Scotland', *Building & Environment*, 42(3): 1391–4.

ATD (2015) *Average Life Span of Homes, Appliances, and Mechanicals*. Available at: www.atdhomeinspection.com/advice/average-product-life/ (accessed 22 April 2015).

Buchanan, A., and Honey, B. (1994) 'Energy and carbon dioxide implications of building construction', *Energy & Building*, 20: 205–17.

Cabeza, L.F., Rincón, L., Vilariño, V., Pérez, G., and Castell, A. (2014) 'Life cycle assessment (LCA) and life cycle energy analysis (LCEA) of buildings and the building sector: A review', *Renewable & Sustainable Energy Reviews*. http://doi.org/10.1016/j.rser.2013.08.037

Census (2014) *Monthly Tourist Record at Jorsalle*, Sagarmatha, Nepal: SNPBZ.

Cuéllar-Franca, R.M., and Azapagic, A. (2012) 'Environmental impacts of the UK residential sector: Life cycle assessment of houses', *Building & Environment*, 54: 86–99.

Fan, C.W., and Junfeng, Z. (2001) 'Characterization of emissions from portable household combustion devices: Particle size distributions, emission rates and factors, and potential exposures', *Atmospheric Environment*, 35(7): 1281–90.

Gentle, P., and Maraseni, T.N. (2012) 'Climate change, poverty and livelihoods: Adaptation practices by rural mountain communities in Nepal', *Environmental Science & Policy*, 21: 24–34.

Huberman, N., and Pearlmutter, D. (2008) 'A life-cycle energy analysis of building materials in the Negev desert', *Energy & Building*, 40: 837–48.

IPCC (2013) *Climate Change 2013: The Physical Science Basis*. Geneva: Intergovernmental Panel on Climate Change. http://doi.org/10.1038/446727a

Knauf, M. (2015) 'Applying opportunity costs to correctly interpret resource efficiency in LCA studies and environmental product declarations', *European Journal of Wood & Wood Products*, 73: 251–7.

Morel, J., Mesbah, C., Oggero, M., and Walker, P. (2001) 'Building houses with local materials: Means to drastically reduce the environmental impact of construction', *Building & Environment*, 36(10): 1119–26.

O'Neill, J., and Spash, C.L. (1999) *Conceptions of Value in Environmental Decision-Making*. Cambridge, UK: CRE. Available at: www.clivespash.org/eve/prb4-edu.pdf (accessed 8 August 2016).

Ortiz, O., Castells, F., and Sonnemann, G. (2008) 'Sustainability in the construction industry: A review of recent developments based on LCA', *Construction & Building Materials*, 23: 28–39.

Pandit, M.K. (2013) 'The Himalayas must be protected', *International Weekly Journal of Science*, 501: 283.

Pittet, D., and Kotak, T. (2010) 'Environmental impact of building technologies, a comparative study in Kutch District, Gujarat State, India', in *Proceedings of the Ecomateriales 4*, Paths towards Sustainability Conference, Bayamo, Cuba.

Salerno, F., Viviano, G., Thakuri, S., Flury, B., Maskey, R.K., Khanal, S.N., Bhuju, D., Carrer, M., Bhochhibhoya, S., Melis, M.T., Gianniono, F., Staiano, A., Carteni, F., Mazzoleni, S., Cogo, A. et al. (2010) 'Energy, forest, and indoor air pollution models for Sagarmatha National Park and Buffer Zone, Nepal', *Mountain Research & Development*, 30(2): 113–26.

Singh, A., Berghorn, G., Joshi, S., Syal, M., and Asce, M. (2011) 'Review of life-cycle assessment applications in building construction', *Journal of Architectural Engineering*, 17(1): 15–23.

Streimikiene, D., and Ciegis, R. (2010) 'The changes of life style: Significant contribution to GHG emission reduction efforts', in *Universities for Climate Change*, ed W.L. Filho, Berlin: Springer-Verlag, pp.167–177

43

INTEGRATED WASTE MANAGEMENT AND RESOURCE RECOVERY IN INDIA

Asha Panwar

Introduction

Managing waste is a perennial concern for both regional administrations and the concerned community across the world. The conventional requirement was for a service sector to uphold public health and maintain civic convenience in the locality concerned. Modern waste management goals are much more complex. We now also link waste and its fate to global objectives of resource security, green economy, and sustainable consumption and production. To achieve this, the planning of waste management and its governance needs to be comprehensive, multi-criteria, and dynamic in nature. It is in this context that the concept of life cycle approaches has unique applications to waste management.

The waste management problem is exacerbated by the increasing complexity of urban and commercial waste that renders its disposal or its recovery complex and perhaps hazardous. Waste management is still commonly considered a local issue, and yet local and regional authorities have mostly been unable to put in place effective management policies, constrained by economic, technological, legal and especially attitudinal and knowledge factors in their communities.

The recent report Global Waste Management Outlook (UNEP–ISWA 2015), like many authoritative reports before it, pointed to the necessity for improved management of (and reduction in) materials across their entire cradle-to-grave cycle, so as to minimize materials going to waste in the first place, and ensuring a safer and more environmentally sound approach to their management.

Waste, its life cycle, and its management

Conventional waste management planning only considers the post-user/consumer life cycle of goods and products, that is, waste generation, its collection, transfer, transportation, and then subsequent disposal. Over time, reuse and recycling stages were added. Life cycle thinking expands the scope of planning to the upstream, pre-user/consumer stages of the concerned product. Focusing on the upstream stages serves a number of purposes.

First, it directs attention to raw materials and also to the suitability of waste-derived secondary materials. A conventional scope tends to limit waste management strategies to consider only safe, cost-effective collection–transfer–transportation–disposal systems. Reuse and recycling may be included for the purpose of diverting waste from the disposal facilities. Life cycle-based strategies, on the other hand, present an opportunity to assess the success of waste management strategies in terms of material circulation, from waste to manufacturing (recycling) or use/consumption stages (reuse). Here, the suitability of secondary materials as substitutes for raw materials for manufacturing, or the ability of recycled products to compete with a similar product, emerges as an important factor in formulating reuse and recycling programmes.

Second, the expanded scope helps waste management planners with identification and assignment of responsibilities attributed to manufacturers, producers, packagers, or retailers. It is in this direction that product design for clean production, and end-of-life options become part of effective waste management strategies. The nature of virgin raw materials used to manufacture the products, the presence of toxic components in products, and dismantling possibilities for composite products all have downstream impacts on waste management planning. Life cycle thinking highlights the multi-stakeholder nature of waste management programmes and facilitates participation of upstream organizations and agencies.

Third, it effectively highlights the role and importance of the users/consumers of the products in question. It is here that consumer choices emerge as a primary component in modern waste management solutions. Procurement policies, product certification, and responsible consumption are seen as having a significant impact on waste management initiatives. This is of importance for countries such as India, where per capita waste generation is expected to increase over the next few years, owing to increasing purchasing power and the expanding urban populace.

Furthermore, life cycle-based thinking is useful for its guidance on impact assessment approaches. Life cycle assessment (LCA) is increasingly used, not only to assess the impacts associated with waste management processes and facilities (see, for example, Chapters 29 and 30), but also to prioritize waste treatment technologies. The addition of more complex sustainability assessment methods that include social and economic factors further increases its applicability for waste management systems. Other life cycle approaches, such as material flow analysis and system dynamics, also have important practical applications, as they help in visualizing the material stocks and flows that lead to waste generation. Application of these comprehensive approaches has provided a strong basis for integrating management systems across waste streams (e.g. household waste, e-waste, construction and demolition waste) and across environment media (e.g. soil, air, water).

The European Union and nations such as Japan have, in varied forms, accepted life cycle thinking as a valuable concept in both waste policies and regional waste management frameworks. Current waste management planning in India has yet to formally utilize life cycle thinking to assist waste management programmes. Still, the Indian waste management debate has changed rapidly in the last decade and is expected to evolve further in coming years. For example, a recent civil awareness campaign on cleanliness and sanitation, titled 'Swachh Bharat Abhiyan', and urban modernization schemes, such as 'Smart Cities Mission', are expected to provide a much-needed impetus at both waste policy and governance level.

Indian scenario

Per capita waste generation in India ranges from 0.17 to 0.67 kg of municipal solid waste per day (MoEF 2010: 12). Current waste collection coverage by municipal bodies, though 70–90 per cent in metropolitan cities, remains below 50 per cent in other urban areas (Annepu 2012: 24). Waste received by municipal bodies typically consists of 51 per cent organic waste, 17 per cent recyclables, 11 per cent hazardous waste, and 21 per cent inert waste (MoEF 2010: 12). The informal waste recycling sector extracts as much as 21 per cent of recyclable material from waste collected by the municipal body (Annepu 2012: 34). Waste reaching disposal sites is largely composed of biodegradable wet waste mixed with recyclables and inert waste. Despite the recyclables harvested by informal waste collectors before collection, the dry recyclables reaching disposal sites could still be a significant amount. This fraction would understandably be higher in cities, where fewer people are associated with informal waste handling. The recyclables harvested by informal waste collectors travel up the hierarchy to small–medium waste traders, product dismantlers, bulk waste traders, and so on, to finally arrive at recycling facilities of different scales, registered officially or otherwise. Such complex, informal systems can be found in a single urban area or are spread over contiguous urban centres, the critical factors being the supply of waste and its market value. Recent estimates on countrywide municipal waste generation suggest around 0.573 million metric tons per day (MoEF 2010: 2). Despite formal and informal recovery, most waste arrives at municipal dumpsites, which, with few exceptions, are essentially dump yards, with no arrangements for material sorting, protective lining, leachate treatment, or energy recovery.

A municipal body typically spends 25–50 per cent of its recurring budget on waste management. In 2009–10, the per capita investment cost for solid waste management was found to be INR 391 (equivalent to US$6), whereas per capita operations and maintenance cost (annual) in the sector was INR 155 (equivalent to US$2.3; MoUD 2011: 74). The services and infrastructure required for waste collection consume a large fraction of the available financial resources, with little left for waste segregation, disposal and processing. Multiple enquiries indicate that, barring a few inspirational cases, municipal bodies generally fall short of achieving the targets set by the central municipal solid waste (management and handling) rules 2000 (Kumar et al. 2009; CPCB 2015). Insufficient waste collection coverage, inadequate source segregation or subsequent sorting of collected waste, and poor management of municipal dump yards are the most common issues. The residents contribute to this state of affairs by littering in public places, not practising waste segregation at bin stage, and dumping waste at undesignated sites. Reasons for unsatisfactory implementation of national guidelines include financial stress on municipal bodies, lack of proper training of personnel, and the attitudes and behaviour of all involved.

The shrinking stock of available land in urban areas and public concerns surrounding dump yards have, in the last decade, compelled municipal bodies to pursue stronger provisions for waste recycling. Annepu (2012: 5) estimates the presence across the country of more than eighty composting plants and seven refuse-derived fuel and waste-to-energy projects. Large-scale projects seeking huge investments are mainly operated on a public–private partnership basis. Many more centralized waste recycling and treatment facilities endorsed through special purpose vehicles are to be set up in the near future. It appears that the municipal bodies are keen to adopt centralized, technology-heavy waste

Figure 43.1 Recycling Tetra Pak, Daman Ganga Board Mills Pvt Ltd, Vapi, Gujarat

Source: Environmental Management Centre, Mumbai, India

management systems that can take care of the bulk quantities of mixed waste that are currently reaching the dump yards. Improving collection and transportation of waste also features in the priorities to help Indian cities appear clean. Many cities, such as Nagpur, Shimla, and Coimbatore, are implementing a zero bin project, aided by smart collection and transportation solutions. An efficient extent of 'integration' between formal and informal waste management systems, centralized and decentralized (and community-level) waste management systems, and ecological, social, and economic aspects nevertheless remain a challenge for the decision-makers.

A preoccupation of regions with direct expenditures has slowed the introduction of systematic waste management systems. Direct costs, that is, financial expenditure by municipal bodies, are rarely seen in the light of the social burdens and impacts that result from unsatisfactory collection and disposal. There are costs associated with constantly cleaning up air, water, land that has been polluted by waste generators, and solid waste management activities and facilities themselves. The negative impacts on the health of waste collectors, recyclers, and people residing in the vicinity of dump yards are other major indirect costs that are rarely calculated, as they are incurred by individuals rather than by the regional administration. They are costs all the same, if only for their impact on economic productivity. Wider indirect costs include those associated with the production of diesel fuel for waste collection vehicles, or the cost of manufacturing and procuring the materials that are needed for waste management systems to function and the secondary wastes resulting from these facilities. A more systematic approach to waste could reduce some of these. On the positive side, a proactive waste management sector creates employment and stimulates innovation. For certain activities (recycling, recovery), waste represents revenues to both informal and formal sectors. Avoided costs accrue to each unit of waste diverted from open burning, undesignated disposal and dump yards. Waste recycling and recovery approaches also reduce the upstream manufacturing costs by replacing the virgin materials. Evaluating these costs and benefits encourages improvements in waste management systems, as well as positively affecting the regional metabolism of materials and energy. It is in these terms that system-based, multi-criteria, cross-boundary considerations could support the formulation of realistic waste management programmes and projects.

A life cycle-based approach highlights the causal linkages across waste streams and management activities – for example, how the absence of a dedicated waste management system for construction and demolition waste affects the landfill's life, or how the absence of a segregation system for hazardous household waste affects the characteristics of compost. Most current waste management policies do not have a provision for life cycle-based decision-making, nor do they make use of complementary programmes such as eco-design, environment labels, sustainable procurement, or producer responsibility. Although a large number of methodologies for applying life cycle thinking and management in the waste sector are available, they have mainly been used in developed countries (Laurent et al. 2014). For developing countries, the conversion of life cycle approaches into practical applications remains a challenge, owing in part to a lack of region-specific waste statistics and life cycle inventories, but also owing to inadequate understanding of their potential benefits.

Current waste management planning in India still appears to be reactionary and lacks a strategic viewpoint on many critical issues. The vision has mostly not moved on from its earlier focus on collection and disposal only. Piecemeal solutions based on uninformed choices prevail owing to limited consultations and inadequate attention to training and capacity building. The success of formal recycling initiatives depends on technologies that are suitable for the waste stream concerned. For example, a municipal body has the choice of sending plastics received in the waste stream to a plastic-to-energy plant, a plastic-to-fuel plant, or a recycling facility that recycles the plastic materials into pellets or some other form. These choices require knowledge of both direct and indirect impacts associated with these recycling facilities. Another important question is whether the recovered energy, fuel, or materials are clean enough for their subsequent utilization, and if the avoided costs associated with recovery or recycling really offset the cost associated with simple disposal. The fitness for recovery or recycling of a waste material is equally important for informal waste recycling. A number of facilities in metropolitan cities, housed in congested residential areas, with no health and safety controls, recycle the waste materials solely on the basis of their market demand. The presence of toxic components in the waste streams affects the recyclers, the neighbouring residents, and finally the end user of the product made from the recycled material. Thoughts on cleaner production and manufacturing, and finding standards for recycled materials and associated product are generally absent. Finding a transition from a simple disposal perspective to preventive waste management approaches is vital to achieve a more sustainable society, as it avoids a huge set of costs. However, it is a challenging task to link a waste management system to the manufacturing and consumption sectors and find a mechanism for cross-sectoral collaboration.

Conclusions

All countries face challenges in dealing with the ever-increasing volumes and complexities of modern waste streams. It is universally agreed that a transition is needed from the current disposal-oriented waste management mindset to a more efficient prevention approach that considers the entire life cycle of materials and goods. Such a change relies on the application of life cycle assessment tools that map the materials and waste flows in society and pinpoint the most effective points of intervention. Developing countries are at a serious disadvantage in that they lack both adequate data, such as consolidated waste statistics, and practitioners trained in applying such tools. Regions and municipalities in India, like those in many other countries, are still taking a piecemeal approach to waste management, often focused on costly technological solutions that may not be completely in tune with waste generation patterns and community attitudes to waste generally.

References

Annepu, R.K. (2012) *Sustainable Solid Waste Management in India*, New York: Department of Earth and Environmental Engineering, Fu Foundation School of Engineering and Applied Science, Columbia University.

CPCB (2015) 'Consolidated Annual Review Report on Municipal Solid Wastes (Management and Handling) Rules, (2000)', *Annual Review Report: 2013–14*, Delhi: Central Pollution Control Board, Ministry of Environment & Forests. Available at http://cpcb.nic.in/annualreport.php (accessed 8 August 2016).

Kumar, S., Bhattacharyya, J.K., Vaidya, A.N., Chakrabarti, T., Devotta, S., and Akolkar, A.B. (2009) 'Assessment of the status of municipal solid waste management in metro cities, state capitals, class I cities, and class II towns in India: An insight', *Waste Management*, 29: 883–95.

Laurent, A., Bakas, I., Clavreul, J., Bernstad, A., Niero, M., Gentil, E., Hauschild, M.Z., and Christensen, T.H. (2014) 'Review of LCA studies of solid waste management systems – Part 1: lessons learned and perspectives', *Waste Management*, 34: 573–88.

MoEF (2010) *Report of the Committee to Evolve Road Map on Management of Wastes in India*, Delhi: Ministry of Environment & Forests (MoEF), Government of India.

MoUD (2011) *Report on Urban Infrastructure and Services*, Delhi: Ministry of Urban Development, Government of India.

UNEP–ISWA (2015) *Global Waste Management Outlook*. Available at: http://web.unep.org/ourplanet/september-2015/unep-publications/global-waste-management-outlook (accessed 8 August 2016).

44

BIODIVERSITY RESTORATION AND SUSTAINABLE TOURISM IN SOUTH-WESTERN AUSTRALIA

Keith Bradby

Introduction

The process of introducing sustainability criteria into regional development can take many directions. This chapter describes how existing work on one key issue – ecological resilience in biodiversity hot spots – can also accommodate and benefit from a broader initiative on tourism development. Such work with multiple outcomes is greatly assisted by modern evaluation techniques such as natural resource accounting, environmental footprints and life cycle costing. These assist scientists, funders, and the broader population to better understand the interconnections between nature conservation and regional development.

The entire region of south-western Australia is classified internationally as an important but ecologically stressed biodiversity hot spot. Over much of the region, the productive phases of resource-based industries are slowing down, and agriculture also faces multiple challenges. The regional community is always hopeful of a more prosperous future, and yet has some economic and social fragility, as the path to other major business activities is not readily apparent. Although tourism appears to many to be a key opportunity, an expansion from its current base into a mainstream economic driver will be short term, unless its resource base is protected and managed well.

Tourism engages many different types of actor throughout its life cycle, requiring a well-thought-out, holistic strategy and co-ordinated implementation. Tourism based on natural values is often not well understood by conventional managers with experience limited to hotels, other service industries, and events.

Context and background

The South West Australia Floristic Region covers an area of more than 300,000 km^2, larger than some countries in other regions (see Figure 44.1: the green line shows the inland limits of the South West Floristic Region, and the swoosh shows the operating area of the Gondwana Link programme). Once part of the ancient supercontinent of Gondwana, it has been geologically stable and continuously vegetated for more than 250 million years. The region enjoys a Mediterranean climate, with annual rainfall ranging from 1,500 to 300 mm. The most extensive land uses are dryland cereal cropping and livestock grazing, some intensive horticulture and grazing in the wetter zone, plus mining and some plantation forestry.

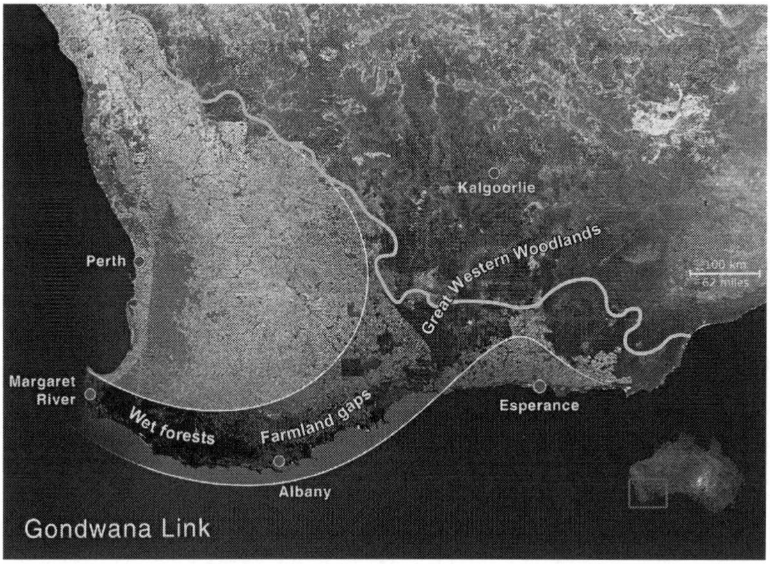

Figure 44.1
Map of Gondwanalink project

The region is recognised as one of the world's top twenty-five biodiversity hot spots, where 'exceptional concentrations of endemic species are also undergoing exceptional loss of habitat' (Myers et al. 2000). This designation provides recognition for the region's ecological values, many of which have only been documented in recent decades. For example, since 1970, the estimated number of identified plant species has more than doubled, from 3,600 to almost 8,000 (Hopper and Gioia 2004).

Significant portions of the region's natural habitats have been removed by activities such as forestry, mining, and farming, with most habitat removal being relatively recent. Large areas were cleared for agriculture from the early 1900s into the late 1990s, driven by government policy, improved fertiliser technologies, favourable commodity prices, and some periods of above-average rainfall.

Even during the development phase, the region's biodiversity values were already recognised. However, protection programmes were somewhat ad hoc, and the dominant conservation response by scientists, government, and community-based conservation bodies was to create protected areas, develop species-specific recovery programmes, and mitigate where possible the impacts of agriculture and other commercial land uses through voluntary programmes. These approaches have had their successes, with some large national parks and reserves having been established, though mostly in areas with poor farming soils and/or high forestry values, some species having been pulled back from near extinction, and voluntary programmes having helped improve land management generally.

However, given the size of the region and the diversity of the ecological systems, none of these efforts, even combined, was ever likely to significantly address the issues of long-term species loss and decline of ecological function. As early as 1982, survey work and analysis, using the emerging principles of island biogeography, determined that reserves larger than 40,000 ha were required to conserve most of the birds and remaining mammals in the inland agricultural area. Only one reserve in that 18 million ha area was within that threshold (Kitchener et al. 1980, 1982).

To change this situation, in 2002, the author and the independent body The Nature Conservancy brought together a range of local and national organisations to establish the Gondwana Link programme.[1] Our vision was of a 1,000-km band of 'reconnected country', from the wet forests to the semi-arid woodlands, where ecosystem function and biodiversity would be restored and maintained. This geographic focus included most of the larger national parks and reserves, where significant habitat still remained, and many of the areas where there were significant concentrations of rare and endemic species.

Two important features of this approach were a lifespan vision of the land and operation at an effective scale. That requires robust situation analysis, a focus on causes rather than on symptoms of ecological decline, and an ambitious intervention programme based on systematic and scientifically rigorous planning.

To implement this approach, increased and diversified funding streams were needed. Previous interventions had been largely based on small-scale donations and constantly changing government grant programmes. The changed thinking to a larger scale, along with a small number of similar initiatives, supported a significant increase in environmentally focused philanthropy, not only in Gondwana Link but nationally.

Our initiative was based on a non-government dynamic. With an avowed aim of operating 'at the scale and speed of business', we have not yet sought formal government endorsement, even though, at operational level, we complement and build on a number of government-funded programmes. Regrettably, our objectives and timescales are not always compatible with those of the authorities (at all levels), and, to achieve our objectives, we have been obliged to formally object to some government policies and development programmes that would damage ecological function and connectivity, reinforcing our independence from government.

We operate as a self-managed, multi-stakeholder initiative driven by enthusiasm and goodwill, with support from a range of locally based landholder groups and large national or global conservation organisations. Co-ordination and support for work across the programme are provided through a parent company, Gondwana Link Ltd (GLL), and the main groups involved are either members or work with us through various arrangements. Although we may have influence, we do not exercise formal control over the actions of groups involved in the programme.

We use planning based on formal Open Standards (Conservation Measures Partnership 2013) to implement a long-term, life cycle land management approach. Used well, this provides a strategic framework focused on achieving those goals of long-term ecological resilience. Planning follows Nature Conservancy guidelines that give priority to underlying system stresses rather than superficial symptoms (Low 2003). Each local plan is owned and managed by the groups working in that area, with each planning team able to incorporate external experts, including from industry and government. Although, initially, the plans were on a 10-year basis, we are steadily lengthening the time frames. Plans are 'bulked up' and integrated into an overall 'Whole of Link' approach by GLL.

Operationally, there has been reasonable success in achieving a large-scale strategic change. We built the initial momentum by modifying current land use so as to strengthen ecological functions in a 72-km-wide farmland gap between two large national parks, with a combination of private, corporate, and some government funding enabling the purchase and restoration of marginal farmlands. Ecological connectivity across this gap has been greatly improved thanks to the purchase of some 13,000 ha of existing habitat and farmland, with around 3,500 ha already restored. Elsewhere, smaller-scale efforts are underway to fill the gaps between other main habitat areas. An additional 16 million ha of habitat area on the eastern end of the Link now receives improved conservation-oriented management. This government-owned land, which was previously largely unprotected and unmanaged, is now known as the Great Western Woodlands.

It is still too early to quantify specific ecological outcomes from these combined efforts. But they have already produced a number of local benefits, such as wildlife use of the restored and protected habitats, improved management of critical threats such as fire and invasive species, and significantly increased management capacity on the ground in key areas. There is now increased public interest in, and support for, reversing ecological decline through large-scale life cycle approaches to land management (see Figure 44.2).

Although there have been some social and economic benefits to the communities involved, these have so far mostly been by-products of the ecological work, rather than an integral component. To address economic outcomes more directly, we are now evolving to a more holistic operating framework where, as outlined by Peter Whitbread-Abrutat in Chapter 35, environmental improvements can also be drivers for socio-economic regeneration.

Figure 44.2
Public interest in conservation measures

The remaining part of this chapter examines one of the initiatives being implemented to achieve simultaneous ecological and economic outcomes.

Broadening the support base by developing sustainable tourism

Central to any life cycle approach is the ability to take the long-term view, link components, and integrate approaches across sectors. A key to operational success in our programme is the ability to recognise and build on opportunities as they arise. Linking tourism development with responsible land use and nature conservation is one of these opportunities.

Considerable experience had already been built through tours of restoration areas run for donors and supporters. These visitors not only wanted to see effective restoration, but also sought interactions with local people and 'in-depth' insights into what biodiversity meant in this region that answered questions such as why a largely flat and dry region was as biologically rich as tropical rainforests. From this evolved the concept of 'restoration tourism' – whereby we could increase tourism and its associated economic benefits, while also increasing broader public understanding and support for land restoration work.

At our 2014 Forum, a small group further developed this concept. In order not to lose sight of the conservation objectives, it was considered important to keep in mind the environmental footprint of current tourism approaches, even those labelled as 'eco-tourism'. This reflection came at a time when the state government was committing substantial resources to exploit the tourism potential of the national park system, including $40 million for road works in one park. We saw this as a costly and risky approach. Providing new infrastructure in our region's large, ecologically important national parks risked degrading natural values and diverting tourists away from commercial infrastructure and attractions. An additional concern was the willingness of government to provide establishment funding for facilities such as interpretive centres, without adequate allowance made for the costs of long-term operation and maintenance. This lack of a life cycle vision means that only a few centres are financially viable, and maintaining some has placed substantial strain on the communities they were meant to benefit.

In late 2014, the Great Southern Development Commission, a state government entity whose regional boundary includes the central section of Gondwana Link, produced a Draft Economic Development Blueprint aimed at achieving 'transformational growth and development'. The blueprint gave environmental values minimal coverage, but did include a proposed 'Biodiversity Drive Trail',[2] designed to entice visitors into the region's hinterland – areas with a pressing need to diversify their agriculture-based economies but lacking the immediate scenic impact of coastal areas.

This was obviously our opportunity. In early 2015, Gondwana Link secured a small contract to develop the trail concept and associated business case. We analysed whether a significant tourist product could be built from the region's biodiversity values and related restoration efforts. This analysis, although focused through the prism of available short-term funding, took a long-term life cycle view of how tourism can grow in an integrated manner and provide longer-term economic returns, while simultaneously supporting ongoing biodiversity management within the region.

Analysis and result

The main driver for international tourism into Australia is the natural environment (GSWESC 2013). Yet the south-west region's tourism sector has had a heavy focus on promoting food, wine, art, and historical events. Environmental appreciation is generally offered as a relatively superficial 'scenic' or 'recreational' experience, largely highlighting the scenic coast, recreational activities such as fishing, or the beauty of the region's wildflowers and forests. It mostly ignores the region's international status as a biodiversity hot spot.

Through wide consultation, it became clear that some inland communities were already engaged in providing good, 'in-depth' local experiences and benefitting from the presence of visitors who were looking beyond superficial visual features.

On the basis that 92 per cent of the existing pattern of visitation is based on 'self-drive', we designed an 800-km-long, region-wide 'Biodiversity Drive Trail' that will deepen the visitor experience through a cohesive regional narrative. Its critical elements are:

- providing depth for the visitor by layering values and experiences, from global significance and the regional narrative through to detailed local trails that highlight local stories;
- using existing infrastructure, such as the existing road network, and appreciating the biological values remaining in the settled landscape;
- strengthening appropriate community efforts in each locality by contributing funding to establish their local trail components;
- providing emphasis on the restoration work already underway across the area; and
- building a social enterprise model for long-term funding to maintain, broaden, and refresh the trail, which includes opportunities for visitors to participate in the local restoration and management efforts, whether with their skills and time or by donations.

Our concept treats people less as 'tourists' and more as 'visitors' – fellow humans interested in our part of the world, open to information and ideas about how to live better with biodiversity, and participating with us in the challenges of living sustainably.

The economics of such a scheme are surprisingly positive. Based on the region's tourist visitation rates of 250,000 persons per year, the analysis showed that a $1.8 million establishment cost would return to regional businesses, over 10 years under a 4 per cent discount rate scenario, $27 for every initial $1 spent. This return was entirely on the basis of immediate expenditure, such as travel and accommodation costs, and did not include additional contributions to restoration programmes.

With the business case for a viable and profitable programme now completed, at the time of writing, we are awaiting confirmation of the core funding from the grant authority. Whatever the outcome, we have already stimulated a positive dialogue across the region and opened doors into other funding streams.

Conclusions

The programme shows that it is possible to combine ecological, economic, and social objectives into a life cycle view of tourism that is based on regional conservation initiatives. The important ingredients include:

- taking a long-term view;
- including upstream and downstream implications;
- identifying linkages between different objectives;
- evaluating costs and benefits across the entire spectrum of inputs and impacts;
- allowing for adaption to the inevitable changes as they occur;
- using evidence to guide decisions, rather than emotive assumptions; and
- actively reducing the footprint of projects and activities.

Our practical working procedures include learning from, building on, and supporting existing efforts, rather than imposing new solutions. We also recognise that, although sustainable solutions require some fundamental changes to 'business-as-usual' economics, we need to build from existing business opportunities rather than try to unilaterally impose new social models.

Adding strategies such as a regional visitor programme to our primary focus of achieving ecological objectives has had its challenges, but it is an essential shift if we are to build mainstream support for the achievement and maintenance of our ecological vision. Clear thinking, as embodied by a life cycle vision, has worked well in both aspects of our work.

Notes

1 So named to emphasise both the connectivity vision and the special nature of southern-hemisphere ecological systems that originated from the ancient continent of Gondwana.
2 The concept of these is basically that tourists self-drive between points of interest, with interpretive material provided at a set number of stops.

References

Conservation Measures Partnership (2013) *Open Standards for the Practice of Conservation: Version 3.0*. Available at: www.conservationmeasures.org (accessed 8 August 2016).

GSWESC (2013) *Great South West Edge National Landscape Experience Development Strategy: Final Report*, Great South West Edge Steering Committee.

Hopper, S.D., and Gioia, P. (2004) 'The southwest Australian floristic region: Evolution and conservation of a global biodiversity hotspot', *Annual Review of Ecology & Systematics*, 35: 623–50.

Kitchener, D.J., Chapman, A., and Muir, B.G. (1980) 'The conservation value for mammals of reserves in the Western Australian wheatbelt', *Biological Conservation*, 18: 179.

Kitchener, D.J., Dell, J., Muir, B.G., and Palmer, M. (1982) 'Birds in Western Australian Wheatbelt Reserves – Implications for conservation', *Biological Conservation*, 22: 127–63.

Low, G. (2003) *Landscape-scale Conservation: A Practitioner's Guide*. The Nature Conservancy.

Myers, N., Mittermeier, R.A., Mittermeier, C.G., da Fonseca, G.A.B., and Kent, J. (2000) 'Biodiversity hotspots for conservation priorities', *Nature*, 403: 803–8.

Part VI

LIFE CYCLE PERSPECTIVES FOR REGIONAL SOCIO-ECONOMIC DEVELOPMENT

45

SYNTHESIS – LIFE CYCLE APPROACHES AND PERSPECTIVES FOR SUSTAINABLE REGIONAL DEVELOPMENT

Fritz Balkau and Guido Sonnemann

General orientation

The chapters in this book have highlighted some of the issues that surround the use of life cycle approaches for sustainable development at the regional level. Their potential application is anchored in long-standing practical experience with the life cycle management (LCM) toolbox that is currently applied to products and materials. They also reflect some of the more recent organisational management procedures that take the longer-term development perspective so fundamental to sustainability. The book has attempted to stay close to the ongoing evolution of life cycle approaches, incorporating recent developments such as organisational LCA, social LCA, and life cycle costing (LCC). Some of these come under the recently coined title of 'life cycle sustainability assessment' (LCSA), a more holistic, multi-criteria approach to assessment. The applicability of such varied approaches is very much influenced by the characteristics of a region – its social, environmental, and economic fabric, as well as its political vision. Regions do not all have the same resource bases, nor the same demographic and trade profiles, nor do their authorities have the same opportunities for action within a national or global policy framework. Each region will have to choose the most appropriate instruments from the available toolbox.

The world is not a static place. Regional leaders have a common desire to move ahead and provide a better life for their citizens. Progress depends on a variety of improvements, reforms, investments, innovations, and supporting actions. To be sustainable, these need to be based on reliable data about the environmental and other sustainability consequences of any change in the region and its global context. Ecological footprints, material flows, resource valuations, industry fabric, and available human resources are just some of the elements of such a database that allow regional administrations to map out a path for measures that can lead to a sustainable future. As this book shows, such a path is infinitely more successful in the long term if a holistic, system-based life cycle approach is used for decision-making on crucial sustainable development challenges. This thinking is reflected in both the title and the structure of the book.

Life cycle thinking blends two ideas – time span and the real sources of impacts. This approach leads to choices that generate economic value over a long period, within the capacities of the region's natural environment, and strengthen the regional social systems. By focusing on asset lifetimes rather than on immediate upfront costs, such approaches achieve a higher degree of system optimisation. A key feature is that a life cycle perspective seeks to identify upstream sources of issues. It tries to avoid 'one-dimensional' decisions that address a particular sustainability challenge (e.g. waste management) but that result in unexpected secondary impacts such as excessive energy use. Life cycle thinking thus helps us to go to the source of the problem and to avoid shifting impacts from one life cycle stage to another; from one geographical area to another; from one generation to another; or from one impact category, and its related environmental medium, to another.

Our times are times of change. Few things remain static when global trade patterns reshape our economies and our social fabric. Scientific discoveries are quickly translated into new products or processes, and this is also true of the way we manage our society – technical progress leads to easier communication, greater transparency, more collaborative ways of doing things, and more accurate measurement and reporting of outcomes. Life cycle thinking is already prominent in modern management practices adopted by leading businesses and by governments, as for example in their adoption of the principles of circular economy. The next step is to extend such thinking to small and medium-sized enterprises and regional administrations. Trade associations and technical centres, public administrations, and local universities can play an important role in encouraging such a trend.

Numerous international standards have been adopted in recent years to harmonise procedures, facilitate comparisons, and agree on common performance levels. Such standardisation may apply equally to a product bought in a store or to the elaborate way in which organisations report to their public stakeholders. International standards also cover management procedures concerning environmental performance, assessment, quality, or risks, among others. Implementation of sustainability programmes by both private and public organisations has thus become substantially codified, together with the life cycle tools on which they depend.

Standardised, yes, but also evolving. The application of life cycle approaches is a dynamic activity where all the boundary conditions seem to be changing simultaneously – methodological, geographical, economic, social, and political. There is no one right procedure, and anyway a different outcome may be expected tomorrow. As standards evolve, further life cycle approaches for the management of sustainability will arise, benefitting from new procedures and new applications.

This is the context in which this book should be viewed – a journey towards sustainability rather than a destination. This also characterises the synthesis below – ideas to be taken on board for consideration, rather than viewed as a rigid methodology to be blindly applied .

Key conclusions for applying life cycle approaches in regions

The contributing authors to this book have approached their subjects from different angles, drawing on their individual experience in a diverse range of professional sectors. The overall message is that the best approach to sustainability is a life cycle approach that takes a system perspective and is long term and holistic. This synthesis draws together some common elements that have been proposed in the different chapters. These elements constitute useful guidance in the further promotion of life cycle approaches for use in regional development.

A number of conclusions that can be drawn from the chapters are developed below.

Clear sustainability objectives are essential in moving forward

The sustainability agenda is broad. Development has many facets. Regional fabrics and dynamics are inevitably complicated. The life cycle of things can be long and complex. The causes of impacts are not always immediately clear, nor may the end-point consequences be well appreciated.

In this complex world, moving forward efficiently requires clear vision and a defined target. It is no accident that an essential early step in many life cycle methodologies is the scoping stage. Although this may appear tedious (after all, we are impatient to act), in fact, it helps to maintain a clear view of the boundaries of our exercise and promotes a more efficient work plan.

Many life cycle methods are limited to well-defined issues or targets (e.g. carbon footprint), but it is important that they continue to be viewed within a wider context of sustainability. LCA has thus traditionally focused on energy and pollution-type issues and has not incorporated, for example, biodiversity or cultural impacts (and has only recently ventured into social impacts). Readers should, therefore, place the applicability of the individual life cycle approaches discussed in this book in the context of the entire set of sustainability objectives that may be adopted by a region or organisation. Not every tool addresses every issue. Many of the tools remain focused on materials flows, waste, and pollution, whereas a few deal with, for example, landscape issues (important for tourism and land management), biodiversity protection (our living space), resource efficiency (of economic interest for doing more with less), and social issues. There are one or more specific tools and related methodologies for each task.

It follows that the application of life cycle approaches requires clear identification of objectives, scoping of the exercise, and an overview of the sustainability context. Just thinking 'life cycle' is already a big step forward, but it is only the first step. The process of defining and elaborating sustainability objectives in detail is beyond the scope of this book; nevertheless, several procedures such as footprinting and LCA already lead us in this direction. Readers seeking further insights into this complex field will find information in the lists of references and sources at the end of each chapter.

There is already a variety of life cycle approaches that can assist regional development

The diversity of development decisions being made by regional administrations and other actors is already opening up major opportunities to apply life cycle tools, procedures, and instruments to enhance the planning process. Relevant approaches include assessment methodologies, resource and materials management procedures, and sustainable infrastructure planning. Their use is enhanced by reference to the additional concepts of circular economy, industrial ecology, environmental resilience and social justice that many regions have already embraced. The use of ecological and other footprinting methods gives regions a clearer view of where they stand on sustainability and identifies items for priority attention. Carbon, water, and waste dominate the sustainability agenda in many regions, and footprinting on these issues is becoming widespread. Materials flow studies (as explained in Chapters 8 and 9) can map resource flows that are generated within a region, but they can also identify materials entering the region that ultimately become part of the waste stream. An understanding of the extended supply chain (or 'value chain') is often necessary for a full assessment. An application of this is in integrated waste and pollution management, which now reaches further and further up the product cycle to try to exercise more influence over the materials and products we are eventually obliged to discard at local level.

All stakeholders benefit from a better understanding of the life cycle consequences of our decisions. Useful tools for such an assessment include LCA itself, especially for studying the future impact of products of importance to regional economies. LCA methodology has broadened in recent years to expand the sustainability factors being evaluated. Several chapters deal explicitly with this. Social LCA allows administrations to map out the human dimension of decision-making and of projects. LCC results in a clearer view of the financial implications at the various stages of a product or a project and is directly relevant to investment and expenditure decisions. A further recent development is organisational LCA (O-LCA), which allows an organisation to evaluate its own relationship with the sustainability agenda by assessing the materials flows and social impacts of its activities and decisions. Taken together, these techniques can be combined into an LCSA (explained in Chapter 14), which gives a clearer picture of where a region stands with respect to various sustainability agenda items. Regional sustainability benchmarking against other territories can be done via the framework of the Global Reporting Initiative (GRI). Although economic comparisons are already common, benchmarking may also apply to natural resources and social capital, materials and product flows, and even to internal management systems.

Few of these tools are novel, although many are still being expanded and improved. Nearly all are subject to standards or guidelines agreed by different stakeholder groups to harmonise their individual applications in a wider national or global arena. Many methodologies are defined under the framework of the International Organization for Standardization (ISO). Although it is true that every region is different, and that many of the LCA tools were first developed to address the product cycle of common materials, their application to regional development issues is both feasible and relevant. The increased use of such tools brings with it a need for appropriate data. Lack of data is currently a handicap for a number of applications, although improved reporting requirements are gradually addressing this issue.

We note that the application of life cycle approaches can be at two levels. Organisations can specify the life cycle performance targets that they expect their clients, suppliers, and contractors to meet. These partners will then apply the appropriate methodologies and tools. But, importantly, regional governments can foster a sustainability mindset in their own administrations through use of life cycle-based administrative procedures by, for example, using LCC to guide procurement. In short, a regional government can take its own internal action and then stimulate others to do the same through policies, planning, and regulations. A fundamental basis for all of this is a coherent regional vision of sustainable development, with clear objectives and targets, and with metrics to allow an evaluation of progress towards these goals. The combined use of O-LCA and sustainability reporting (whether GRI framework or other) can be a powerful first step for organisations to enhance their sustainability efforts.

Some life cycle approaches are still evolving

Although the methodologies and examples presented by authors can already contribute in their present form to enhanced sustainability planning and management, some are less mature than others. In part, this is because expectations have evolved faster than the methodologies, pushing the boundaries of the exercise into new areas. There is less practical experience with LCA in new areas such as social issues or ecosystems than in the traditional product-focused assessments. The new application areas include the use of input–output tables, extensions into O-LCA, deeper applications of social impact assessment, and attempts to apply LCA to water use (water footprinting) and land use. Attempts are also being made to cover biodiversity in LCA applications. Experience gained will of course greatly benefit future users, but first movers will have the deepest learning experiences.

With such a dynamic evolution of methodologies, it is important for regions to keep up to date with evolving practices, tracking real-life applications elsewhere to see what will be most relevant to them. For example, social LCA is building up a good body of experience. However, consensus has not yet been reached on how to deal with land use and the balance for biogenic carbon, although the way this balance is eventually handled can have enormous consequences for a region's forests. There are sure to be further results here soon, and the concept of benchmarking evoked earlier can be usefully applied to enhance the learning experience.

Combine assessment type tools with intervention processes

Sustainable regional development will not be achieved without some changes in the way we do things. Assessment tools remain essential in bringing forth information and defining priorities. However, regions will need to transform assessment into intervention by defining new targets, changing management procedures, repositioning the role of certain materials and products in society, and directing finance to the key sectors that stimulate innovation.

This book shows that a number of these intervention processes are already in regular use in many places and sectors. These include the use of eco-labels, environmental product declarations (EPDs), eco-design, supply-chain management, product–service systems, eco-efficiency measures, and application of environmental standards.

A number of holistic concepts help to provide a proper management framework for such intervention: for example, LCM for industrial clusters, industrial ecology applied to business estates, circular economy as a basis for integrated waste management, landscape management and restoration for natural resource conservation, corporate social responsibility to guide community programmes of resource companies, and an integrated approach to sustainable energy to provide a foundation for renewables. Several life cycle communication and information tools are included in the book (EPD for example), and others, such as public sustainability reporting (perhaps under the GRI framework), could easily be added.

A choice of appropriate intervention tools leads to a more systematic application of LCM. This structures our interventions according to an agreed collective action by all the value-chain stakeholders. Much development is currently under way in the world of LCM, and regional entities would benefit from becoming involved in this evolution by bringing forward their needs and objectives. A faster shift from LCM's present product-centric focus to one that also incorporates the needs of the public sector is one of the developments that regions should welcome and work towards.

Address multiple application areas

Many of the methodologies described in this book have very specific applications, often aimed at addressing one or two critical sustainability issues such as energy, water, waste, and so on. Even the holistic concepts of industrial ecology and circular economy are heavily oriented towards considering materials and waste flows. These remain valuable tools for resolving specific concerns, but, at regional level, a wider range of sustainability issues usually comes into play, and life cycle tools should ideally be capable of addressing a broader spectrum of outcomes. For example, public procurement criteria based only on, say, a product's carbon footprint will entirely ignore its waste disposal implications. Here, it is not necessarily the life cycle tool that is deficient: it is the limited objectives of the procurement criteria (perhaps also reflecting limited sustainability objectives in the first place). Similar considerations apply to other areas where a broader vision is needed at the intervention stage if the sustainability agenda as a whole is to move forward.

Even within the broader context, many of the issue-specific life cycle tools presented in this book retain an important place when certain issues reach a critical level (for example, water or waste). It is, nevertheless, important to maintain a 'sustainability overview' of their application to avoid possible burden shifting and other secondary consequences. In any case, few of the tools discussed are capable of addressing the entire sustainability agenda at one time. LCSA attempts this at the assessment level. LCM could, in theory, be used to guide the subsequent intervention in a more holistic way, but LCM is not yet sufficiently structured to easily permit this in general application. For the moment, an issue-based approach to life cycle intervention is the best we can hope for, and is certainly preferable to no action at all.

A number of chapters deal with a more life cycle view of natural resource management, covering issues of (landscape) assessment, stewardship, sustainable exploitation, and land care and remediation. This sector includes service industries, such as tourism, that benefit from a sustainable environment but need themselves to operate according to responsible social and environmental criteria. Lastly, it should be highlighted that regional administrations will, in any case, benefit from greater resource efficiency, through, for example, more efficient buildings, less wasteful bureaucracies, lower operating costs, and more sustainable financial practices (including investments). Given the substantial budgets required for infrastructure development and maintenance, a life cycle view of such investments avoids the frequent past practice of shifting the burden of renovation and end-of-life impacts to future generations.

Move up the LCM maturity curve; 'mainstream' the intervention phase

It is well understood that sustainability initiatives within and by an organisation are not fully successful if they are simply grafted on to a current programme structure as an 'extra' function. It is unrealistic to expect a special department to take responsibility for sustainability while the remainder of an organisation continues to do business as usual. Change, and the life cycle thinking that underlies it, is most successful when both are built into – or, to use the management term, 'mainstreamed' – into all areas of management and administration. Mainstreaming is important for all organisations, whether corporate or public administration. The use of formal life cycle methodologies and instruments, as outlined in this book, gives an organisation some tools to assist in such change and to measure its progress, but they are not sufficient on their own. The UNEP/SETAC Life Cycle Initiative has developed a '*life cycle management maturity matrix*' that can be helpful in this regard. The use of O-LCA (see Chapter 15) and public sustainability reporting (not covered in this book) can be useful first steps in the change management process, without unduly perturbing the existing administrative structure. Chapter 4 also provides some pointers on sustainable financing. The development and formal adoption of a broad-based sustainability strategy and objectives are an important subsequent step. Footprinting and materials flow studies are able to indicate the current regional position on sustainability and point to priority areas where change is needed by the various operational divisions within an organisation.

Learn from experience, introduce benchmarking, and study feasibility and prerequisites

The case studies in Parts III–V show a number of applications of life cycle approaches within regions, either by a regional administration itself or by business entities or clusters. The case studies show how the concept of LCM can function in practice if it is appropriately integrated into regional decision-making. They also reveal some of the success factors (or challenges!) for such initiatives. Few of the outcomes described in the case studies arose by accident – most were the result of deliberate change management in accordance with a clearly defined purpose. Many involved consultation with stakeholders in the product life cycle, including sustainability experts. In particular, the overall sustainability objectives needed to be clear, including consideration of geographical (supply-chain issues and global trade) and societal boundaries (for instance, workplace, life style, health and safety).

The case studies well illustrate the importance of using available assessment methodologies as a prerequisite for decision-making. Clear identification of current resource bases, material flows, and footprints was needed in these case studies to provide adequate base-line data.

The major case studies (as well as the smaller examples quoted throughout the text) can inspire others to take action and intervene appropriately to move forward on sustainability objectives. Although a few cases limited themselves to single sustainability criteria, many were based on an integrated approach that covers a combination of environmental, social, and economic criteria (e.g. the chapters on Hungary, Chile, and India; the operation of eco-parks; sectors such as mining, energy, tourism; and so on). Some studies demonstrate how mainstreaming is important, and that all levels of organisation along the value chain need to be involved in implementation (e.g. the chapters on industrial estates, landscapes, industrial innovation).

Challenges to an effective use of life cycle approaches in regions

The preceding section highlighted the potential of applying life cycle approaches successfully, but there are inevitably some challenges to their implementation. Three important ones are discussed below.

Different levels of scale and maturity of various tools

Although a number of life cycle methodologies have been standardised under the ISO 14020 and 14040 series, others remain independent concepts, without a clear operational definition. Some terminology is still evolving, as the profession grapples with new challenges and expanded objectives. There has been considerable progress in social life cycle assessment in recent years, and, as we have seen, a new initiative has developed in ISO with support from the UNEP/SETAC Life Cycle Initiative on O-LCA. However, progress is still slow on natural resource-based issues such as biodiversity, landscape and cultural and heritage impacts. The major handicap of the existing life cycle instruments is that they were originally developed with a focus on materials and products and with predominantly private-sector companies as client users. The structure of most of the approaches (with the probable exception of EEIO and MFA and of the ecological footprint) is less appropriate for dealing with multiple sustainability issues, in an extended geographical area, under a public administration. An advantage of the life cycle impact assessment framework is that the impact criteria, once developed, are scalable and can be used to assess the potential impacts related to the economic activity of a region and of its administration as an organisation.

Moving from assessment to intervention and mainstreaming

As a category, the assessment methodologies we have discussed in this book are more mature, and more standardised, than the tools for taking action. However, the intervention stage is where the future is built, where actual changes are made, and where the new ways are found to move more rapidly towards sustainable development, including at regional level. The intervention stage is concerned with the application of LCM.

It has already been mentioned that, to be effective, LCM needs to be built in, not added on, to day-to-day organisational procedures. It is, however, clear that for this to be achieved, a clearer LCM framework for regional application would be beneficial, comparable to but different from, the approach being used for implementing LCM in big companies. A number of LCM components have already been described in this book, but these represent a starting point rather than a complete toolbox. The challenge is that LCM, like LCA itself, has evolved around an industrial clientele interested in optimising its products. Its application to the public administration and infrastructure sector is less well described and there is less experience to share. Nevertheless, as LCM is a holistic concept rather than a standardised tool, the range of definitions and practices currently in use also spills over into applications that can be useful at regional level.

Unconscious use of incomplete sets of key sustainability performance indicators

We have already mentioned the importance of broadening the scope of LCA techniques beyond energy and materials flows so as to be able to address a fuller range of sustainability concerns. Beyond the selection of suitable techniques, it is important that users of LCA and other life cycle approaches re-examine their initial choice of sustainability. Although such a choice remains the prerogative of the client region, recent years have seen a sustainability dialogue that focuses mainly on energy and carbon density, replacing to some extent the earlier preoccupation with the waste and pollution agendas, and not at all incorporating other important environmental and social priorities. Depending on the assessment and management processes chosen, the single-issue approach leads all too easily either to a neglect of other important actions or to a cross-media transfer of impacts, or both. Such a focus is sometimes referred to as 'the partial solution that blocks the way to a complete outcome'. A common example in regions is an aggressive recycling policy that has unfortunate consequences for waste management by withdrawing wanted components from the waste stream, leading, for example, to a malfunction of a waste incinerator that previously depended on the paper waste to maintain its combustion temperature. Other single-issue examples are also known, such as forest clearing for biofuels that damages biodiversity and is thus counterproductive for climate change. A further example is the installation of solar PV in remote regions without incorporation of a management system for battery recovery. Although single-issue action is necessary to get some things done, it is important not to lose sight of the bigger picture. Each region must define the breadth of its sustainability agenda, its priority issues, and its objectives.

Perspectives

The discussion elements above flow naturally from the contributions of the authors of this book. They can be seen as a central rallying point for attempts to apply life cycle approaches to sustainable regional development, in the same way that LCM evolved as a valuable tool for leading businesses and industry.

Challenges remain. We have to be aware that to date most life cycle practitioners have not seen regions as a potential new client base. Practitioners have developed in-depth expertise in many assessment procedures, often standardising methodologies along the way, but a strategic view of their importance in regional development planning is still missing. The evolution of life cycle approaches to address additional ecological, social, and economic issues in a consensus-based way is not rapid enough to keep pace with the growing need for such tools, which continue to have a mainly product-centric focus.

On the other hand, most regional organisations have little awareness or practical knowledge of LCM approaches that could assist them in meeting their sustainability objectives. They are not benefitting from the rich experience of the growing community, in particular of LCA/LCM practitioners, working with the private sector, and, more importantly, they seem to be unable to articulate their needs or give appropriate guidance to possible life cycle initiatives in their regional operations.

We hope that, in the future, there will be more opportunities for these two worlds to meet, so that the use of life cycle approaches to sustainable regional development becomes mainstream.

INDEX

Note: italics denote figures; bold denotes tables and boxes.

Kolibri 360 GmbH
Ferdinand-Jühlke-Straße 7
99095 Erfurt, Deutschland
produktsicherheit@kolibri360.de